AF166358

# Communications
# in Computer and Information Science 2165

## Rationale

The CCIS series is devoted to the publication of proceedings of computer science conferences. Its aim is to efficiently disseminate original research results in informatics in printed and electronic form. While the focus is on publication of peer-reviewed full papers presenting mature work, inclusion of reviewed short papers reporting on work in progress is welcome, too. Besides globally relevant meetings with internationally representative program committees guaranteeing a strict peer-reviewing and paper selection process, conferences run by societies or of high regional or national relevance are also considered for publication.

## Topics

The topical scope of CCIS spans the entire spectrum of informatics ranging from foundational topics in the theory of computing to information and communications science and technology and a broad variety of interdisciplinary application fields.

## Information for Volume Editors and Authors

Publication in CCIS is free of charge. No royalties are paid, however, we offer registered conference participants temporary free access to the online version of the conference proceedings on SpringerLink (http://link.springer.com) by means of an http referrer from the conference website and/or a number of complimentary printed copies, as specified in the official acceptance email of the event.

CCIS proceedings can be published in time for distribution at conferences or as post-proceedings, and delivered in the form of printed books and/or electronically as USBs and/or e-content licenses for accessing proceedings at SpringerLink. Furthermore, CCIS proceedings are included in the CCIS electronic book series hosted in the SpringerLink digital library at http://link.springer.com/bookseries/7899. Conferences publishing in CCIS are allowed to use Online Conference Service (OCS) for managing the whole proceedings lifecycle (from submission and reviewing to preparing for publication) free of charge.

## Publication process

The language of publication is exclusively English. Authors publishing in CCIS have to sign the Springer CCIS copyright transfer form, however, they are free to use their material published in CCIS for substantially changed, more elaborate subsequent publications elsewhere. For the preparation of the camera-ready papers/files, authors have to strictly adhere to the Springer CCIS Authors' Instructions and are strongly encouraged to use the CCIS LaTeX style files or templates.

## Abstracting/Indexing

CCIS is abstracted/indexed in DBLP, Google Scholar, EI-Compendex, Mathematical Reviews, SCImago, Scopus. CCIS volumes are also submitted for the inclusion in ISI Proceedings.

## How to start

To start the evaluation of your proposal for inclusion in the CCIS series, please send an e-mail to ccis@springer.com.

Ngoc-Thanh Nguyen · Bogdan Franczyk ·
André Ludwig · Manuel Nunez · Jan Treur ·
Gottfried Vossen · Adrianna Kozierkiewicz
Editors

# Advances in Computational Collective Intelligence

16th International Conference, ICCCI 2024
Leipzig, Germany, September 9–11, 2024
Proceedings, Part I

 Springer

*Editors*
Ngoc-Thanh Nguyen ⓘ
Wrocław University of Science
and Technology
Wrocław, Poland

André Ludwig ⓘ
University of Leipzig
Leipzig, Germany

Jan Treur ⓘ
Vrije Universiteit Amsterdam
Amsterdam, The Netherlands

Adrianna Kozierkiewicz ⓘ
Wrocław University of Science
and Technology
Wrocław, Poland

Bogdan Franczyk ⓘ
University of Leipzig
Leipzig, Germany

Manuel Nunez ⓘ
Universidad Complutense de Madrid
Madrid, Spain

Gottfried Vossen ⓘ
University of Münster
Münster, Germany

ISSN 1865-0929          ISSN 1865-0937 (electronic)
Communications in Computer and Information Science
ISBN 978-3-031-70247-1          ISBN 978-3-031-70248-8 (eBook)
https://doi.org/10.1007/978-3-031-70248-8

This Springer imprint is published by the registered company Springer Nature Switzerland AG
The registered company address is: Gewerbestrasse 11, 6330 Cham, Switzerland

If disposing of this product, please recycle the paper.

# Preface

This volume contains the first part of the proceedings of the 16th International Conference on Computational Collective Intelligence (ICCCI 2024), held in Leipzig, Germany from 9–11 September 2024. The conference was organized in a hybrid mode which allowed for both on-site and online paper presentations. The conference was hosted by Leipzig University, Germany and jointly organized by Wrocław University of Science and Technology, Poland in cooperation with IEEE SMC Technical Committee on Computational Collective Intelligence, European Research Center for Information Systems (ERCIS), and International University-VNU-HCM (Vietnam).

Following the successes of the 1st ICCCI (2009), held in Wrocław, Poland, the 2nd ICCCI (2010) in Kaohsiung - Taiwan, the 3rd ICCCI (2011) in Gdynia - Poland, the 4th ICCCI (2012) in Ho Chi Minh City - Vietnam, the 5th ICCCI (2013) in Craiova - Romania, the 6th ICCCI (2014) in Seoul - South Korea, the 7th ICCCI (2015) in Madrid - Spain, the 8th ICCCI (2016) in Halkidiki - Greece, the 9th ICCCI (2017) in Nicosia - Cyprus, the 10th ICCCI (2018) in Bristol - UK, the 11th ICCCI (2019) in Hendaye - France, the 12th ICCCI (2020) in Da Nang - Vietnam, the 13th ICCCI (2021) in Rhodes - Greece, the 14th ICCCI (2022) in Hammamet - Tunisia, and the 15th ICCCI (2023) in Budapest - Hungary, this conference continued to provide an internationally respected forum for scientific research in computer-based methods of collective intelligence and their applications.

Computational collective intelligence (CCI) is most often understood as a subfield of artificial intelligence (AI) dealing with soft computing methods that facilitate group decisions or processing knowledge among autonomous units acting in distributed environments. Methodological, theoretical, and practical aspects of CCI are considered as the form of intelligence that emerges from the collaboration and competition of many individuals (artificial and/or natural). The application of multiple computational intelligence technologies such as fuzzy systems, evolutionary computation, neural systems, consensus theory, etc., can support human and other collective intelligence, and create new forms of CCI in natural and/or artificial systems. Three subfields of the application of computational intelligence technologies to support various forms of collective intelligence are of special interest but are not exclusive: the Semantic Web (as an advanced tool for increasing collective intelligence), social network analysis (as a field targeted at the emergence of new forms of CCI), and multi-agent systems (as a computational and modeling paradigm especially tailored to capture the nature of CCI emergence in populations of autonomous individuals).

The ICCCI 2024 conference featured a number of keynote talks and oral presentations, closely aligned to the theme of the conference. The conference attracted a substantial number of researchers and practitioners from all over the world, who submitted their papers for the main track and nine special sessions.

The Main Track, covering the methodology and applications of CCI, included: collective decision-making, data fusion, deep learning techniques, natural language processing, data mining and machine learning, social networks and intelligent systems, optimization, computer vision, knowledge engineering and application, as well as Internet of Things: technologies and applications. The Special Sessions, covering some specific topics of particular interest, included: cooperative strategies for decision making and optimization, security and reliability of information, networks and social media, anomalies detection, machine learning, deep learning, digital image processing, artificial intelligence, speech communication, IOT applications, natural language processing, and innovative applications in data science.

We received 234 papers submitted by authors coming from 45 countries around the world. Each paper was reviewed by at least three members of the international Program Committee (PC) and by the Meta-reviewing Committee. The final decision regarding acceptance for all submissions to the Main Track and Special Sessions was made by the ICCCI Program Chairs to select the best papers and ensure the highest quality of the chosen papers. Our review model ensured that papers from both the Main Track and Special Sessions were held to an equally high standard.

Finally, we selected 59 papers for oral presentation and publication in two volumes of the Lecture Notes in Artificial Intelligence series and 67 papers for oral presentation and publication in two volumes of the Communications in Computer and Information Science series.

We would like to express our thanks to the keynote speakers: Jarosław Jankowski from West Pomeranian University of Technology in Szczecin (Poland), Klaus Solberg Söilen from Halmstad University (Sweden), Sören Auer from Leibniz University Hannover (Germany), and Krzysztof Czarnecki from University of Waterloo (Canada)

Many people contributed to the success of the conference. First, we would like to recognize the work of the PC co-chairs and Special Sessions organizers for taking good care of the organization of the reviewing process, an essential stage in ensuring the high quality of the accepted papers. The Special Session chairs deserve a special mention for the evaluation of the proposals and the organization and coordination of the work of nine Special Sessions. In addition, we would like to thank the PC members for performing their reviewing work with diligence. We thank the Local Organizing Committee chairs, Publicity chairs, Web chair, and Technical Support chairs for their fantastic work before and during the conference. Finally, we cordially thank all the authors, presenters, and delegates for their valuable contribution to this successful event. The conference would not have been possible without their support.

Our special thanks are also due to Springer for publishing the proceedings and to all the other sponsors for their kind support.

It is our pleasure to announce that the ICCCI conference series continues to have a close cooperation with the Springer journal Transactions on Computational Collective Intelligence, and the IEEE SMC Technical Committee on Transactions on Computational Collective Intelligence.

Finally, we hope that ICCCI 2024 contributed significantly to the academic excellence of the field and will lead to the even greater success of ICCCI events in the future.

September 2024

Ngoc Thanh Nguyen
Bogdan Franczyk
André Ludwig
Manuel Núñez
Jan Treur
Gottfried Vossen
Adrianna Kozierkiewicz

# Organization

## Organizing Committee

### Honorary Chairs

Eva Inés Obergfell      Leipzig University, Germany
Arkadiusz Wójs      Wrocław University of Science and Technology, Poland
Piotr Jędrzejowicz      Gdynia Maritime University, Poland

### General Chairs

Ngoc Thanh Nguyen      Wrocław University of Science and Technology, Poland
Bogdan Franczyk      Leipzig University, Germany

### Program Chairs

André Ludwig      Leipzig University, Germany
Manuel Núñez      Universidad Complutense de Madrid, Spain
Jan Treur      Vrije Universiteit Amsterdam, The Netherlands
Gottfried Vossen      University of Münster, Germany

### Steering Committee

Ngoc Thanh Nguyen (Chair)      Wrocław University of Science and Technology, Poland
Piotr Jędrzejowicz      Gdynia Maritime University, Poland
Shyi-Ming Chen      National Taiwan University of Science and Technology, Taiwan
Kiem Hoang      University of Information Technology, VNU-HCM, Vietnam
Dosam Hwang      Yeungnam University, South Korea
Lakhmi C. Jain      University of South Australia, Australia
Geun-Sik Jo      Inha University, South Korea
Janusz Kacprzyk      Polish Academy of Sciences, Poland
Ryszard Kowalczyk      University of South Australia, Australia

| | |
|---|---|
| Yannis Manolopoulos | Open University of Cyprus, Cyprus |
| Toyoaki Nishida | Kyoto University, Japan |
| Manuel Núñez | Universidad Complutense de Madrid, Spain |
| Klaus Solberg Söilen | Halmstad University, Sweden |
| Khoa Tien Tran | International University-VNUHCM, Vietnam |

## Organizing Chairs

| | |
|---|---|
| Philippe Krajsic | Leipzig University, Germany |
| Marcin Pietranik | Wrocław University of Science and Technology, Poland |

## Special Session Chairs

| | |
|---|---|
| Adrianna Kozierkiewicz | Wrocław University of Science and Technology, Poland |
| Paweł Sitek | Kielce University of Technology, Poland |
| Patrick Zschech | Leipzig University, Germany |

## Doctoral Track Chairs

| | |
|---|---|
| Marek Krótkiewicz | Wrocław University of Science and Technology, Poland |
| Rainer Unland | University of Duisburg-Essen, Germany |

## Publicity Chairs

| | |
|---|---|
| Andreas Barton | Leipzig University, Germany |
| Marcin Jodłowiec | Wrocław University of Science and Technology, Poland |
| Rafal Palak | Wrocław University of Science and Technology, Poland |

## Webmaster

| | |
|---|---|
| Marek Kopel | Wrocław University of Science and Technology, Poland |

# Local Organizing Committee

| | |
|---|---|
| Martin Schieck | Leipzig University, Germany |
| Martin Max Röhling | Leipzig University, Germany |
| Christian Alverman | Leipzig University, Germany |
| Patient Zihisire Muke | Wrocław University of Science and Technology, Poland |
| Thanh-Ngo Nguyen | Wrocław University of Science and Technology, Poland |
| Jose Fabio Ribeiro Bezerra | Wrocław University of Science and Technology, Poland |

# Keynote Speakers

| | |
|---|---|
| Jaroslaw Jankowski | West Pomeranian University of Technology in Szczecin, Poland |
| Klaus Solberg Söilen | Halmstad University, Sweden |
| Krzysztof Czarnecki | University of Waterloo, Canada |
| Sören Auer | University of Hannover, Germany |

# Special Session Organizers

### ADMDL2024: Special Session on Anomalies Detection using Machine and Deep Learning

| | |
|---|---|
| Yousra Chabchoub | Institut Supérieur d'Electronique de Paris, France |
| Maurras Togbe | Institut Supérieur d'Electronique de Paris, France |

### AISC2024: Special Session on AI and Speech Communication

| | |
|---|---|
| Ualsher Tukeyev | Al-Farabi Kazakh National University, Kazakhstan |
| Orken Mamyrbayev | Institute of Information and Computational Technologies, Kazakhstan |

*CCINLP2024: Special Session on Computational Collective Intelligence and Natural Language Processing*

Ismail Biskri                University of Québec a Trois-Rivieres, Canada
Nadia Ghazzali              University of Québec a Trois-Rivieres, Canada

*CSDMO2024: Special Session on Cooperative Strategies for Decision Making and Optimization*

Piotr Jędrzejowicz          Gdynia Maritime University, Poland
Dariusz Barbucha            Gdynia Maritime University, Poland
Ireneusz Czarnowski         Gdynia Maritime University, Poland

*DICV2024: Special Session on Recent Advances of Deep Learning and Internet of Things in Computer Vision-Related Applications*

Wadii Boulila               Prince Sultan University, KSA
Maha Driss                  Prince Sultan University, KSA
Anis Koubaa                 Prince Sultan University, KSA
Jawad Ahmad                 Edinburgh Napier University, UK
Faisal Saeed                Birmingham City University, UK

*DIPMAI2024: Special Session on Digital Image Processing for Medical and Automotive Industry*

Debora Gil                  Universitat Autònoma de Barcelona, Spain
Mihail Gaianu               West University of Timisoara, Romania

*IADS2024: Special Session on Innovative Applications in Data Science*

Małgorzata Przybyła-Kasperek   University of Silesia in Katowice, Poland
Agnieszka Wosiak            Lodz University of Technology, Poland
Agnieszka Duraj             Lodz University of Technology, Poland
Rafał Skinderowicz          University of Silesia in Katowice, Poland
Wiesław Paja                University of Rzeszów, Poland

*MLRWD2024: Special Session on Machine Learning in Real-World Data*

Jan Kozak                   University of Economics in Katowice, Poland
Mikhail Moshkov             King Abdullah University of Science and
                            Technology, KSA
Artur Kozłowski             Łukasiewicz Research Network, Poland

Przemysław Juszczuk                   Polish Academy of Sciences, Poland
Barbara Probierz                      University of Economics in Katowice, Poland

*SIRENE2024: Special Session on Security and Reliability of Information, Networks and Social Media*

Rafal Kozik                           Bydgoszcz University of Science and Technology,
                                      Poland
Adrianna Kozierkiewicz                Wroclaw University of Science and Technology,
                                      Poland
Marcin Pietranik                      Wroclaw University of Science and Technology,
                                      Poland
Marek Pawlicki                        Bydgoszcz University of Science and Technology,
                                      Poland
Wojciech Mazurczyk                    Warsaw University of Science and Technology,
                                      Poland
Michal Choraś                         Bydgoszcz University of Science and Technology,
                                      Poland

## Senior Program Committee

Plamen Angelov                        Lancaster University, UK
Costin Badica                         University of Craiova, Romania
Nick Bassiliades                      Aristotle University of Thessaloniki, Greece
Maria Bielikova                       Slovak University of Technology in Bratislava,
                                      Slovakia
Abdelhamid Bouchachia                 Bournemouth University, UK
David Camacho                         Universidad Autónoma de Madrid, Spain
Richard Chbeir                        University of Pau and Pays de l'Adour, France
Shyi-Ming Chen                        National Taiwan University of Science and
                                      Technology, Taiwan
Paul Davidsson                        Malmö University, Sweden
Mohamed Gaber                         Birmingham City University, UK
Daniela Godoy                         ISISTAN Research Institute, Argentina
Manuel Grana                          University of the Basque Country, Spain
William Grosky                        University of Michigan, USA
Francisco Herrera                     University of Granada, Spain
Tzung-Pei Hong                        National University of Kaohsiung, Taiwan
Dosam Hwang                           Yeungnam University, South Korea
Lazaros Iliadis                       Democritus University of Thrace, Greece
Mirjana Ivanovic                      University of Novi Sad, Serbia

| | |
|---|---|
| Piotr Jędrzejowicz | Gdynia Maritime University, Poland |
| Geun-Sik Jo | Inha University, South Korea |
| Kang-Hyun Jo | University of Ulsan, South Korea |
| Janusz Kacprzyk | Polish Academy of Sciences, Poland |
| Ryszard Kowalczyk | Swinburne University of Technology, Australia |
| Ondrej Krejcar | University of Hradec Kralove, Czech Republic |
| Hoai An Le Thi | University of Lorraine, France |
| Edwin Lughofer | Johannes Kepler University Linz, Austria |
| Yannis Manolopoulos | Aristotle University of Thessaloniki, Greece |
| Grzegorz J. Nalepa | AGH University of Science and Technology, Poland |
| Toyoaki Nishida | Kyoto University, Japan |
| Manuel Núñez | Universidad Complutense de Madrid, Spain |
| George A. Papadopoulos | University of Cyprus, Cyprus |
| Radu-Emil Precup | Politehnica University of Timişoara, Romania |
| Leszek Rutkowski | Częstochowa University of Technology, Poland |
| Tomasz M. Rutkowski | University of Tokyo, Japan |
| Ali Selamat | Universiti Teknologi Malaysia, Malaysia |
| Edward Szczerbicki | University of Newcastle, Australia |
| Ryszard Tadeusiewicz | AGH University of Science and Technology, Poland |
| Muhammad Atif Tahir | National University of Computer and Emerging Sciences, Pakistan |
| Jan Treur | Vrije Universiteit Amsterdam, The Netherlands |
| Serestina Viriri | University of KwaZulu-Natal, South Africa |
| Bay Vo | Ho Chi Minh City University of Technology, Vietnam |
| Gottfried Vossen | University of Munster, Germany |
| Lipo Wang | Nanyang Technological University, Singapore |
| Michał Woźniak | Wrocław University of Science and Technology, Poland |

## Program Committee

| | |
|---|---|
| Muhammad Abulaish | South Asian University, India |
| Sharat Akhoury | University of Cape Town, South Africa |
| Stuart Allen | Cardiff University, UK |
| Ana Almeida | GECAD-ISEP-IPP, Portugal |
| Bashar Al-Shboul | University of Jordan, Jordan |
| Adel Alti | University of Setif, Algeria |
| Taha Arbaoui | University of Technology of Troyes, France |

| | |
|---|---|
| Mehmet Emin Aydin | University of the West of England, Bristol, UK |
| Thierry Badard | Laval University, Canada |
| Amelia Badica | University of Craiova, Romania |
| Hassan Badir | École Nationale des Sciences Appliquées de Tanger, Morocco |
| Paulo Batista | Universidade de Évora, Portugal |
| Khalid Benali | University of Lorraine, France |
| Morad Benyoucef | University of Ottawa, Canada |
| Szymon Bobek | Jagiellonian University, Poland |
| Leon Bobrowski | Bialystok University of Technology, Poland |
| Grzegorz Bocewicz | Koszalin University of Technology, Poland |
| Urszula Boryczka | University of Silesia, Poland |
| Mariusz Boryczka | University of Silesia, Poland |
| János Botzheim | Eötvös Loránd University, Hungary |
| Peter Brida | University of Žilina, Slovakia |
| Ivana Bridova | University of Žilina, Slovakia |
| Krisztian Buza | Budapest University of Technology and Economics, Hungary |
| Aleksander Byrski | AGH University of Science and Technology, Poland |
| Alberto Cano | Virginia Commonwealth University, USA |
| Roberto Casadei | Università di Bologna, Italy |
| Amine Chohra | Paris-East Créteil University (UPEC), France |
| Kazimierz Choros | Wrocław University of Science and Technology, Poland |
| Robert Cierniak | Częstochowa University of Technology, Poland |
| Mihaela Colhon | University of Craiova, Romania |
| Antonio Corral | University of Almería, Spain |
| Jose Alfredo Ferreira Costa | Universidade Federal do Rio Grande do Norte, Brazil |
| Rafal Cupek | Silesian University of Technology, Poland |
| Ireneusz Czarnowski | Gdynia Maritime University, Poland |
| Camelia Delcea | Bucharest University of Economic Studies, Romania |
| Shridhar Devamane | Global Academy of Technology, India |
| Muthusamy Dharmalingam | Bharathiar University, India |
| Tien V. Do | Budapest University of Technology and Economics, Hungary |
| Márk Domonkos | Eötvös Loránd University, Hungary |
| Nadia Essoussi | University of Tunis, Tunisia |
| Rim Faiz | University of Carthage, Tunisia |
| Marcin Fojcik | Western Norway University of Applied Sciences, Norway |

| | |
|---|---|
| Anna Formica | IASI-CNR, Italy |
| Bogdan Franczyk | University of Leipzig, Germany |
| Dariusz Frejlichowski | West Pomeranian University of Technology in Szczecin, Poland |
| Naoki Fukuta | Shizuoka University, Japan |
| Mauro Gaspari | University of Bologna, Italy |
| K. M. George | Oklahoma State University, USA |
| Janusz Getta | University of Wollongong, Australia |
| Daniela Gifu | Romanian Academy - Iaşi Branch, Romania |
| Arkadiusz Gola | Lublin University of Technology, Poland |
| Foteini Grivokostopoulou | University of Patras, Greece |
| László Gulyás | Eötvös Loránd University, Hungary |
| Petr Hajek | University of Pardubice, Czech Republic |
| Kenji Hatano | Doshisha University, Japan |
| Marcin Hernes | Wrocław University of Economics, Poland |
| Huu Hanh Hoang | Hue University, Vietnam |
| Jeongky Hong | Yeungnam University, South Korea |
| Frédéric Hubert | Laval University, Canada |
| Zbigniew Huzar | Wrocław University of Science and Technology, Poland |
| Fethi Jarray | University of Gabes, Tunisia |
| Joanna Jedrzejowicz | University of Gdansk, Poland |
| Gordan Jezic | University of Zagreb, Croatia |
| Ireneusz Jóźwiak | Wrocław University of Science and Technology, Poland |
| Przemysław Juszczuk | University of Economics in Katowice, Poland |
| Arkadiusz Kawa | Poznań School of Logistics, Poland |
| Attila Kiss | Eötvös Loránd University, Hungary |
| Marek Kopel | Wrocław University of Science and Technology, Poland |
| Petia Koprinkova-Hristova | Bulgarian Academy of Sciences, Bulgaria |
| Ivan Koychev | University of Sofia "St. Kliment Ohridski", Bulgaria |
| Jan Kozak | University of Economics in Katowice, Poland |
| Dalia Kriksciuniene | Vilnius University, Lithuania |
| Stelios Krinidis | Centre for Research and Technology Hellas (CERTH), Greece |
| Dariusz Krol | Wrocław University of Science and Technology, Poland |
| Marek Krotkiewicz | Wrocław University of Science and Technology, Poland |
| Jan Kubicek | VSB - Technical University of Ostrava, Czech Republic |

| | |
|---|---|
| Elzbieta Kukla | Wrocław University of Science and Technology, Poland |
| Julita Kulbacka | Wrocław Medical University, Poland |
| Marek Kulbacki | Polish-Japanese Academy of Information Technology, Poland |
| Piotr Kulczycki | Polish Academy of Science, Systems Research Institute, Poland |
| Kazuhiro Kuwabara | Ritsumeikan University, Japan |
| Florin Leon | "Gheorghe Asachi" Technical University of Iaşi, Romania |
| Doina Logofatu | Frankfurt University of Applied Sciences, Germany |
| Aphilak Lonklang | Eötvös Loránd University, Hungary |
| Juraj Machaj | University of Žilina, Slovakia |
| George Magoulas | Birkbeck, University of London, UK |
| Bernadetta Maleszka | Wrocław University of Science and Technology, Poland |
| Marcin Maleszka | Wrocław University of Science and Technology, Poland |
| Adam Meissner | Poznań University of Technology, Poland |
| Manuel Méndez | Universidad Complutense de Madrid, Spain |
| Jacek Mercik | WSB University in Wrocław, Poland |
| Radosław Michalski | Wrocław University of Science and Technology, Poland |
| Peter Mikulecky | University of Hradec Kralove, Czech Republic |
| Miroslava Mikusova | University of Žilina, Slovakia |
| Jean-Luc Minel | Université Paris Ouest Nanterre La Défense, France |
| Javier Montero | Universidad Complutense de Madrid, Spain |
| Dariusz Mrozek | Silesian University of Technology, Poland |
| Manuel Munier | University of Pau and Pays de l'Adour, France |
| Phivos Mylonas | Ionian University, Greece |
| Anand Nayyar | Duy Tan University, Vietnam |
| Filippo Neri | University of Napoli Federico II, Italy |
| Linh Anh Nguyen | University of Warsaw, Poland |
| Sinh Van Nguyen | International University – Vietnam National University, HCMC, Vietnam |
| Loan Thuy Thi Nguyen | International University Ho Chi Minh City, Vietnam |
| Adam Niewiadomski | Lodz University of Technology, Poland |
| Adel Noureddine | University of Pau and Pays de l'Adour, France |
| Alberto Núñez | Universidad Complutense de Madrid, Spain |
| Mieczyslaw Owoc | Wrocław University of Economics, Poland |

| | |
|---|---|
| Marcin Paprzycki | Systems Research Institute, Polish Academy of Sciences, Poland |
| Marek Penhaker | VSB -Technical University of Ostrava, Czech Republic |
| Isidoros Perikos | University of Patras, Greece |
| Elias Pimenidis | University of the West of England, Bristol, UK |
| Nikolaos Polatidis | University of Brighton, UK |
| Piotr Porwik | University of Silesia, Poland |
| Paulo Quaresma | Universidade de Évora, Portugal |
| David Ramsey | Wrocław University of Science and Technology, Poland |
| Mohammad Rashedur Rahman | North South University, Bangladesh |
| Ewa Ratajczak-Ropel | Gdynia Maritime University, Poland |
| Virgilijus Sakalauskas | Vilnius University, Lithuania |
| Ilias Sakellariou | University of Macedonia, Greece |
| Khouloud Salameh | University of Pau and Pays de l'Adour, France |
| Imad Saleh | Université Paris 8, France |
| Sana Sellami | Aix-Marseille University, France |
| Yeong-Seok Seo | Yeungnam University, South Korea |
| Andrzej Sieminski | Wrocław University of Science and Technology, Poland |
| Dragan Simic | University of Novi Sad, Serbia |
| Stanimir Stoyanov | University of Plovdiv "Paisii Hilendarski", Bulgaria |
| Grażyna Suchacka | University of Opole, Poland |
| Piotr Sulikowski | West Pomeranian University of Technology, Poland |
| Libuse Svobodova | University of Hradec Kralove, Czech Republic |
| Martin Tabakov | Wrocław University of Science and Technology, Poland |
| Yasufumi Takama | Tokyo Metropolitan University, Japan |
| Joe Tekli | Lebanese American University, Lebanon |
| Trong Hieu Tran | VNU-University of Engineering and Technology, Vietnam |
| Maria Trocan | Institut Superieur d'Electronique de Paris, France |
| Krzysztof Trojanowski | Cardinal Stefan Wyszyński University in Warsaw, Poland |
| Ualsher Tukeyev | al-Farabi Kazakh National University, Kazakhstan |
| Olgierd Unold | Wrocław University of Science and Technology, Poland |
| Serestina Viriri | University of KwaZulu-Natal, South Africa |
| Adam Wojciechowski | Lodz University of Technology, Poland |

# Contents – Part I

**Natural Language Processing**

## Data Mining and Machine Learning

## Social Networks and Intelligent Systems

# Contents – Part II

**Computational Intelligence for Digital Content Understanding**

## Collective Intelligence in Healthcare

# Collective Intelligence and Collective Decision-Making

# Formalization of Agent-Based Model of Group Learning

Barbara Wędrychowicz[(✉)] and Marcin Maleszka

Wroclaw University of Science and Technology, 27 wybrzeże Stanisława Wyspiańskiego st., 50-370 Wroclaw, Poland
{barbara.wedrychowicz,marcin.maleszka}@pwr.edu.pl

**Abstract.** A formalisation of agent-based model of group learning is introduced in this paper. The idea of multi-agent model of knowledge diffusion during cooperation in small heterogeneous groups has been inspired by observations of human cooperation and characteristics. Most research, on how humans learn and cooperate in groups, comes from the psychology and pedagogy literature. Therefore, the model environment is inspired by the surroundings in which learning takes place while people cooperate: school and company. Individual agents possess features that influence communication, cooperation, and knowledge diffusion. These features can either facilitate or hinder the acquisition of knowledge. As shown in the previous paper [6], a multi-agent model can be prepared to accurately simulate a specific group's work process, considering students' characteristics and behaviours. The current paper focusses on generalising and formalising the model. The basic postulates and assumptions are described. Postulates on dividing agents into groups, groups size, and agents' knowledge change are defined. The main postulates about the impact of agents' characteristics on cooperation, communication and knowledge gain have been settled and the example of the model was discussed.

**Keywords:** Agent based modelling · Teaching model · Group modelling

## 1 Introduction

The study of the teaching and learning process and the acquiring knowledge by students date back to the 1940s, when Dale proposed the cone of experience (Fig. 1) which is widely used today [1]. Even then, psychologists and educators noticed that teaching only from books and lectures was not the most optimal way of imparting knowledge to students. Then, they have begun to consider and study new approach to education, the process of learning and teaching. Their research shows that learners effectively acquire much more knowledge through "doing" rather than "listening" and "watching" [1]. The term currently used to describe this phenomenon is active learning. According to researchers considering this topic, active learning allows faster knowledge acquisition [2,3]. It also

N.-T. Nguyen et al. (Eds.): ICCCI 2024, CCIS 2165, pp. 3–15, 2024.
https://doi.org/10.1007/978-3-031-70248-8_1

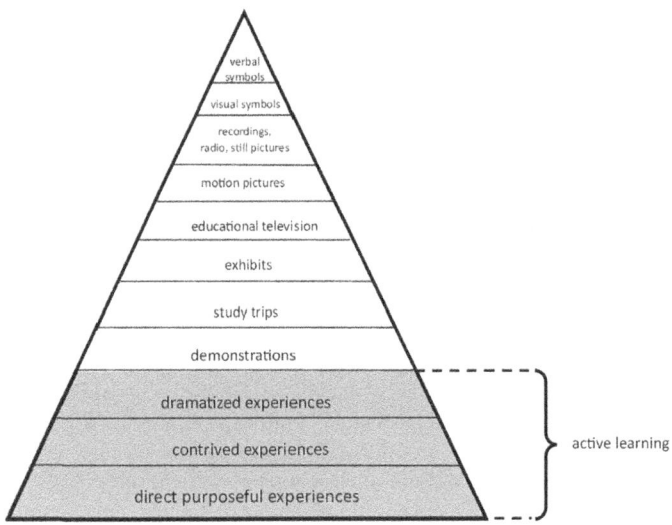

**Fig. 1.** Dale's Cone of Experience

enables more effective development of abilities and skills and allows to change or consolidate one's personal values that motivate or discourage learning [2].

Current technological developments force society to think even more about ways to improve the teaching and learning process. Methods that engage students in active learning are constantly being refined. Nowadays, students have difficulty concentrating throughout the entire duration of the lecture. Taking notes and studying from books also cause quite a lot of problems. Therefore, psychologists and educators are looking for alternative teaching methods. Based on Dale's cone of experience, they have noted that the most knowledge is acquired through active, dynamic, and engaging activities. Participating in discovering new knowledge and imparting one's knowledge to others may also be part of active learning.

One of the most frequently used active teaching methods is group work. It encourages students to interact among peers, communicate, cooperate, and exchange knowledge - share existing knowledge, absorb knowledge provided by peers, and agree on a common version of knowledge if any inaccuracies or contradictions arise. Salvin suggests that cooperatively developed solutions are usually more valuable than those invented by a single person [4].

Some experts [5, 7] claim that placing students in feature-based groups usually has a minor but positive impact on their educational achievements. In rare circumstances, inaccurate group composition may even cause learning to decline in terms of quality and speed. Depending on the outcome of the students' division into groups, the characteristics of individuals may have a positive or negative impact on the entire groups work. The combined students' characteristics can help the group quickly absorb all the knowledge provided by individuals and easily discover new knowledge. However, an incorrect division may be the reason that, despite the individual's great potential, the increase in knowledge will be small or nonexistent.

The above observations became an inspiration to prepare a model for knowledge diffusion during cooperation in small heterogeneous groups. Individual agents have features corresponding to human characteristics. Those features selected from pedagogical literature were suggested by researcher as having an impact on communication, cooperation, and knowledge diffusion. Depending on the intensity of a given features, they may facilitate and advance the acquisition of knowledge, facilitate and advance the transfer of knowledge, or hinder both processes.

Our previous research [6] has shown that it is possible to prepare a multi-agent model that is able to simulate and reflect the work process of a specific group quite accurately. It takes into account the characteristics and behaviors of students, as well as the characteristics of the modeled group. Currently, our work concentrate on generalizing and formalizing the previously prepared model. The current paper proposes basic postulates and assumptions about the model. It describes agents representing students, groups and knowledge.

The composition of this paper is as follows: next section contains the literature overview of crucial elements of the environment on which the model is based: school environment, students and teachers characteristics and phases of learning. Then the model is described by formalizing agents, groups of agents and knowledge, and appropriate criteria and postulates are added to each part.

## 2   Environment

As the interactions and learning process between members of a school environment and the teams within companies have been very thoroughly described in the psychological and pedagogical literature, it serves as a model environment for our model. Each agents that learns (student-agent) and each teaching agent (teacher-agent) can be described by different characteristics and behaviors. According to the pedagogical literature, the set of features that interest us and have an impact on learning process will be similar for all of them. All student-agents function, communicate, and cooperate with each other to achieve a common goal. In the case of our model, it is to acquire as much knowledge as possible.

The above view of the problem makes the multi-agent system an ideal tool for modeling the indicated environment.

### 2.1   Students and Teachers

Each student-agent can be represented by a set of characteristics, and their intensity, level of knowledge and kinds of behavior. A teacher-agent can be described in a similar way. The research shows that the characteristics and behaviors of the student and teacher may influence the effectiveness of learning [8]. When modeling the above-mentioned environment, the actions performed by individual units - interactions between the different type of agents, interactions between the same type of agents, as well as independent learning and forgetting the knowledge acquired by the student-agent, should be taken into account. The concepts of student and teacher used in this article do not always refer to the classically understood concepts in which the student is a younger person who needs to learn specified subject, and the teacher is an experienced and pedagogically-oriented educator. In the case of this article, each individual can be both a student-agent and a teacher-agent. We assume that a student-agent is an agent that in the current moment acquires and gains some knowledge, and a teacher-agent is an agent that shares its knowledge. In the model presented in this paper the learning and teaching process takes place at the same time, through group work and communication.

## 2.2   Phases of Learning

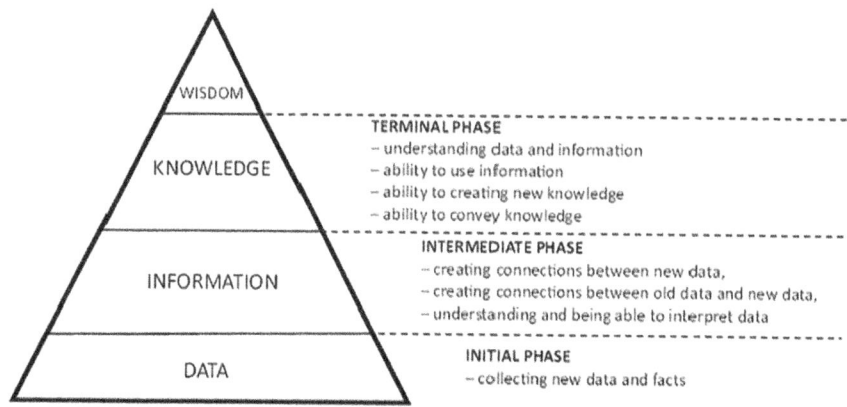

**Fig. 2.** Comparison of the knowledge pyramid (DIKW) and the three phases of learning

According to Shuell [9], each learning process can be divided into three main phases: the initial phase, the intermediate phase, and the terminal phase. Each phase is characterized by a different way of acquiring data, using it and converting into knowledge. First, when the student starts learning, he collects data and facts. Typically, each of these data is a separate unit, part of the information, and is not connected to others. The student has only facts that he is unable to use them. By devoting more time to learning, he enters the next phase - the intermediate phase. The student begins to see connections between the data received, creates links, and later also sees connections between the new data and the knowledge he had before. He starts to interpret and understand the received data and transform them into information. By dedicating even more time to learning, the student enters the third, final phase of learning - terminal phase.

In the last part, the student is able to use new information, effectively combine it with previously acquired knowledge, create new knowledge based on the existing knowledge, and clearly communicate and impart it to others. We can see that the subsequent phases perfectly coincide with the first levels of the knowledge pyramid DIKW (Fig. 2).

## 3 Model

As mentioned in the previous section, because most research and data in the topic of human learning comes from the field of pedagogy and psychology, the school environment has become the basis on which the agent environment is modelled. In the initial proposed model, we only include part of it. The model presents the last two learning phases (intermediate and terminal) combined in group work. We deliberately omit the first phase - data collection - data and fact acquisition from the teacher-educator. We assume that when starting cooperation, each student has some data, and subsequent learning phases are carried out through communication and enabling the acquisition and transfer of knowledge between peers while working in a group. In the proposed model, we consider a set of n agents representing students. All agents are divided into m groups within which they cooperate - communicate, transfer knowledge, combine knowledge and, in some cases, create new knowledge.

### 3.1 Agents

By $A$, we denote the set of all agents, and by $A_i$, we denote a single agent. Each agent is represented by a set that contains basic information about the agent - an ordered tuple $C_i$ representing the agent's characteristic features and a variable $K_i$ representing the agent's current state of knowledge:

$$A = \{A_1, A_2, ..., A_n\}$$

and

$$A_i = \{C_i, K_i\} \qquad for \quad i = 1, 2, ..., n$$

According to the previously cited work [6], part of each agent description is represent-ed by an ordered tuple of several features that were found in the pedagogical literature, where it is indicated that they have some impact on the process of knowledge acquisition, knowledge transfer, and interaction and communication with peers. In the initial model four main features were selected:

- *Communicativeness*- communication skills - $c_1$,
- *Leader potential* - ability to lead the group in some tasks - $c_2$,
- *Interest* - interest in the topic - $c_3$,
- *Ease* - ease of learning - $c_4$.

In this model, the value of each feature is in the range $[0, 1]$ where 0 means that the student does not have the described skills, and 1 is the maximum value, suggesting that the student has mastered a given skill or feature perfectly.

$$C = (c_1, c_2, c_3, c_4) \quad and \quad \forall c \in C : c \in [0, 1]$$

*Example 3.1-1* the ordered set $C = (1, 0.1, 0.5, 0)$ represents a student whose characteristics are distributed as follows:

- *Communicativeness*
  $c_1 = 1$ - means that the student very easily establishes contact with others, is open and has no problems with transmitting information
- *Leader potential*
  $c_2 = 0.1$ - means that the student has little ability to be a leader, it is unlikely that he will become the leader of the group;
- *Interest*
  $c_3 = 0.5$ - means that an intermediate student's interest in the subject
- *Ease*
  $c_4 = 0$ - means that learning is difficult for him, the remembering new knowledge is usually small.

## 3.2   Groups

Let us denote the set of all agents' subgroups by $G$: $G = \{G_1, G_2, ..., G_m\}$ which are

- nonempty:
$$\forall G_i \in G : G_i \neq \varnothing \quad for \ i = 1, 2, ..., m$$
- pairwise disjoint:
$$\forall G_i, G_j \in G : G_i \cap G_j = \varnothing \quad for \ i = 1, 2, ..., m; \ i \neq j$$
- the union of the subsets $G_i$ is equal to $G$:
$$\bigcup_{i=1}^{m} G_i = A$$

Below we define a set of postulates related to the division of agents into groups and the size of groups.

1. The number of groups should be not less than two,
2. Each group should have at least two agents:

$$m \in \{2, 3, ..., \lfloor \frac{n}{2} \rfloor\}$$

3. The size of groups should not differ by more than 1:

$$\forall G_i, G_j \in G : ||G_i| - |G_j|| \leq 1 \quad for \ i = 1, 2, ..., m;$$

4. The optimal number of groups is such that each group can contain the same number of agents:

$$m : \frac{n}{m} \in \mathbb{Z}^+$$

### 3.3   Agent's Knowledge

The second component of the description of the agent $A_i$ is the $K_i$ parameter representing its knowledge. The knowledge of each agent varies over time, and its level depends on the agent's characteristics $(C_i)$, the composition of the group and the communication process between agents. Let us denote the knowledge of the agent in a given iteration of the simulation of the ongoing $T$ iterations by $K_i$.

**Initial Knowledge** is the knowledge with which an agent starts communicating and cooperating in a group. With the current assumptions, only the second and third phases of learning are included in the model. These phases are represented by group working and the introductory knowledge is the knowledge at the beginning denote by $K_{i,0}$. The basic postulates regarding agents' introductory knowledge are:

1. Each agent can start working in a group with the knowledge:
   (a) equal to 0 (has no knowledge),
   (b) greater than 0 (has some knowledge before starting cooperation):

$$\forall A_i \in A(K_{i,0} \geq 0)$$

2. Introductory knowledge of individual agent may be different from others:

$$\forall A_i, A_j \in A(K_{i,0} = K_{j,0} \lor K_{i,0} \neq K_{j,0})$$

**The Increase in Knowledge** during group work in the preliminary model is described by several assumptions. In the preliminary model, we assume that the agents' knowledge is consistent, we assume there is no inconsistent knowledge. The process of forgetting knowledge by students is deliberately omitted, and thus the agent's knowledge never decreases. By $\Delta K_{i,t1,t2}$ we denote the increase in the agent's knowledge between time $t1$ and $t2$, such that $t1 < t2$:

$$\forall t1, t2 \in \{0, 1, ..., T\} : t1 < t2 \quad (\Delta K_{i,t1,t2} = K_{i,t2} - K_{i,t1})$$

1. The agent's knowledge increases or remains unchanged during working in a group:

$$\forall t1, t2 \in \{0, 1, ..., T\} : t1 < t2 \quad \forall A_i \in A(\Delta K_{i,t1,t2} \geq 0)$$

2. The increase of the agent's knowledge in subsequent units of time may vary:

$$\forall t_1, t_2, t_3 v \in \{0, 1, ..., T\} : t_1 < t_2 < t_3 \quad (\Delta K_{i,t1,t2} = \Delta K_{i,t2,t3} \lor \Delta K_{i,t1,t2} \neq \Delta K_{i,t2,t3})$$

3. At each moment of cooperation, the knowledge of individual agents may be different from others:

$$\forall t_1, t_2, t_3 \in \{0, 1, ..., T\} \quad \forall A_i, A_j \in A(K_{i,t} = K_{j,t} \lor K_{i,t} \neq K_{j,t})$$

4. The increase in knowledge of individual agents per unit of time may be different:

$$\forall t_1, t_2 \in \{0, 1, ..., T\} \quad \forall A_i, A_j \in A(\Delta K_{i,t1,t2} = \Delta K_{j,t1,t2} \lor \Delta K_{i,t1,t2} \neq \Delta K_{j,t1,t2})$$

**Terminal Knowledge** of the agent $A_i$ is the knowledge with which the agent ends cooperation and communication in the group. Under the current assumptions, this is the final knowledge that the agent acquired on a selected topic while working in a group for a specified period of time $T$ and we denote it as $K_{i,T}$.

1. Each agent can finish working in a group with the knowledge:
   (a) equal to introductory knowledge,
   (b) greater than introductory knowledge

$$\forall A_i \in A(K_{i,T} \geq Ki, 0)$$

2. Terminal knowledge of individual agent may be different from others:

$$\forall A_i, A_j \in A(K_{i,T} = K_{j,T} \ \lor \ K_{i,T} \neq K_{j,T})$$

### 3.4   Agent Characteristics

According to the researcher, a student's and teacher's features and behaviours might affect how they learn. As shown in the previous section, we represent the agent $i$ by a set $A_i = \{C_i, K_{i,t}\}$ where $C_i = (c_{1i}, c_{2i}, c_{3i}, c_{4i})$ is agent's characteristic and $K_{i,t}$ is agent's state of knowledge in time $t$. Let assume that the agent's initial knowledge depends on its characteristics:

$$K_{i,0} = k(c_{1i}, c_{2i}, c_{3i}, c_{4i})$$

where $k$ is some knowledge function. In that preliminary model, we propose several postulates for the range of agent's characteristics and their impact on the level of individual agent's knowledge. It is assumed that the same features provide the same initial knowledge:

$$(c_{1i} = c_{1j} \ \land \ c_{2i} = c_{2j} \ \land \ c_{3i} = c_{3j} \ \land \ c_{4i} = c_{4j}) \quad \Rightarrow \quad K_{i,0} = K_{j,0}$$

**Communicativeness** - communication skills ($c_1$). Some study indicates that communication quality significantly impacts group working [10]. It is assumed that the better communication inside the group, the easier the transfer and diffusion of knowledge, while the lack of communication prevents the transfer of knowledge. A few intuitive postulates regarding that feature of the agent were assumed:

1. The higher the level of the feature, the easier it is to gain knowledge:

$$(c_{1i} > c_{1j} \ \land \ c_{2i} = c_{2j} \ \land \ c_{3i} = c_{3j} \ \land \ c_{4i} = c_{4j}) \quad \Rightarrow \quad K_{i,0} > K_{j,0}$$

2. The lower the level of the feature the more difficult it is to gain knowledge:

$$(c_{1i} < c_{1j} \ \land \ c_{2i} = c_{2j} \ \land \ c_{3i} = c_{3j} \ \land \ c_{4i} = c_{4j}) \quad \Rightarrow \quad K_{i,0} < K_{j,0}$$

The communicativeness may also have impact on communication and knowledge transfer between agents. Informal assumptions have been proposed as follow:

1. The higher the level of the feature of agents:
   - the easier it may be to establish communication,
   - the communication may be more stable,
   - the communication may be faster,
   - the conflict resolving may be more effective.
2. The lower the level of the feature of agents:
   - the more difficult it may be to establish communication,
   - the communication may be unstable,
   - the communication may be slower,
   - the conflict resolving may be less effective.
3. The zero rate or very low rate of the feature in all agent makes it impossible to communicate in group.

**Leader Potential** - ability to lead the group in some tasks $(c_2)$. The leadership skills of individual agents primarily influence group cooperation and, to a lesser extent, affect the individual learning. Depending on the composition of the group and the number of agents with strong or weak leadership skills, communication within the group may be easier or more difficult [11]. This feature has no effect on the agent's initial knowledge:

1. $(c_{1i} = c_{1j} \land c_{2i} > c_{2j} \land c_{3i} = c_{3j} \land c_{4i} = c_{4j}) \quad \Rightarrow \quad K_{i,0} = K_{j,0}$
2. $(c_{1i} = c_{1j} \land c_{2i} < c_{2j} \land c_{3i} = c_{3j} \land c_{4i} = c_{4j}) \quad \Rightarrow \quad K_{i,0} = K_{j,0}$

The impact on communication and knowledge transfer between agents. Informal assumptions have been proposed as follow:

1. The more agents in the group have high leadership potential:
   - the more difficult it may be to establish communication,
   - the communication may be unstable,
   - the communication may be slower,
   - the conflict resolving may be less effective.
2. The lower the level of the feature of agents:
   - the easier it may be to establish communication,
   - it may be easy to maintain communication stable,
   - the communication may be quicker,
   - the conflict resolving may be more effective.
3. No one with have high leadership potential in group may not have significant impact on knowledge acquisition and communication.

**Interest** - interest in the topic ($c_3$). The results of the study [12] suggests that student's interest and enthusiasm for a subject might boost their academic performance. There appears to be a reciprocal link: students who are more interest in the subject gain knowledge quicker, and students who gain knowledge quicker also grow more interested in the subject. A few intuitive postulates regarding that feature of the agent were assumed:

1. The higher the level of the feature, the easier it is to gain knowledge:

$$(c_{1i} = c_{1j} \wedge c_{2i} = c_{2j} \wedge c_{3i} > c_{3j} \wedge c_{4i} = c_{4j}) \quad \Rightarrow \quad K_{i,0} > K_{j,0}$$

2. The lower the level of the feature the more difficult it is to gain knowledge:

$$(c_{1i} = c_{1j} \wedge c_{2i} = c_{2j} \wedge c_{3i} < c_{3j} \wedge c_{4i} = c_{4j}) \quad \Rightarrow \quad K_{i,0} < K_{j,0}$$

The impact on communication and knowledge transfer between agents. Informal assumptions have been proposed as follow:

1. The higher the level of the feature of agents:
   – the easier it may be to establish communication,
   – the communication may be more stable,
   – the communication may be faster,
   – the possibility of generating new knowledge is higher.
2. The lower the level of the feature of agents:
   – the more difficult it may be to establish communication,
   – the communication may be ineffective,
   – the communication may be slower,
   – the possibility of generating new knowledge is smaller.
3. All agents, or almost all agents high rate of the feature makes the possibility of generating new knowledge very high.
4. All agents, or almost all agents low or zero rate of the feature makes the possibility of generating new knowledge very low.

**Ease** - ease of learning ($c_4$). It is assumed that the individual's abilities and capabilities have impact on learning process and knowledge acquisition:

1. The higher the level of the feature, the easier it is to gain knowledge:

$$(c_{1i} = c_{1j} \wedge c_{2i} = c_{2j} \wedge c_{3i} = c_{3j} \wedge c_{4i} > c_{4j}) \quad \Rightarrow \quad K_{i,0} > K_{j,0}$$

2. The lower the level of the feature the more difficult it is to gain knowledge:

$$(c_{1i} = c_{1j} \wedge c_{2i} = c_{2j} \wedge c_{3i} = c_{3j} \wedge c_{4i} < c_{4j}) \quad \Rightarrow \quad K_{i,0} < K_{j,0}$$

The impact on communication and knowledge transfer between agents. Informal assumptions have been proposed as follow:

1. The more agents have high ease of learning:
   – the communication may be faster,

– the possibility of generating new knowledge is higher.
2. The less agents have high ease of learning:
   – the communication may be ineffective,
   – the possibility of generating new knowledge is smaller.
3. All agents, or almost all agents high rate of the feature makes the possibility of generating new knowledge very high.
4. All agents, or almost all agents low or zero rate of the feature makes the possibility of generating new knowledge very low.

## 4    Example

Let's analyze a simple example. Suppose we have a group of five agents with different characteristics: $G = \{A_1, A_2, A_3, A_4, A_5\}$. The first one, described as in *Example 3.1-1* in Sect. 3.1. The others are described as follows:
$A_2 = \{(0.8,\ 1,\ 0.2,\ 1), K_2\};\ A_3 = \{(0.7,\ 1,\ 0.8,\ 0.1), K_3\};$
$A_4 = \{(0.7,\ 0.3,\ 0.8,\ 0.5), K_4\};\ A_5 = \{(0.4,\ 0,\ 0.6,\ 1), K_5\}.$
For the purposes of the analyzed example, let's assume that $K_{i,0} - k(c_1, c_2, c_3, c_4)$ such that $k(c_1, c_2, c_3, c_4) = c_1 + c_3 + c_4$ then:

$$K_{1,0} = 1.5;\quad K_{2,0} = 2.0;\quad K_{3,0} = 1.6;\quad K_{4,0} = 2.0;\quad K_{5,0} = 2.0$$

Analyzing the attributes of individual agents, we can see that all the postulates from the previous section regarding the agent's initial knowledge have been satisfied. Agents with one different level of feature have proportionally different knowledge and two agents with different characteristics have the same knowledge level.

Taking into account the postulate regarding the number of groups and the number of agents in a group, we see that agents can only be divided into two groups $G = \{G_1, G_2\}$ such that $(|G_1| = 2 \land |G_2| = 3) \lor (|G_1| = 3 \land |G_2| = 2)$. Let's consider an example that $G_1 = \{A_1, A_2, A_3\}$, $G_2 = \{A_4, A_5\}$ . It's obvious that all postulates regarding groups are satisfy. $G_1, G_2$ are: nonempty, pairwise disjoint and the union of the sets $G_1$ and $G_2$ is equal to $G$.

Considering the presumptions about how agents' characteristics affect group work, we notice that this divide results in a group $G_1$ in which:

1. all agents have high communicativeness:
   (a) it may be easy to establish communication,
   (b) communication may be faster,
   (c) conflict resolving may be more effective;
2. two agents have high leadership potential:
   (a) it may be difficult it may be to establish communication,
   (b) the communication may be unstable,
   (c) conflict resolving may be less effective;
3. all agents has some interest:
   (a) it is hard to determine how difficult it will be to establish communication,
   (b) it is hard to determine how stable the communication will be,

    (c) it is hard to determine if generating new knowledge is possible;
4. all agents has some interest:
    (a) it may be on average difficult to establish communication,
    (b) it may be on average difficult to determine how stable the communication will be,
    (c) it may be possible to generating small amount of new knowledge;
5. all agents ease of learning if different, one has very high, one low and one equal to 0:
    (a) it is hard to determine how difficult it will be to establish communication,
    (b) it is hard to determine how stable the communication will be,
    (c) it is hard to determine if generating new knowledge is possible.

Analyzing the above postulates, it is easy to notice that the quality of communication and knowledge growth in the entire group is easy to determine if all agents have a similar level of all characteristics. If the features are different, it is difficult to decide what the quality of communication will be. Some characteristics are mutually exclusive: 1a–c and 2a–c. In the future, it is necessary to determine the significance of the impact of individual features on the quality of communication or to propose functions that resolve the conflict and allow for mitigating the selected feature of the agent for the duration of group work.

## 5   Conclusion

In that paper we presented a proposal to formalize the basic features and a set of postulates regarding a multi-agent model of group learning in a small heterogeneous group, in which the features of agents are inspired by the features of students and teachers. The work is a preliminary definition and formalization that allows us to construct initial conditions, postulates and assumptions for the simulation of the complex process of knowledge diffusion in small group using a multi-agent systems. In future works, we plan to formally define knowledge representation, the method of communication between agents, and propose methods for resolution of the inconsistent knowledge. Given that the communication has the ability to affect and direct any learning process, especially within a group, some research on its impact on knowledge diffusion and acquisition is planned. The influence of the characteristics of individual agents and the composition of characteristics in the group will be taken into account when describing and simulating communication within the group.

**Acknowledgment.** The work was supported by the project Minigrants for doctoral students of the Wroclaw University of Science and Technology.

# References

1. Dale, E.: AudioVisual Methods in Teaching. Dryden Press, New York (1946)
2. Bonwell, C.C., Eison, J.A.: Active learning: creating excitement in the classroom. 1991 ASHE-ERIC higher education reports. ERIC Clearinghouse on Higher Education, The George Washington University, One Dupont Circle, Suite 630, Washington, DC 20036-1183 (1991)
3. Prince, M.J.: Does active learning work? A review of the research. J. Eng. Educ. **93**(3), 223–231 (2004)
4. Slavin, R.: Cooperative Learning: Theory, Research and Practice. Allyn and Bacon, Boston (1995)
5. Dolata, R.: Czy segregacja uczniow ze wzgledu na uprzednie osiagniecia szkolne zwieksza efektywnosc nauczania mierzona metoda EWD?. In: XIV Konferencja Dia-gnostyki, Opole (2008)
6. Wędrychowicz, B., Maleszka, M.: Agent based model of elementary school group learning - a case study. In: Nguyen, N.T., et al. (eds.) ICCCI 2023. LNCS, vol. 14162, pp. 56–67. Springer, Cham (2023). https://doi.org/10.1007/978-3-031-41456-5_5
7. Berliner, D.C., Rosenshine, B.: The acquisition of knowledge in the classroom. In: Schooling and the Acquisition of Knowledge, pp. 375–396. Routledge (2017)
8. Houser, M.L., Frymier, A.B.: The role of student characteristics and teacher behaviors in students' learner empowerment. Commun. Educ. **58**(1), 35–53 (2009). https://doi.org/10.1080/03634520802237383
9. Shuell, T.J.: Phases of meaningful learning. Rev. Educ. Res. **60**(4), 531–547 (1990)
10. Lam, C.: The role of communication and cohesion in reducing social loafing in group projects. Bus. Prof. Commun. Q. **78**(4), 454–75 (2015)
11. Wu, J.B.: Consequences of differentiated leadership in groups. Acad. Manag. J. **53**, 90–106 (2010)
12. Hasan, R., Lukitasari, M., Darmayani, O., Santoso, S.: The variation pattern of cooperative learning models implementation to increase the students creative thinking and learning motivation. J. Phys. Conf. Ser. **1157**(2), 022075 (2019)

# Music Genre Classification Using Hybrid Committees and Voting Mechanisms

Daniel Kostrzewa$^{(\boxtimes)}$ (iD), Piotr Berezka, and Robert Brzeski (iD)

Department of Applied Informatics, Silesian University of Technology, Gliwice,
Poland
{daniel.kostrzewa,robert.brzeski}@polsl.pl

**Abstract.** In this paper, using the example of music genre classification, seven voting methods for classifier committees are presented and used. This work focuses on improving the results of classification quality mainly by using different voting methods in different classifier committees. Examination of these methods was carried out on four classifier committees created from a total of 31 individual models. Classification quality is crucial in many machine learning applications. Therefore, methods for improving this quality may be of great practical value. The use of classifier committees and the proposed voting methods significantly improved the results comparing to base models.

**Keywords:** Voting system · Music Information Retrieval · Audio Features · Classifier Committee · MFCC · FMA · Convolutional Neural Network

## 1 Introduction

In the age of digital streaming platforms and massive music libraries, effective music classification is crucial for creating personalized playlists and recommendations [23]. Machine learning algorithms based on neural networks and deep learning have significantly improved the accuracy and efficiency of music classification systems. These mechanisms analyze large data sets, identifying patterns and similarities in songs, enabling the automation of genre labeling and recommendation processes.

This work explores the use of hybrid committees in music classification, where different algorithms are combined to create a more comprehensive structure. Using the strengths of individual algorithms, such as decision trees, neural networks, support vector machines, and deep convolutional networks, hybrid ensembles enable more precise categorization of musical genres. The work examines

This work was supported by the Rector's pro-quality grant, Silesian University of Technology, Gliwice, Poland, grant number 02/100/RGJ24/0034, and by the Statutory Research funds of Department of Applied Informatics, Silesian University of Technology, Gliwice, Poland, grant number 02/100/BK_24/0035.

various committee voting systems and proposes new, different approaches to this problem.

In a hybrid committee, each model in the ensemble is trained independently, often using different subsets of the data or using different data engineering techniques. Once individual models are trained, their predictions are combined using techniques such as comparison, voting, or weighted averaging, in which the contribution of each model is weighted based on its performance or expertise in specific areas of the feature space. The purpose of creating a hybrid committee is to increase prediction accuracy. By combining models that excel at different aspects of a problem, hybrid committees can provide more accurate and reliable predictions compared to a single model. They are particularly useful in complex tasks in which algorithms capture various aspects of underlying patterns in the data. Hybrid committees are used in applications where achieving high prediction accuracy is crucial and the data is diverse and multi-faceted.

Hybrid ensembles in situations where individual models may fail due to noise or outliers in the input, the committee approach ensures that errors made by one model are compensated by correct predictions from the others, leading to a more stable and reliable final result.

The remainder of the paper is as follows. Section 1 introduces the undertaken issue, provides an overview of the classifiers committee and related works, and highlights the contribution. Section 2 describes the used database and built system. Section 3 provides detailed information on used voting systems. The conducted experiments and received results are presented in Sect. 4. The paper is concluded in Sect. 5, where the future work is also outlined.

## 1.1   Related Work

The rising popularity and practical significance of machine learning methodologies have had a profound impact on the domain of acoustic signal analysis and processing. There are multiple research areas that are affected by those methods, like Indian musical type classification [1], music genre classification utilizing convolutional neural networks [3], DenseNet and data augmentation [11], residual networks [26], attributes relevance exploration for music recommendation systems [8], and architecture with attention mechanism with usage of different features for music recommendation [22] to name a few. Convolutional Neural Networks (CNNs), in particular, have emerged as highly effective tools in the realm of audio processing [5,20,25,28].

The domain of music classification [12,18,19,21] utilizing machine learning techniques has witnessed a substantial surge in research interest over the past decade. This increased attention can be attributed to the expanding availability of digital music data, advancements in machine learning algorithms, and the broadening applications of automated music analysis across various industries.

Within the scientific literature describing hybrid ensembles, many papers are available [4,27]. In [6], the authors prepared a decent review of ensemble mechanisms in deep learning. They highlighted different classifications of ensemble learning, including a set of fusion strategies, which are closely related to the topic of this work. The voting ensemble (both weighted and unweighted) can be constructed as a committee of classical classifiers [2,14] or a set based on deep learning classifiers [15]. The ultimate classification result is determined through the fusion mechanism (voting method) applied to individual components of the committee [17]. In his work titled "A Theoretical Analysis of Why Hybrid Ensembles Work", auhor delves into the concept of hybrid ensembles in machine learning [7], exploring how combinations of diverse classifiers contribute to performance improvement.

### 1.2   Contribution

The contribution of the paper mainly focuses on the creation and examination of different voting systems of different hybrid committees for musical genre classification. For this purpose, seven different voting systems with four different committees using a total of 31 independent classifiers were used.

## 2   Outline of the System

### 2.1   Used Dataset

In the field of Music Information Retrieval (MIR), many publicly available datasets have emerged over the years. Currently, the most important repositories containing data describing musical works are, for example, MSD - Million Song Dataset, FMA - Free Music Archive or GTZAN Dataset. These sets differ from each other primarily in the size and structure of the data available.

The FMA dataset was used in the research environment created for this work [13]. Free Music Archive, created by Michaël Defferrard, Kirell Benzi, Pierre Vandergheynst and Xavier Bresson, is an open and easily accessible collection of both numerical data of musical songs and their audio tracks.

Moreover, to ensure the repeatability of research using FMA, the entire set is initially divided into training, validation and test subsets in the following proportions: 80% - training set, 10% - validation set and 10% - test set.

In addition to audio tracks, FMA also contains computed numerical values that describe the characteristics of music tracks. Each item present in the set corresponds to a total of 350 values from nine different features (Table 1). These are statistical values of the described features, such as minimum, maximum, mean, median, standard deviation, skewness, and kurtosis. This data was used to train classical classifiers (Sect. 2.2).

For the training of deep classifiers (Sect. 2.2) the graphic data in the form of Mel-spectrograms where used. Mel-spectrogram is a type of spectrogram created from the soundtrack of a musical piece. Mel-spectrograms used for the training of deep classifiers are widely used in audio processing and machine learning due to their ability to offer a perceptually meaningful representation of sound. Unlike traditional spectrograms, which display the amplitude of different frequencies over time, Mel-spectrograms are based on the Mel scale, which is close to human auditory perception.

**Table 1.** Values available for each song contained in FMA.

| Feature | Value quantity |
| --- | --- |
| Chroma | 84 |
| Tonnetz | 42 |
| MFCC | 140 |
| Spectral centroid | 7 |
| Spectral bandwidth | 7 |
| Spectral contrast | 49 |
| Spectral rolloff | 7 |
| Mean square energy | 7 |
| Zero-crossing rate | 7 |
| Total | 350 |

Mel-spectrograms are created on the basis of a method for mapping a parametric acoustic signal - the Mel-Frequency Cepstral Coefficients (MFCC). They are often used to determine the genre and timbre of a musical piece [24]. The effectiveness of MFCC has also been confirmed in terms of music recommendations [10, 18].

## 2.2   Used Classifiers

All of the 31 base models used in the research described in this work are classic classifiers or based on deep networks. Despite differences in architecture, operation and training, classical and deep classifiers with very similar results on the same test set when trying to categorize on the same test data can often give different class predictions, which favors the results obtained by the hybrid committee.

The following 18 classical classifiers trained on numerical data were used in the research:

- Logistic Regression (LR),
- K Nearest Neighbors (kNN),
- Support Vector Machine (SVCrbf, SVCsigmoid and SVCpoly1),
- Linear Support Vector Machine (linSVC1 and linSVC2),
- Decision Tree (DT),
- Random Forest (RF),
- Algorithm AdaBoost (AdaBoost),
- Multilayer Perceptron (MLP1 - MLP6),
- Naive Bayes classifier (NB).
- Quadratic Discriminant Analysis (QDA)

For the purposes of this work, 13 different deep classifiers trained on Mel-spectrograms belonging to two types of convolutional neural networks were implemented:

– Convolutional Neural Networks (CNN1, CNN2, CNN3),
– Convolutional Recurrent Neural Networks (CRNN1 - CRNN10).

Four statistics of the quality of the classification were determined, which allowed to evaluate the results of the base classifiers according to the following quality metrics:

– Accuracy,
– Precision,
– Sensitivity,
– F1-score.

With the use of the mentioned eighteen classical classifiers and thirteen deep classifiers, which gives a total of 31 different models the 4 hybrid committees research environment where implemented (Sect. 4.2). Then, 7 voting systems (Sect. 3) were tested using these 4 hybrid committees.

The test software was created in Python using libraries such as Keras-Tensorflow, Librosa and Scikit-Learn.

## 3    Voting Systems

For the purposes of this work, the following voting systems were implemented in the research environment:

– **Simple Solution, SS** - each individual base classifier assigns a probability to each of the possible classes. Then the probabilities of all classifiers are summed, the class that has the highest sum is selected as the winner of the committee vote [9].
– **Majority Voting, MV** - the classifier votes only for one class, the one with the highest probability. The class with the most votes wins [9].
– **K Approval Voting, KAV** - each base model selects the three classes it considers most likely. In the next step, the classifier assigns points. The most likely class will receive three points, the next will receive two points, and the least likely of the three will receive one point. The class with the highest total points is the result of the committee's vote [9].
– **Highest Probability Voting, HPV** - the input data is classified using one model - the one with the highest confidence.
– **Threshold Based Voting, TBV** - an acceptance threshold is introduced into the probability lists. The threshold can take on different values and should be tuned appropriately to the problem being solved. The single prediction of each base classifier is checked against the threshold and if it is lower than the threshold, it is completely rejected and is not taken into account any further. The remaining probabilities are averaged within the class. The class with the highest average probability wins the vote. This approach ensures that only votes with appropriate certainty, determined by the acceptance threshold, are considered.

– **Rank Based Voting, RBV** - based on probability, classifiers assign a rank to each of the possible classes. The higher the probability, the lower the rank. If two classes have identical probabilities assigned, they receive the same rank. Then the ranks of each class are averaged. The class with the lowest average rank becomes the committee's classification. In the event that two or more classes have the lowest value, each average rank is multiplied by the lowest rank assigned to that class. If the committee is still unable to determine one smallest value, the existing ranks are multiplied by the normalized probability determined as $p_z = 1 - p_n$, where $p_n$ - the highest probability of the considered class. If there is no winner of the vote, the committee's result is selected at random.

For better explain how this voting system works, an example is provided. Rank voting for example, for eight classes K1 to K8 and four classifiers would proceed as follows.

Base classifiers assign a rank to each class:

|               | $K1$ | $K2$ | $K3$ | $K4$ | $K5$ | $K6$ | $K7$ | $K8$ |
|---------------|------|------|------|------|------|------|------|------|
| $Classifier\,1:$ | 1 | 6 | 5 | 7 | 3 | 8 | 2 | 4 |
| $Classifier\,2:$ | 7 | 1 | 5 | 6 | 4 | 8 | 3 | 2 |
| $Classifier\,3:$ | 1 | 7 | 3 | 5 | 8 | 4 | 6 | 2 |
| $Classifier\,4:$ | 1 | 8 | 3 | 4 | 5 | 6 | 7 | 2 |

Average ranks are determined:

| $K1$ | $K2$ | $K3$ | $K4$ | $K5$ | $K6$ | $K7$ | $K8$ |
|------|------|------|------|------|------|------|------|
| 2, 5 | 5, 5 | 4, 0 | 5, 5 | 5, 0 | 6, 5 | 4, 5 | 2, 5 |

Both the first and last classes have the same average rank. To eliminate conflict, the previously obtained ranks are multiplied by the set of the best ranks of each class:

| $K1$ | $K2$ | $K3$ | $K4$ | $K5$ | $K6$ | $K7$ | $K8$ |
|------|------|------|------|------|------|------|------|
| 1 | 1 | 3 | 4 | 3 | 4 | 2 | 2 |

receiving:

| $K1$ | $K2$ | $K3$ | $K4$ | $K5$ | $K6$ | $K7$ | $K8$ |
|------|------|------|------|------|------|------|------|
| 2, 5 | 5, 5 | 12, 0 | 22, 0 | 15, 0 | 26, 0 | 9, 0 | 5, 0 |

This allows us to determine the final classification - the first class wins with a score of 2,5.

- **Diversity Driven Voting, DDV** - a diversity score is determined for each of the classifiers, which is based on the Jensen-Shannon divergence [16]. On its basis, weights are calculated to select the committee's final predictions. For better explain how this voting system works, an voting example is provided below. Diversity Driven Voting, for example, for eight classes K1 to K8 and four classifiers would proceed as follows:

Determining diversity scores for base models:

$$
\begin{aligned}
Classifier\,1 \quad & W_1 = 0.095 \\
Classifier\,2 \quad & W_2 = 0.180 \\
Classifier\,3 \quad & W_3 = 0.090 \\
Classifier\,4 \quad & W_4 = 0.090
\end{aligned}
$$

Sum of Diversity Scores - Total Diversity Ratio:

$$
w_{cr} = \sum_{i=1}^{n} W_i = 0.455
$$

where $n$ - the number of classifiers in the committee. Final classifier weights:

$$
\begin{aligned}
Classifier\,1 \quad & W_{o1} = W_1/w_{cr} = 0.209 \\
Classifier\,2 \quad & W_{o2} = W_2/w_{cr} = 0.396 \\
Classifier\,3 \quad & W_{o3} = W_3/w_{cr} = 0.198 \\
Classifier\,4 \quad & W_{o4} = W_4/w_{cr} = 0.198
\end{aligned}
$$

The probabilities determined by the classifiers are multiplied by the appropriate final weight. The obtained results are summed for each class:

| $K1$ | $K2$ | $K3$ | $K4$ | $K5$ | $K6$ | $K7$ | $K8$ |
|------|------|------|------|------|------|------|------|
| 0.334 | 0.200 | 0.078 | 0.064 | 0.072 | 0.050 | 0.078 | 0.124 |

The highest value is selected from the obtained weighted probabilities - the first class wins the vote.

First 3 of voting systems (SS, MV, KAV) ware taken from other works. The proposed solutions are last 4 voting systems (HPV, TBV, RBV, DDV).

# 4    Experimental Results

In this section received results of each individual classifier (base models) as well as received results of created from these models 4 voting committees are presented.

## 4.1    Classifiers

The obtained classification quality values for all of 31 individual classifiers (including eighteen classical classifiers and thirteen deep classifiers) are summarized in Tables 2 and 3. Those models that achieved an accuracy level equal to or higher than 40% are marked in bold.

Among classical classifiers, the best results were achieved by models based on support vector machines (SVC) and traditional neural networks - multi-layer perceptrons (MLP). The highest accuracy value among convolutional neural networks was achieved by networks containing recurrent layers (CRNN). The highest result (0.471) of all base classifiers was given by SVCrbf.

**Table 2.** Statistics of implemented classical classifiers.

| Classifier | Accuracy | Precision | Sensitivity | F1 |
|---|---|---|---|---|
| **LR** | **0.432** | **0.434** | **0.432** | **0.432** |
| kNN | 0.336 | 0.392 | 0.336 | 0.299 |
| **SVCrbf** | **0.471** | **0.473** | **0.471** | **0.469** |
| **SVCsigmoid** | **0.431** | **0.436** | **0.431** | **0.43** |
| **SVCpoly1** | **0.470** | **0.468** | **0.470** | **0.467** |
| **linSVC1** | **0.415** | **0.425** | **0.415** | **0.415** |
| **linSVC2** | **0.419** | **0.414** | **0.419** | **0.414** |
| DT | 0.330 | 0.299 | 0.330 | 0.298 |
| RF | 0.325 | 0.306 | 0.325 | 0.293 |
| AdaBoost | 0.364 | 0.368 | 0.364 | 0.338 |
| **MLP1** | **0.451** | **0.445** | **0.451** | **0.447** |
| **MLP2** | **0.452** | **0.451** | **0.452** | **0.451** |
| **MLP3** | **0.452** | **0.454** | **0.452** | **0.453** |
| **MLP4** | **0.450** | **0.441** | **0.450** | **0.444** |
| **MLP5** | **0.446** | **0.444** | **0.446** | **0.444** |
| **MLP6** | **0.456** | **0.454** | **0.456** | **0.454** |
| **NB** | **0.408** | **0.405** | **0.408** | **0.393** |
| **QDA** | **0.402** | **0.431** | **0.402** | **0.405** |

Table 3. Statistics of implemented deep classifiers.

| Classifier | Accuracy | Precision | Sensitivity | F1 |
|---|---|---|---|---|
| CNN1 | 0.394 | 0.389 | 0.394 | 0.388 |
| **CNN2** | **0.409** | **0.400** | **0.409** | **0.402** |
| **CNN3** | **0.404** | **0.404** | **0.404** | **0.402** |
| **CRNN1** | **0.409** | **0.453** | **0.409** | **0.414** |
| **CRNN2** | **0.415** | **0.429** | **0.415** | **0.413** |
| CRNN3 | 0.361 | 0.369 | 0.361 | 0.339 |
| **CRNN4** | **0.401** | **0.414** | **0.401** | **0.395** |
| **CRNN5** | **0.405** | **0.406** | **0.405** | **0.404** |
| **CRNN6** | **0.410** | **0.412** | **0.410** | **0.403** |
| **CRNN7** | **0.424** | **0.442** | **0.424** | **0.424** |
| **CRNN8** | **0.410** | **0.414** | **0.410** | **0.395** |
| **CRNN9** | **0.424** | **0.414** | **0.424** | **0.411** |
| **CRNN10** | **0.401** | **0.419** | **0.401** | **0.401** |

## 4.2  Classifiers Committee

Four different hybrid committees were created to conduct the research:

The first one (HKW) consists of all 31 classifiers implemented in the research environment.

The second committee (HK40) was formed from 25 models with an accuracy of at least 0.40 (which are: LR, SVCrbf, SVCsigmoid, SVCpoly1, linSVC1, linSVC2, MLP1, MLP12, MLP3, MLP4, MLP5, MLP6, NB, QDA, CNN2, CNN3, CRNN1, CRNN2, CRNN4, CRNN5, CRNN6, CRNN7, CRNN8, CRNN9, CRNN10).

The third committee (HK42) contains 12 classifiers with an accuracy score greater than or equal to 0.42 (which are: LR, SVCrbf, SVCsigmoid, SVCpoly1, MLP1, MLP12, MLP3, MLP4, MLP5, MLP6, CRNN7, CRNN9).

The last committee created for the purposes of research is the HK44 committee, composed of only 8 models with a minimum accuracy of 0.44 (which are: SVCrbf, SVCpoly1, MLP1, MLP12, MLP3, MLP4, MLP5, MLP6).

All of the created committees were examined for quality metrics for each of the available voting systems.

Table 4 summarizes the accuracy metrics for all hybrid committees by voting systems - simple solution (SS), majority voting (MV), k approval voting (KAV), highest probability voting (HPV), threshold based voting (TBV), rank based voting (RBV) and diversity driven voting (DDV). For the HKW, HK40 and HK42 committees, the best results were achieved by a rank based voting system (RBV). The HK44 committee recorded the best accuracy result for the diversity driven voting system (DDV).

**Table 4.** Summary of committees' Accuracy results for all votes.

| Vote | Committee HKW | Comm. HK40 | Comm. HK42 | Comm. HK44 | average |
|---|---|---|---|---|---|
| SS | 0.519 | 0.520 | 0.496 | 0.478 | 0.503 |
| MV | 0.530 | 0.526 | 0.496 | 0.479 | 0.508 |
| KAV | 0.524 | 0.529 | 0.500 | 0.478 | 0.508 |
| HPV | 0.415 | 0.416 | 0.474 | 0.463 | 0.442 |
| TBV | 0.520 | 0.520 | 0.495 | 0.474 | 0.502 |
| RBV | *0.536* | **0.530** | **0.504** | 0.475 | **0.511** |
| DDV | 0.518 | 0.525 | 0.495 | **0.481** | 0.505 |
| average | 0.509 | 0.509 | 0.494 | 0.497 | 0.497 |

Table 5 shows the F1 metric results for all committees according to the voting systems used. The HKW, HK40 and HK42 committees recorded the best results for rank based voting. Committee HK44 achieved the highest F1 score for diversity driven voting.

**Table 5.** Summary of committees' F1 results for all votes.

| Vote | Committee HKW | Comm. HK40 | Comm. HK42 | Comm. HK44 | average |
|---|---|---|---|---|---|
| SS | 0.515 | 0.417 | 0.493 | 0.476 | 0.475 |
| MV | 0.527 | 0.524 | 0.494 | 0.479 | 0.506 |
| KAV | 0.521 | 0.528 | 0.497 | 0.477 | 0.506 |
| HPV | 0.411 | 0.413 | 0.469 | 0.458 | 0.438 |
| TBV | 0.516 | 0.516 | 0.491 | 0.472 | 0.499 |
| RBV | *0.537* | **0.531** | **0.502** | 0.473 | **0.511** |
| DDV | 0.513 | 0.521 | 0.493 | **0.480** | 0.502 |
| average | 0.506 | 0.493 | 0.491 | 0.474 | 0.491 |

Simple solution, k approval voting, threshold based voting and diversity driven voting recorded the best results for a committee composed of classifiers with an accuracy of at least 0.4 (HK40). For these systems, as the committees were limited to a minimum accuracy of 0.42 (HK42) and then 0.44 (HK44), they achieved increasingly lower accuracy values. These results were caused by reducing the number of base models in the ensembles. Analyzing the obtained data, it can be seen that the results of the SS, KAV, TBV and DDV voting systems depend not only on the average accuracy of the classifiers, but also rely heavily on the number of different models included in the committee.

Systems such as majority voting (MV) and rank based voting (RBV) have certain noticeable characteristics. Using this scheme, the committee's output was proportional to the number of base models it included - the best accuracy was achieved by the HKW committee. This reveals the characteristics of these systems, which are low sensitivity to accuracy and benefiting from a larger number of base classifiers.

Of all the tested hybrid committees and voting types, the highest accuracy and F1-score metrics result was given by the HKW committee composed of all classifiers voting according to RBV with a result of 0.536 and 0.537, respectively. In this way, we achieve an accuracy of over 13% better than the best base classifier (0.471).

Rank based voting provides an evaluation of each prediction made by all models, favoring internal rankings of committee members and indirectly considering probabilities. These features ensure high accuracy metric scores.

# 5    Conclusions and Future Work

The aim of the work was to create and examine 4 hybrid committees in terms of known and proposed voting systems. The implemented 31 base models, which are classic classifiers and deep convolutional networks, allowed the creation of sets of classifiers - voting committees. Four different committees were tested for each voting system. Voting patterns taken from other works are simple solution (SS), majority voting (MV), and k approval voting (KAV). The proposed solutions are highest probability voting (HPV), threshold based voting (TBV), rank based voting (RBV) and diversity driven voting (DDV).

The committees studied voted through seven different methods. One of the possible directions of development of the work may be the creation of further voting systems. The voting system gives the committee its unique character and greatly influences its accuracy.

The proposed hybrid committee solutions assume the determination of the final ensemble prediction only on the basis of the votes of individual models. This solution could be extended to include external data to assist the committee in classification. At the time of voting, in addition to the prediction matrix, the ensemble may have specific data at its disposal to improve it. Such data may include, for example, weights determined based on the previously known results of each classifier. Weights would be calculated for each class so that during classification, a model that previously had high precision for a certain class would be more important during voting. Moreover, it is possible to use evolutionary algorithms to improve the efficiency of committees. This approach assumes tuning the weights of the examined hybrid committee by applying appropriate algorithms to it. This should result in an increase in the hybrid committees' quality metrics and performance.

The presented methods for improving the quality of classification are so universal that they can be used beyond music genre classification in other systems. The application may be limited primarily by the increased demand for computing power and the increased implementation time of the learning process.

# References

1. Aswale, S.P., Shrivastava, P.C., Bhagat, R., Joshi, V.B., Shende, S.M.: Multilingual Indian musical type classification. In: Nagaria, R.K., Tripathi, V.S., Zamarreno, C.R., Prajapati, Y.K. (eds.) VCAS 2022. LNEE, vol. 1024, pp. 419–430. Springer, Singapore (2022). https://doi.org/10.1007/978-981-99-0973-5_31
2. Bahuleyan, H.: Music genre classification using machine learning techniques. arXiv preprint arXiv:1804.01149 (2018)
3. Choudhury, N., Deka, D., Sarmah, S., Sarma, P.: Music genre classification using convolutional neural network. In: 2023 4th International Conference on Computing and Communication Systems (I3CS), pp. 1–5. IEEE (2023)
4. Costa, Y.M., Oliveira, L.S., Silla, C.N., Jr.: An evaluation of convolutional neural networks for music classification using spectrograms. Appl. Soft Comput. **52**, 28–38 (2017)
5. Elbir, A., Aydin, N.: Music genre classification and music recommendation by using deep learning. Electron. Lett. **56**(12), 627–629 (2020)
6. Ganaie, M.A., Hu, M., Malik, A., Tanveer, M., Suganthan, P.: Ensemble deep learning: a review. Eng. Appl. Artif. Intell. **115**, 105151 (2022)
7. Hsu, K.W.: A theoretical analysis of why hybrid ensembles work. Comput. Intell. Neurosci. **2017**, 1–12 (2017)
8. Kostrzewa, D., Chrobak, J., Brzeski, R.: Attributes relevance in content-based music recommendation system. Appl. Sci. **14**(2), 855 (2024)
9. Kostrzewa, D., Ciszynski, M., Brzeski, R.: Evolvable hybrid ensembles for musical genre classification. In: Proceedings of the Genetic and Evolutionary Computation Conference Companion, pp. 252–255 (2022)
10. Kostrzewa, D., Mazur, W., Brzeski, R.: Wide ensembles of neural networks in music genre classification. In: Groen, D., de Mulatier, C., Paszynski, M., Krzhizhanovskaya, V.V., Dongarra, J.J., Sloot, P.M.A. (eds.) ICCS 2022, Part II. LNCS, vol. 13351, pp. 64–71. Springer, Cham (2022). https://doi.org/10.1007/978-3-031-08754-7_9
11. Le Thuy, D.T., Van Loan, T., Thanh, C.B., Cuong, N.H.: Music genre classification using densenet and data augmentation. Comput. Syst. Sci. Eng. **47**(1), 657–674 (2023)
12. Lukashevich, H., Grollmisch, S., Abeßer, J.: Quantifying uncertainty in music genre classification. In: Proceedings of The 49th Annual Conference on Acoustics DAGA, Hamburg, Germany, pp. 1378–1381 (2023)
13. Michaël, D., Kirell, B., Pierre, V., Xavier, B.: FMA: a dataset for music analysis, 5 September 2017
14. Nanni, L., Costa, Y.M., Lumini, A., Kim, M.Y., Baek, S.R.: Combining visual and acoustic features for music genre classification. Expert Syst. Appl. **45**, 108–117 (2016)
15. Nanni, L., Maguolo, G., Brahnam, S., Paci, M.: An ensemble of convolutional neural networks for audio classification. Appl. Sci. **11**(13), 5796 (2021)
16. Nielsen, F.: Jensen-Shannon divergence and diversity index: origins and some extensions, Sony Computer Science Laboratories Inc., Tokyo, Japan, April 2021
17. Onan, A., Korukoğlu, S., Bulut, H.: A multiobjective weighted voting ensemble classifier based on differential evolution algorithm for text sentiment classification. Expert Syst. Appl. **62**, 1–16 (2016)
18. Van den Oord, A., Dieleman, S., Schrauwen, B.: Deep content-based music recommendation. Adv. Neural Inf. Process. Syst. **26** (2013)

19. Parmezan, A.R.S., Silva, D.F., Batista, G.E.: A combination of local approaches for hierarchical music genre classification. In: ISMIR, pp. 740–747 (2020)
20. Pons, J., Serra, X.: Randomly weighted CNNs for (music) audio classification. In: ICASSP 2019-2019 IEEE International Conference on Acoustics, Speech and Signal Processing (ICASSP), pp. 336–340. IEEE (2019)
21. Ramírez, J., Flores, M.J.: Machine learning for music genre: multifaceted review and experimentation with audioset. J. Intell. Inf. Syst. **55**(3), 469–499 (2020)
22. Sachdeva, N., Gupta, K., Pudi, V.: Attentive neural architecture incorporating song features for music recommendation. In: Proceedings of the 12th ACM Conference on Recommender Systems, pp. 417–421 (2018)
23. Soares, A.C., Marco, C., Rafael, G.: Predicting music popularity on streaming platforms. In: Simpósio Brasileiro de Computação Musical (2019)
24. Tzanetakis, G., Cook, P.: Musical genre classification of audio signals. IEEE Trans. Speech Audio Process. **10**(5), 293–302 (2002). https://doi.org/10.1109/TSA.2002.800560
25. Vall, A., Dorfer, M., Eghbal-Zadeh, H., Schedl, M., Burjorjee, K., Widmer, G.: Feature-combination hybrid recommender systems for automated music playlist continuation. User Model. User-Adap. Inter. **29**, 527–572 (2019)
26. Xu, Z., et al.: Research on music genre classification based on residual network. In: Chenggang, Y., Honggang, W., Yun, L. (eds.) MobiMedia 2022. LNCISD, vol. 451, pp. 209–223. Springer, Cham (2022). https://doi.org/10.1007/978-3-031-23902-1_16
27. Zhang, L., Lim, C.P., Yu, Y., Jiang, M.: Sound classification using evolving ensemble models and particle swarm optimization. Appl. Soft Comput. **116**, 108322 (2022)
28. Zhang, Y., et al.: Music recommendation system and recommendation model based on convolutional neural network. Mobile Inf. Syst. **2022** (2022)

# Towards Practical Large Scale Traffic Model of Electric Transportation

Marcin Maleszka[1,2,3(✉)] 

[1] Wroclaw University of Science and Technology, st. Wyspianskiego 27,
50-370 Wroclaw, Poland
marcin.maleszka@pwr.edu.pl
[2] School of Computer Science and Engineering, International University,
Ho Chi Minh City, Vietnam
[3] Vietnam National University, Ho Chi Minh City, Vietnam

**Abstract.** There is currently a transition from internal combustion engine vehicles to various kinds of green transportation, including battery electric vehicles. Multiple generations of such vehicles had many serious limitations, which leads to multiple concerns being raised about their viability and impact on driver behavior. Literature mostly studies the problem in small scale, or from the point of view of electric grid. In this paper we conduct a large scale simulation of driver behavior, using an agent model of traffic on a simplified map of Poland. We simulate several generations of existing and announced future electric vehicles, as well as different levels of charging infrastructure adopted, to look for the point where the driver behavior is not impacted at all, or only slightly impacted. The move to a larger scale requires adoption of some modification to the agent model, in order to decrease the computational requirements.

**Keywords:** Agent-based simulation · Traffic simulation · Battery electric vehicle · Charging behavior

## 1 Introduction

The ongoing transition to green transport poses many challenges. On the most general level, there is the economic and ecological decision on which approach is the best: synthetic fuels, hydrogen-electric, or battery-electric. There are also challanges are on the engineering, or even optimization level. For example, multiple research papers use the battery electric vehicle (BEV) approach as a basis for a new variant of the classical optimization problem: the electric traveling salesman problem [8]. The problems are usually considered from the point of view of decision makers, transportation companies, infrastructure, or car producers. In this paper we consider the green transition challanges (specifically for battery electric vehicles) from the point of view of a typical driver in a country in the early stages of the change. We also consider large scale adoption, as compared to many previous papers limited to city-wide environment. Both those changes lead to unique challanges in the model.

N.-T. Nguyen et al. (Eds.): ICCCI 2024, CCIS 2165, pp. 29–43, 2024.
https://doi.org/10.1007/978-3-031-70248-8_3

The purpose of this research is to create a traffic model geared for BEV simulation purposes, that will be better suited to large-scale simulations in terms of computational resource requirements. The traffic model itself will be used to determine the threshold of BEV parameters, when the travel times are indistinguishable from other vehicles. To optimize the performance, the model does not attempt to realistically replicate the full driving situation, but rather the most important factors influencing the travel process.

For purposes of simulation we narrow the applicability of the model to a specific subtype of applications. We focus on typical drivers in Poland, to represent countries without wide-scale adoption of electric vehicles (in which different perspective needs to be considered). Statistics show [14], that such a driver makes only around 12.000 km per year in their car, but an anecdotal anti-EV argument is "making an 800 km drive without stopping". A recent study of different groups of drivers show this type of thinking as common also in other countries [9]. The road network in Poland is lacking in terms of both highways, but also in infrastructure needed for BEV cars. On the other hand most drivers have access to overnight charging [13]. These hypothetical requirements for BEVs to operate similar to traditional cars (Internal Combustion Engine Vehicle, ICEV) are related to a difference between fueling and charging. Fueling an ICEV requires centralized distribution and does not differentiate fueling speed, or amount of fuel loaded. Charging a BEV may be done using alternate or direct current, where the second is usually much faster. In fact AC charging may be done in any electrical outlet (Level 1 charging) facilitating overnight charging, in a three-phase socket at home or at a charging point (Level 2 charging, usually simply called AC), and the DC charging is done only at fast charging points (Level 3 charging, usually simply called DC). Additionally, due to charging curve of the battery, the speed of charging is dynamic for DC and drops significantly when battery is more than 80% full (charging from 10 to 80% usually takes the same amount of time *or less* than charging from 80 to 100%). Thus a longer route starts with a battery charged to full using AC, but once it drops to ca. 10%, it is only charged up to 80%. Interestingly, while this values are well known to EV users and manufacturers, they are almost never taken into account in research papers.

The simulation model constructed in this paper represents various types of cars driving along major routes in Poland. We consider several types of BEV vehicles, with parameters based on real vehicles as used by drivers (we do not use manufacturer data, but data from [12]), and several levels of EV infrastructure adoption. We also simulate ICEV cars with their existing infrastructure. The various mixes of BEV vehicles and infrastructure adoption levels are tested to determine the specific situations, when the driver behavior may not be distinguished between BEV and ICEV.

This paper is organized as follows: Sect. 2 presents an overview of various works describing BEV behavior, modifications of existing vehicle-related computer science problems, and other related works. Section 3 provides details of the basic concepts and notions used to build the simulation model. Section 4 describes the details of the simulation model and its parameters using ODD methodology, while Sect. 5 provides the results of the simulational experiments, both in terms of application, and model performance. Finally, Sect. 6 provides general discussion of the results and their applicability.

## 2  Related Works

While the literature on electric vehicles is very rich in the engineering aspect, it is much rarer when considered as a computer science problem. The most prominent are the various paper that consider the Electric Traveling Salesman Problem (ETSP) and Electric Vehicle Routing Problem (EVRP). In some early papers the problem was considered for hybrid electric approaches, for example in [3] the classic TSP problem is enhanced with some routes requiring additional *boost* – in that case the electric engine is used in addition to the combustion one, and it may later require charging (or the charging may be done during driving). In a full ETSP the first major change is that the distance between two cities is supplemented or supplanted by energy consumption along the route – interestingly, limited models assume no recharging and only minimize the battery use [4]. In turn, as the charging process takes time, so with multiple recharging stations visited, it is the travel-and-recharge time that needs to be minimized [2]. When the problem is discussed in terms of heavy transport, the weight of the vehicle is another variable that needs to be considered [1] – the heavier the vehicle, the higher the energy consumption.

The transition to electric vehicles was also considered from the point of view of the energy grid. A complex model that includes energy infrastructure, EV vehicles, different types of charging, and drivers traveling on a *map*, was proposed in [6]. The model is more complex than what is proposed in the current paper, but it does not reflect driving and charging in real road conditions. The authors assume DC charging up to 100% and estimate driving range of theoretical vehicles based on aerodynamics instead of using real data. The *map* used is also a small graph consisting of 10 vertices. The authors of [5] build a traffic model for Beijing, and consider charging behaviour divided between three types of location: home, work, and public. They base their research on real data about taxi drivers, which seems to introduce a bias to the distribution of remaining charge at the start of the charging process – it follows normal distribution in all cases, instead of public (DC) charging only being done when the remaining range is withing 10–20%. Additionally, while real car data is considered, it is only official ratings, and the vehicles mostly correlate to the group we consider *Obsolete* in this paper. Still, the shown results bear high similarity to those in the current paper, especially for the aforementioned group. In [7] a similar division to home, work and public chargers is made, but the authors additionally consider the ownership of garage (home charging location). The aim of the work is to distribute public chargers in a small city, and authors simulate vehicle traffic to find the locations with most demand. The vehicle data is taken from official statistics, not real driving data, and the vehicles correspond to *Obsolete* group in the current research. While the rest of the paper is outside the scope of current research, it must be noted that the charger data is also discussed in abstract, which lowers the applicability of the results (e.g., public AC charger uses speed of 15 kWh/h, this value is possible device-side, but most BEV will only use 11 kWh/h).

## 3  Basic Concepts in Simulation

In the presented model we use a static map with a simplified representation of major cities and major roads in Poland. For better representation of real driving conditions, the roads allow different average speed of driving (e.g., while the allowed maximum speed on Polish highways is 140 km/h, the traffic situation over a longer route leads to an average of ca. 120 km/h). In total there is ca. 20% of roads with speed average 120 km/h (highways), 50% of roads with speed average of 80 km/h (country roads), and the remaining roads are urban and suburban areas with average speed of 40 km/h. With three categories of roads the car ranges need to be described only in terms of those three average speeds, as $R_{120}, R_{80}, R_{40}$. The last, and largest, of those values, usually is around or even exceeds official range rating of the vehicle in the WLTP range measure procedure. It is instead taken from real vehicle data in [12]. We also assume full availability of fuel stations and a partial access to AC charging stations in cities (residential areas), while along the roads the availability of fuel stations and DC charging points is scarce (one per a given distance). Additional simplifying assumption is the lack of queues at fueling stations and charging points. The used map is based on OpenStreetMap [11] data of Poland. An example of the source data limited to a small region is given in Fig. 1a), while Fig. 1b) shows the miniature of the used simplified map. The areas in the simplified map (buildings in the base one) have an assigned type, with around 90% representing residential areas, and 10% representing industrial ones. We assign the type randomly, due to lack of such labels in OpenStreetMap and the size of the data preventing manual annotation.

As noted, vehicle agents move periodically between two areas along different types of roads. Initially, the vehicles are fully charged (fueled), and use up charge (fuel) as they travel, based on the distance and speed. Once the charge (fuel) drops to 10% of full, the vehicle will stop to charge at a DC charging point (fuelling station) until it is 80% full (100% in case of fuel). If the vehicle would not be able to make it to the next charging point, it will instead stop earlier. The charging takes $T_{10-80}$ minutes. BEV additionally charge at night, to full if there is enough time, or as much as possible. Charging from 0 to full takes $T_{AC}$ hours.

Using the abovementioned parameters, we define different populations of possible vehicles, with parameters presented in Table 1:

- The first group of parameters represents cars with older technology, that were produced before 2013 (e.g. Nissan Leaf with 24 kWh battery). This group uses

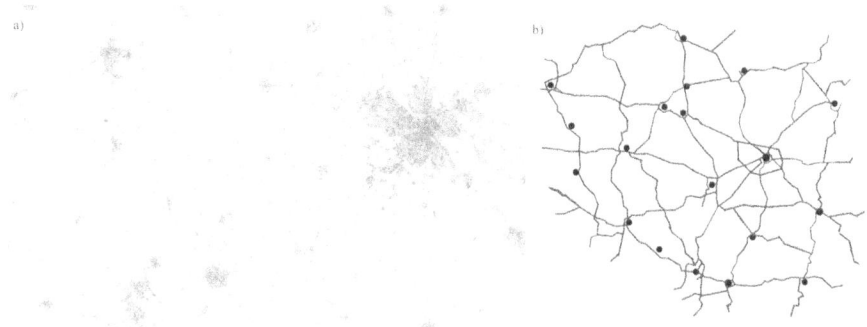

**Fig. 1.** a) Sample of source map data from OpenStreetMap [11], representing a single small part of Poland. b) Major roads in Poland as used in the simplified map in simulation.

the shorthand *Obsolete*. This is also the only group where we consider AC charging as slower than 11 kWh/h.

- Parameters for *City* represents group of small electric cars available around 2020 with average parameters (e.g. VW ID.3).
- Parameters for *Current* represents group of electric cars available around and before 2020, optimized for long range travel (e.g. Tesla 3 LR).
- The next group of parameters represents a recommended driving approach, that is a short (15 min) break after every three hours of driving. As of the date of publication of this paper, there exists a single car model operating within these parameters (Lucid Air), with more having been announced. While charging time is longer, it also allows for longer driving time, i.e., when charging time is reduced to 15 min, the range allows for 3 h of driving. Parameters for *Long Range*.
- Finally, we use the anecotical problem that is raised as a point against EV transition, i.e. "800 km drive on a motorway with no stops", as the basis for the last group of parameters for the experiment in *Future* group. It also corresponds to some marketing materials of "future electric vehicles", as presented by major car manufacturers.

**Table 1.** Parameters of vehicle groups used in the model.

| Agent group | $R_{120}$ | $R_{80}$ | $R_{40}$ | $T_{10-80}$ | $T_{AC}$ |
|---|---|---|---|---|---|
| Obsolete | 120 | 155 | 205 | 27 | 8 |
| City | 325 | 405 | 520 | 32 | 6 |
| Current | 480 | 590 | 730 | 27 | 8 |
| Long Range | 655 | 790 | 960 | 33 | 11 |
| Future | 800 | 1100 | 1400 | 5 | 15 |

In simulation we consider ICEV with $R$ identical to Long Range group (80% of cases), or Extreme group (remaining 20%), but with fuelling time of 5 min.

The EV infrastructure level is considered at three levels of adoption:

- *Low* level represents situation typical to countries with less that 5% total sales volume for factory new EV. This is represented with a charging point with a minimum time of 50 min every 200 km of major roads.
- *Medium* adoption represents a variant of proposed requirements for charging infrastructure, here generalized to a charging point with minimum time of 15 min located every 60 km of major roads.
- *Full* adoption represents a situation when the charging may be done at any random location within the standard time of a given vehicle group.

In contrast the ICEV infrastructure is similar to the Medium and Extreme level of adoption. The fuel stations are located only in select places, and the process of fueling up the car takes approx. 5 min.

## 4   Agent Model

The simulation of the vehicle behavior was prepared in GAMA [10] environment, and the model was constructed to make best use of this tools features. As such, there are some superficial similarities to other models deployed in GAMA: there is an underlying map that includes residential areas, industrial areas and roads; and there are vehicle (people) agents that regularly move between a selected place in the residential area to a selected place in the industrial one. This serves as a comparison basis for the optimizations done in the proposed model, that allow it to better represent large-scale BEV traffic.

There are two types of entities considered in the model: the agents that represent a combined vehicle-driver, and the global environment that represents the real map (as described in previous section) and other global variables, e.g., time.

The map in the environment uses graph representation, specifically it is a multi-labeled undirected graph $G = (V, E)$, where:

- Each node $v \in V$ representing a point of interest, has a list of descriptors that include at least one of {residential, industrial, charging, crossroads}. As a single node represents a larger location (in source GIS map: a node is a single building; in the model it corresponds in size to a city quarter) multiple such labels are allowed.
- Each edge $e \in E$ representing a road fragment, has two labels assigned: one of the allowed values of speed $s \in \{40, 80, 120\}$, which corresponds to road types in the model, and numerical value of length $d$. A single edge represents a fragment of a road of a given length $d$, where the average traveling speed is $s$. Only a limited number of road-types (speeds) are allowed, instead of full range of speeds – this has a large impact in limiting the number of necessary calculations in the model.

In terms of used data, initially many nodes labeled only as *crossroads* did not have more than two edges – as smallest roads were removed from the simplified map, the additional edges that would create a real crossroads were not present. In further preparation of data these nodes were deleted and the edges merged, all in order to limit the number of calculations required.

The environment defines also the current time (time moment $\tau$ (numerical) and time of the day $t$ (date format)). The nodes with *charging* descriptor all have an identical limitation: charging time in minutes $C_{DC}$ (10% to 80% for BEV, but 0 to 100% when simulating ICEV). This is another simplification in the model used to limit computational costs – only one type (speed) of DC chargers is present.

Every agent in the simulation is of the same type, but some of their parameters are generated at random. An agent represents a single vehicle-driver traveling along the roadmap. A single agent is described by the following parameters:

- Maximum range of the agent $R_{max}$ (range at $s = 40$), which indirectly represents the battery size (maximum charge) of BEV. Numerical value depending on the vehicle group of the agent.
- Current remaining range of current agent $R_c$, a numerical value that represents remaining battery charge in BEV. In simulation, instead of calculating charge used per distance, the remaining range is decreased directly, proportionally more so for $s = 80$ and $s = 120$ – this approach, combined with limiting the roads to three possible speeds, significantly reduces the number of required calculations.
- Charging time (AC and DC) $T_{AC}, T_{DC}$ are two numerical parameters, identical for all agents in the same vehicle group. The AC (slow) charging time is measured in hours required to increase the charge level from 0 to 100%, while the DC (fast) charging time is measured in minuted required to increase the charge level from 10 to 80%. As the charging station also has a charging time, the process takes the *maximum* of the two ($max(T_{DC}, C_{DC})$).
- Home node $V_H$ in $G$ (node with a descriptor: residential) and work node $V_W$ in $G$ (node with a descriptor: industrial). All agents start in the home node and travel to work nodes, then travel back.
- A Boolean parameter $H_C$ determining if the given agent has access to a home charger, allowing for slow charging the car when the location is *home* (during the night).
- To-work time $T_W$ and to-home time $T_H$, determine the time of day when the agents leave their current location to travel to work/home.
- Current location $v_c$ and current target $v_t$ are variables used for determining how the agent moves in the graph.

The simulation is initialized at time moment representing midnight. All agents start at home location with full range. After that each agent operates on a cycle, testing its internal and environmental variables, to determine:

- If time of day is $t = T_W$ then it sets the target to work node $v_t = V_W$, and if $t = T_H$ then it sets the target to home node $v_t = V_H$.

- If the target is different than the current node ($v_c \neq v_t$) then the agent moves one node towards the former (if it takes more than one time moment, it is immobilized at the target node for the appropriate time). After that the remaining range is recalculated, proportionally to the speed on the traveled edge.
- Fast charging – while the agent is not at home node ($v_c \neq V_H$), then at each node with the descriptor *charging* it observes its remaining range. If it is lower than 10% or lower than the distance to the next charging station, then the vehicle will stops to charge. The remaining range increases to 80% of maximum, but the vehicle is immobilized for $max(T_{DC}, C_{DC})$.
- Night charging – while the agent is at home node ($v_c = V_H$) it will increase its charge each hour by $\frac{R_{max}}{T_{AC}}$, until it has restored the full range.

## 5   Results

The simulations are run on the simplified map of Poland, as described in Sect. 3, with 10000 agents, for a period of time corresponding to one week. We consider all combinations of BEV groups and infrastructure adoptions levels, as well as the ICEV agents with traditional fuel stations. During the experiments we observe the remaining range of all agents (vehicles) in kilometers: maximum value, minimum value, average value; as well as the number of vehicles currently fast charging/refueling (i.e. stopped during travel) and number of vehicles with more than 80% of range remaining. The simulations were repeated in the same setup, but with a random seed of agents, multiple times for each configuration. Some specific examples are shown in Figs. 2, 3, 4, 5.

There are several general observations that can be made on the basis of the simulations. On the most general level, all groups of vehicles are proportionally similar: there is a always a group of cars near the full range, a group of cars near the minimum range, and a group of cars currently charging (refueling); the average remaining range for the whole group is also always more than half. In terms of range in kilometers, this leads to major differences between each group, but the proportions remain similar.

The first clear difference between ICEV and BEV is the number of vehicles that are at a maximum charge during different parts of the cycle (times of day), which is a direct consequence of how recharging/refueling works. As it is preferable to fast charge BEVs only to 80% and slow charge them to full at night (and the model reflects that), this type of vehicles usually begin the cycle with full range and if recharge is needed – it only brings them back to 80%. ICEVs can only be refueled at stations and they have no practical limitations on the process, therefore they are at full range at random points of the day, but always less when returning home at night. A surprising case is the hypothetical *Future* group – due to the requirement for large battery, more time is needed for slow charging – and in consequence it in some limited cases it may not charge up to full at night. Still, even in that case it will have larger remaining range than any other group, either BEV or ICEV. This ties directly to the frequency of fast charging – as BEVs use their idle time at night for charging, most of them use public infrastructure (fast charging) rarer than ICEVs refuel. The only exception is the *Obsolete* group, which has too low base maximum range, and need to recharge more often.

Interpreting the results when the remaining range is expressed in terms of kilometers instead of percentage (battery charge level), we can make additional comparisons between BEV and ICEV. The map used in simulation is deliberately large scale, so the beforementioned pattern of night charging is not sufficient for *all* travel needs. For a randomized sample, even in the *Current* group, between 20% and 30% of vehicles use only night charging, while additional 5–10% only need to fast recharge occasionally (once per simulation corresponding to a single week) – these values are out of the total, and only 60% of agents have access to this option. For vehicles with smaller maximum range these numbers decrease, but it increases only marginally for vehicles with larger range (as AC charging requires more time due to larger battery). Taking this into account, the theoretical *Future* group, combined with maximum level of infrastructure, leads to less time recharging than refueling ICEV (i.e., they are statistically indistinguishable with no night charging). With this level of infrastructure, the *Current* and *Long Range* are within 10% additional time required for travel in *all* cases, while the *City* group is within this threshold on *Average*. For *Medium* level of infrastructure, the *Future* and *Long Range* fall within this threshold (the former always, and the latter   on average), while for the *Low* Infrastructure the smallest extension of travel time, for the *Future* group, is 12% on average, while all the other groups achieve much worse results.

**Table 2.** Simulation of traveling on the longest route. Time in hours: minutes for all BEV agent groups with all levels of Infrastructure.

| Agent group | Low infrastructure | Medium Intrastructure | Full Infrastructure |
|---|---|---|---|
| Obsolete | exceeds time limit | exceeds time limit | exceeds time limit |
| City | exceeds time limit | 11:57 | 11:52 |
| Current | 11:40 | 11:15 | 11:12 |
| Long Range | 11:17 | 10:29 | 10:23 |
| Future | 10:01 | 9:46 | 9:40 |

A separate consideration needs to be done for singular cases of agents with very long daily drives, which correspond to the anecdotal arguments against electric vehicles. In the simulated map the maximum distance between two points is 713 km, which could be travel with no stops in 9 h 22 min using the fastest route. In real-world short stops are recommended every 3 h, which brings the travel time to around 10 h (the stops could be calculated in such a manner, that

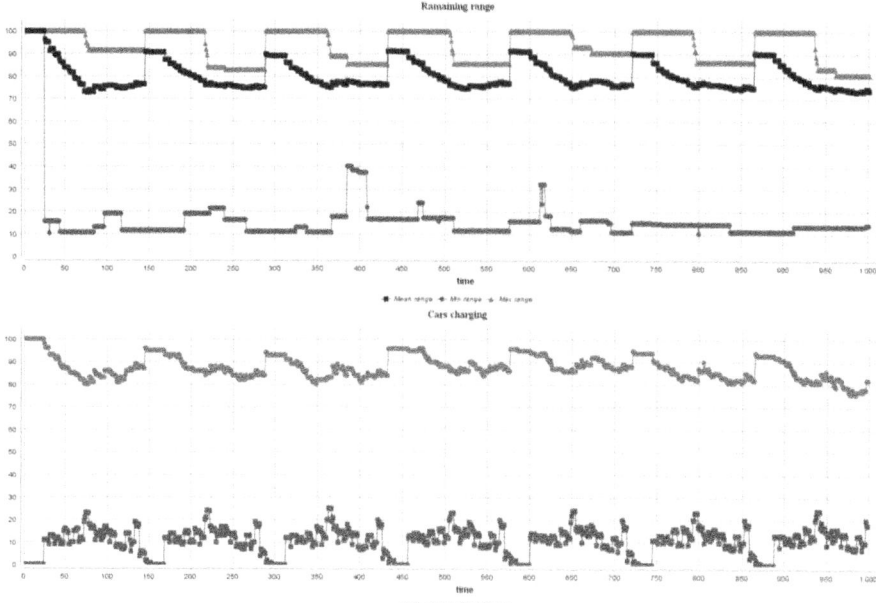

**Fig. 2.** Simulation of the BEV group *Obsolete* with *Low* infrastructure level, with the full range of 205 km (at speed 40 km/h). Top chart shows the percentage of remaining range: maximum, minimum and mean value. Bottom chart shows the number of vehicles currently fast charging, and the number of vehicles that currently have more than 80% of range remaining.

it takes exactly 10 h), but in the proposed model such stops are not present. One may note that in such situation, the agent spends more time traveling than waiting at a location (i.e., working or resting), but this borderline case is the basis for some of more important results. In the base simulations such a situation had a negligible probability of occurring, therefore a new setup was created, in which only this longest route was observed. The baseline of ICEV takes on average 9 h 41 min to travel and otherwise the results do not differ from the base simulation. The time taken for all groups of BEV vehicles with different levels of infrastructure is shown in Table 2. As may be noted the *Obsolete* group and the *City* group in case of *Low* infrastructure have a travel time in one direction that exceeds 12 h, and the agents become stuck between locations. This is a limitation of the simulation, where the vehicles travel between two locations within a 24 h cycle – in real world scenario the driver would switch travel method, or allocate more time for travel. An opposite situation occurs with the theoretical *Future* group of agents – here the difference between BEV and ICEV vehicle is statistically insignificant: the base range is larger, therefore even charging up to 80%, combined with nightly slow charging, allows for time

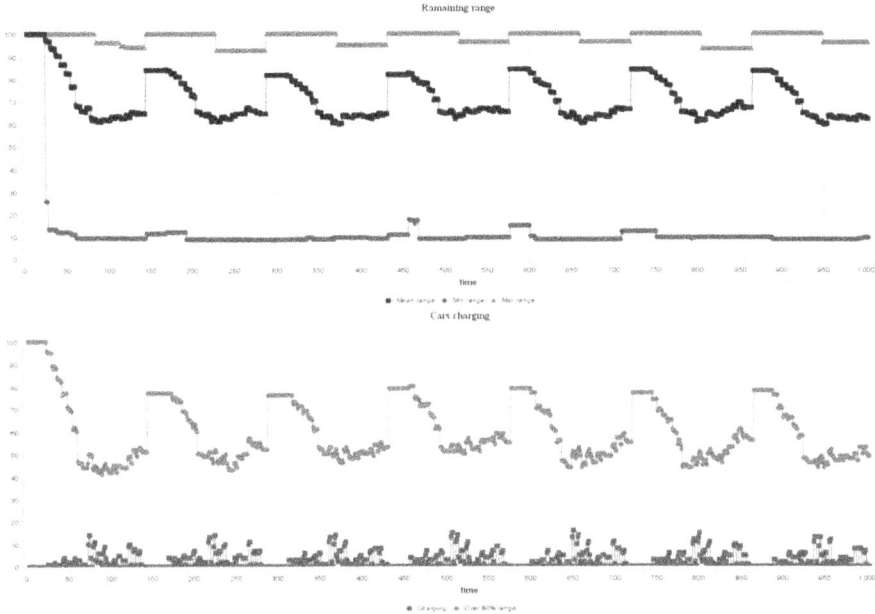

**Fig. 3.** Simulation of the BEV group *Current* with *Medium* infrastructure level, with the full range of 730 km (at speed 40 km/h). Top chart shows the percentage of remaining range: maximum, minimum and mean value. Bottom chart shows the number of vehicles currently fast charging, and the number of vehicles that currently have more than 80% of range remaining.

parity with ICEV refueling. Similar situation occurs for *Long Range* group, where the electric vehicles take longer to travel, but the difference is not large unless the infrastucture is on the lowest level. In that case, as well as for all remaining combinations (in practical terms: real-world vehicles as of year 2024), the travel time is over 11 h. There is only a very small time window available for slow charging at night, so the charging behavior becomes more similar to refueling. This increases the travel time, as the vehicles need more time for recharge than to refuel. Note that for a singular travel (in simulation: first home/work/home cycle) the BEV agents start from 100% charge and the travel time is further reduced by a shorter charging stop required.

The model in the presented final form was also compared to others, the base model utilizing the full real-world data and various influence on range/charge, and the intermediary models with different non-key features simplified. Each next step had lower computational requirements in terms of memory and/or processing power, but results were rarely fully comparable. We started with a baseline model that was a combination of both [7] and one of build-in models

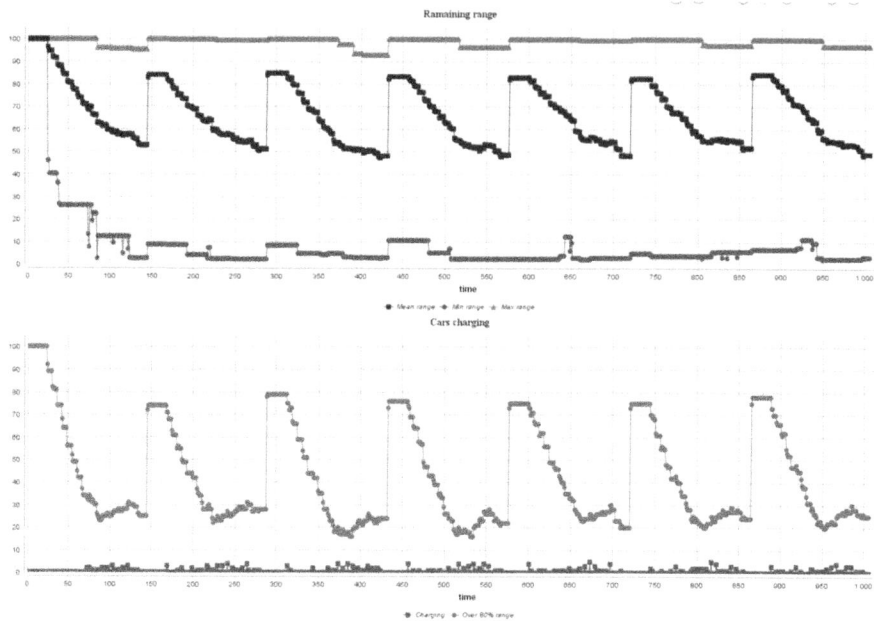

**Fig. 4.** Simulation of the BEV group *Future* with *Full* infrastructure level, with the full range of 1400 km (at speed 40 km/h). Top chart shows the percentage of remaining range: maximum, minimum and mean value. Bottom chart shows the number of vehicles currently fast charging, and the number of vehicles that currently have more than 80% of range remaining.

[10], with additional options that reflect driver perspective on EV travel, as to our knowledge there is solution that relates exactly to the problem considered in this paper. Full vehicle and environmental details were used, including detailed influence on vehicle range and charging curve. We considered a fully-detailed map first, but due to memory requirements had to limit simulations to a single metropolitan area. Limiting the map details to main routes, as detailed previously, allowed expanding the map area ca. 800 times to country-wide map. We then reduced the details of the model, with only three allowed speeds remaining (using real user data instead of calculating all values using equations) and using a proportion of 10–80% charge time instead of full charging curve. This reduced computational requirements of the model by ca. 75% as compared to the baseline model. Finally, switching from using battery charge to remaining range as an agent variable allowed for another ca. 12% decrease in computational requirements.

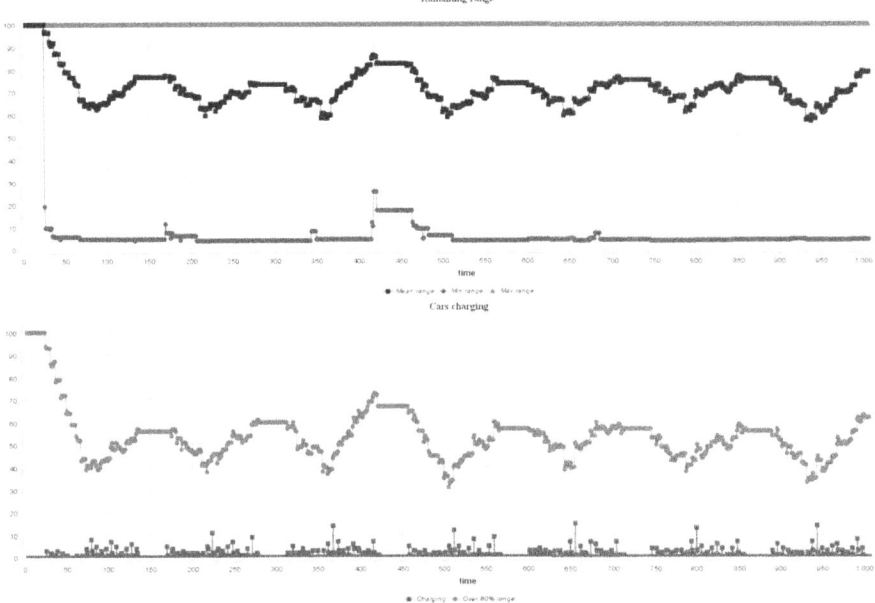

**Fig. 5.** Simulation of the ICE vehicles with current refueling infrastructure, with the full range of 960 km for 80% of agents and 1400 km for the rest (at speed 40 km/h). Top chart shows the percentage of remaining range: maximum, minimum and mean value. Bottom chart shows the number of vehicles currently refueling, and the number of vehicles that currently have more than 80% of range remaining.

## 6    Conclusions

The simulational model is focused on solving a different problem than many present in literature, including more realistic agent behaviour and data. We discuss a traffic model for electric vehicles on a simplified map of a country. Unsurprisingly, if we assume that some *Future* group of vehicles will have parameters of ICEV, we observe that it does not impact the driver behaviour negatively. In fact, for a significant part of the population the time required for public charging will be lower than the time required for public refueling. In more realistic situation of high-end vehicles existing these days, the switch from ICEV to BEV leads to an increase of travel times by less than 10% due to additional charging stops. This still requires the infrastructure to be at least on the level described here as *Medium* – in real-world as planned on major routes in European Union.

Another result of the paper is the approach used to decrease the computational requirements and allow for larger simulation, while not influencing the quality of the simulation. The vehicle charging is non-linear in terms of kWh/h, the another the charging points may be currently used, or may divide available power between vehicles. These components, among others, were considered in initial iterations of the model, but were removed as their impact on the simulation is miniscule.

The considered problem is part of a larger situation caused by the move from fossil fuels to battery electric vehicles, and other problems need also be considered. In a parallel line of research we are currently studying is the problem of charging station location – these days it is mostly considered from a theoretical perspective in graphs, or inside cities with high-powered electric grid already existing, instead of taking into account costs of creating new electrical infrastructure to connect key locations along major travel routes.

# References

1. Baek, D., Chen, Y., Macii, E., Poncino, M., Chang, N.: Battery-aware electric truck delivery route planner. In: 2019 IEEE/ACM International Symposium on Low Power Electronics and Design (ISLPED), pp. 1–6. IEEE (2019)
2. Ceselli, A., Righini, G.: The Electric Traveling Salesman Problem: properties and models. Technical report 2434/789142-University of Milan (2020). https://doi.org/10.13140/RG.2.2.17712. 99848
3. Doppstadt, C., Koberstein, A., Vigo, D.: The hybrid electric vehicle-traveling salesman problem. Eur. J. Oper. Res. **253**(3), 825–842 (2016)
4. Erdogdu, K., Karabulut, K.: Distance and energy consumption minimization in electric traveling salesman problem with time windows. In: 2020 7th International Conference on Electrical and Electronics Engineering (ICEEE), pp. 160–164. IEEE (2020)
5. Hu, L., Dong, J., Lin, Z.: Modeling charging behavior of battery electric vehicle drivers: a cumulative prospect theory based approach. Transp. Res. C Emerg. Technol. **102**, 474–489 (2019)
6. Marmaras, C., Xydas, E., Cipcigan, L.: Simulation of electric vehicle driver behaviour in road transport and electric power networks. Transp. Res. C Emerg. Technol. **80**, 239–256 (2017)
7. Pagani, M., Korosec, W., Chokani, N., Abhari, R.S.: User behaviour and electric vehicle charging infrastructure: an agent-based model assessment. Appl. Energy **254**, 113680 (2019)
8. Roberti, R., Wen, M.: The electric traveling salesman problem with time windows. Transp. Res. E Logist. Transp. Rev. **89**, 32–52 (2016)
9. Sprei, F., Kempton, W.: Mental models guide electric vehicle charging. Energy **292**, 130430 (2024)
10. Taillandier, P., et al.: Building, composing and experimenting complex spatial models with the GAMA platform. GeoInformatica **23**(2), 299–322 (2019). https://doi.org/10.1007/s10707-018-00339-6
11. OpenStreetMap in GIS format. https://download.geofabrik.de/europe/poland.html. Accessed 29 Jan 2024
12. Source of real use data for BEV, EV-Database. https://ev-database.org/. Accessed 29 Jan 2024

13. Polish Statistics Portal, Home Infrastructure Data. https://stat.gov.pl/obszary-tematyczne/infrastruktura-komunalna-nieruchomosci/nieruchomosci-budynki-infrastruktura-komunalna/gospodarka-mieszkaniowa-i-infrastruktura-komunalna-w-2022-roku,13,17.html. Accessed 29 Jan 2024
14. Polish Statistics Portal, Driver Data. https://stat.gov.pl/obszary-tematyczne/transport-i-lacznosc/transport/transport-drogowy-w-polsce-w-latach-2020-i-2021,6,7.html. Accessed 29 Jan 2024

# A Systematic Literature Review on Affective Computing Techniques for Workplace Stress Detection

## Challenges, Future Directions, from Data Collection to Stress Detection

Iris Mezieres[1], Abir Gorrab[2], Rébecca Deneckère[1(✉)], Nourhène Ben Rabah[1], and Bénédicte Le Grand[1]

[1] Centre de Recherche en Informatique, Université Paris 1 Panthéon-Sorbonne, Paris, France
{Rebecca.Deneckere,Nourhene.Ben-Rabah,
Benedicte.Le-Grand}@univ-paris1.fr
[2] RIADI Laboratory, National School of Computer Science, University of Manouba, Manouba, Tunisia
Abir.Gorrab@riadi.rnu.tn

**Abstract.** In a world where work significantly impacts daily life, its influence on well-being is undeniable. The prevalence of stress at work has gain an increased attention due to its profound effects on both individual health and corporate performance. Addressing and detecting stress has become an essential challenge in fostering a healthy work environment. Technological innovations, especially in the field of affective computing, which involves various IT resources for analyzing human behavior and emotions, offer promising solutions for measuring employee stress. This paper provides a Systematic Literature Review (SLR) focusing on the existing scientific research in stress at work assessment using affective computing technologies. What distinguishes our work from others is that we deeply focus on each phase of the stress quantification process. We start by reviewing application contexts, before detailing the data collection process, including data sources and collection devices. We then highlight data analysis techniques used in the literature. Finally, we emphasize the challenges discussed by researchers during their work and give insight into future work.

**Keywords:** Affective Computing · Stress · Workplace · SLR

## 1 Introduction

Stress is a natural response of the human body to a feeling of physical, or mental tension or anxiety to enable rapid reaction [2]. However, in the workplace where life is -generally- not at risk, stress has counterproductive consequences. The European Agency for Safety and Health at Work (EU-OSHA)[1] reports that 51% of European workers consider stress

---

[1] EU-OSHA Homepage, https://osha.europa.eu/fr. Accessed 10 Feb 2024.

© The Author(s), under exclusive license to Springer Nature Switzerland AG 2024
N.-T. Nguyen et al. (Eds.): ICCCI 2024, CCIS 2165, pp. 44–56, 2024.
https://doi.org/10.1007/978-3-031-70248-8_4

a common workplace issue. Despite the consequences on health, such as professional burnout or illnesses, according to the European ESENER survey[2], only 34% of continental businesses had adopted an action plan to prevent stress at work in 2019. Identifying stress indicators and allowing proactive intervention significantly increases employees' commitment, job satisfaction, and loyalty, enhancing their productivity [3]. Affective computing may help in this regard, by gathering data for emotion identification and analysis with devices such as smartwatches, phones, or webcams [5]. Data types vary from physiological (heart rate), and physical (body movements) to behavioral characteristics. Adopting affective computing seems promising but some challenges must be addressed for a secure and successful application in real-world environments. In this work, we propose an SLR to answer the following research issue: *'How are affective computing technologies used to improve the measurement of employee stress at work?'*.

We first provide background information about causes and consequences of stress at work, and related works in Sect. 2, before detailing our methodology for this SLR in Sect. 3. In Sect. 4, we present the results of our literature review and then discuss the challenges associated with this research question in Sect. 5. Finally, we draw our conclusions in Sect. 6.

## 2  Background and Related Works

In this section, we start by presenting a background on stress at work: its causes and consequences, before discussing related works and initiating our contributions.

### 2.1  Stress at Work

**Causes.** We propose in Fig. 1 a classification of the causes of stress at work, which may vary according to individuals, their personalities, histories, or perceptions.

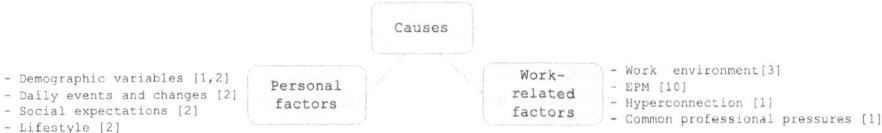

**Fig. 1.** Classification of Stress at Work Causes.

First, there are several ***personal factors*** [2], including *demographic variables* such as age, gender, education, or nationality. For instance, women tend to be more affected by stress due to evolutionary differences in neurobiology and hormones [1]. Another personal factor is *lifestyle*, such as lack of physical activity or consumption of alcohol or drugs. *Daily events and changes* can also influence stress, such as relationships with friends or colleagues, family problems, the death of a loved one, etc. Finally, there are

---

[2] ESENER Homepage, https://osha.europa.eu/fr/facts-and-figures/esener. Accessed 10 Feb 2024.

demands and *expectations of society*, impacting people differently depending to their personalities.

Next, we find a set of ***work-related factors***. What happens in the *work environment* greatly affects our emotions and feelings. Noise or over-illumination may impact our daily lives even outside of work [3]. *Hyperconnection* also makes it challenging to completely detach from work [1]. *Electronic Performance Monitoring* (EPM) systems provide managers with indicators of employees' activity (e.g., work pace, connection and disconnection time). The authors of [10] report that these tools increase stress for 81% of employees. Finally, we identified *common professional pressure* as potential causes of stress, such high workload, tight deadlines, long working hours, low salary, and unattainable goals [1].

**Consequences.** We classified the consequences of stress at work in Fig. 2. On the *Personal level*, we note that both mental and physical health may be impacted.

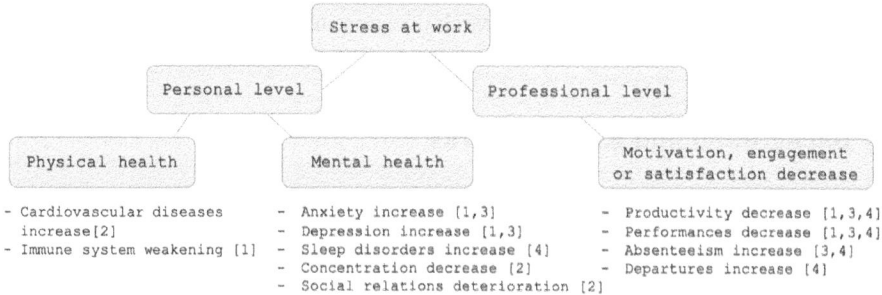

**Fig. 2.** Classification of Stress at Work Consequences.

Stress also has repercussions on a professional level, through its impact on motivation, engagement, and satisfaction. This results in significant financial losses for companies. Recent studies estimate stress at work to cost the United States more than 300 billion dollars, due to absenteeism, accidents, and reduced productivity[3], All of this underscores the need for organizations to prevent, or at least recognize and manage employee stress, to foster a healthy -and therefore constructive- work environment.

## 2.2 Related Works

Other works have studied the existing literature on stress measurement in professional contexts using affective computing [3, 5, 8, 9]. In [3], the authors highlighted the evolution of methods used to measure the stress of workers. They also compared existing solutions based on cost and privacy impact. In the same context, authors of [5] focused on identifying how affective computing technologies could be used to improve workers' daily lives. Furthermore, authors of [9] conducted a systematic literature review by identifying the way wearable devices could be used in the workplace for stress measurement.

---

[3] The American Institute of Stress https://www.stress.org/workplace-stress. Accessed 1 March 2024.

They aimed at determining sensors that could be used, their eventual contributions, and the challenges arising from them. Moreover, authors of [8] studied the way wearable devices and phones could be used to detect stress at work, in education, and transport. They attempted to analyze challenges related to data collection and how contexts could impact the stress measure process. These existing works present some specificities. Some works focus on measuring stress at work [3, 8, 9] while few other works assess stress and emotional states in different aspects of daily life [7]. Some of them focus specifically on studying the use of wearable devices to measure stress [9], while others investigate solutions [3, 7]. In our work, we followed a rigorous research protocol and carefully collected and analyzed research articles to conduct a systematic literature review dedicated to stress assessment at work, using affective computing technologies. What distinguishes our work from others is that we deeply focus on each phase of the stress quantification process. We start by reviewing application contexts, before detailing the data collection process, including data sources and collection devices. We then highlight data analysis techniques used in the literature. Furthermore, we emphasize the challenges discussed by researchers during their work and give insight into future work.

## 3   Systematic Research Papers Collection Methodology

A SLR [6] is a scientific method employed to investigate a precisely defined issue. It entails employing well-defined methods to select a body of scientific papers sharing a common theme, conducting an in-depth analysis, extracting pertinent information, and synthesizing these findings to address the initial research question comprehensively.

**Research Questions.** We have structured our initial research issue into 4 Research Questions (RQ):

–   **RQ1**: In which contexts is affective computing used to assess stress at work?
–   **RQ2**: What data collection methods are used to assess stress at work?
–   **RQ3**: How is collected data analyzed?
–   **RQ4**: What are the challenges related to the use of affective computing to measure stress at work and how are they addressed in existing works?

**Search Query.** The following search query was executed on the Scopus library:
    ("affective computing" OR "sentiment analysis" OR "emotion recognition" OR "artificial intelligence" OR "human-computer interaction" OR "sensors").
    AND (workers OR workplace OR offices OR employees OR professionals OR company OR enterprise) AND stress.

**Papers Selection.** Figure 3 illustrates the steps of the selection process among the 162 papers retrieved automatically at step 1. We note that N refers to the number of papers retained at each step.
    We applied *inclusion and exclusion criteria* in step 2. Our Inclusion criteria are: (a) Paper proposing a solution to measure employees' stress at work, using affective computing technologies, (b) Paper published after 2018 (to focus on recent studies), and (c) Paper accessible online. Our exclusion criteria are: (a) Paper not written in English,

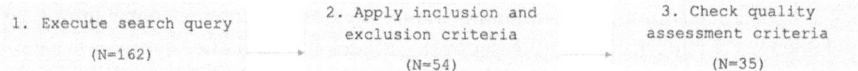

**Fig. 3.** Scientific Papers Selection Process.

and (b) Paper not peer-reviewed (such as abstract, poster, proposal, technical report, thesis). We obtained 54 papers after this first selection. At step 3, the 54 remaining papers were checked against the following *quality assessment criteria*: (a) The way data is collected and analyzed is clearly described (yes/no/partially), (b) The context of the study is defined (yes/no/partially), and (c) The study evaluates the performance of the proposed solution (yes/no). Each "yes" is assigned a score of 1, "no" a score of 0 and "partially" a score of 0,5. Papers that scored less than 2 were excluded, resulting in 35 papers for the SLR.

## 4  SLR Results

### 4.1  RQ1: Which Context for Stress at Work Assessment?

Table 1 shows that most research works for stress assessment at work using affective computing took place in a static activity context (25 papers), among which 24 in an office. Only 7 papers reported experiments in dynamic environments such as healthcare, industry, or emergency services, while 3 papers on our corpus did not specify the work context.

**Table 1.** Applications Domains.

|  | Applications domain | Example of references |
|---|---|---|
| Static 25 sources | Office | [16, 20, 35] |
|  | Customer service | [37] |
| Dynamic 7 sources | Healthcare – paramedical | [24 30, 31] |
|  | Industry | [14, 33, 42] |
|  | Security - Emergency services | [27] |

### 4.2  RQ2: Which Method of Data Collection for Stress Assessment?

A categorization of collected **data sources** is proposed in Table 2.

**Physiological data** consists of biometric data, which may be related to the nervous system, the respiratory and circulatory system, or the integumentary system (skin conductance response and skin temperature).

**Physical data** reflect the movement of a part of an individual's body. We distinguish body and facial movement.

**Table 2.** Collected Data Sources.

| Data type | | References |
|---|---|---|
| Physiological data | Nervous system | [13, 26] |
| | Respiratory and circulatory system | [14, 17, 19, 21, 24, 26, 27, 30, 31, 34, 38, 40, 43, 45] |
| | Integumentary system | [14, 15, 22, 24, 27, 29, 32, 38, 40, 43, 45] |
| Physical data | Body movements | [11, 17, 20, 28, 36, 45] |
| | Facial expressions and eye blinks | [18, 19, 26, 33] |
| Behavioural data | Vocal characteristics and speech like voice pitch and speech content | [26, 30, 37, 40] |
| | lexical analysis and punctuation usage | [25] |
| | mouse clicks speed while using a computer | [12, S6, S9] |
| Others | personal data | [S9, 30, 44] |
| | environmental data | [11, 26, 27, 30, 39, 40, 44] |
| | work-related data | [11, 19, 21, 39, 44] |

**Behavioral data** characterize the reactions of an individual to his/her emotions. In the context of workplace stress, this includes voice, writing, or interactions with work tools such as computers or phones.

**Other data** types may be personal data (gender, age, sleep quality), environmental data (temperature, light intensity), or work-related data (schedule, break number). However, this data remains secondary in these works. We note that physiological data are mostly utilized in our corpus, in 19 papers among 35; 10 papers incorporate physical data, while 8 articles use behavioral data. We also observe that different data types are combined in some works [11, 19, 30, 40], enriching data to be analyzed.

We categorize the **devices** used for data collection into four categories:

**a) Sensors worn or manipulated by users**, such as smartwatches, glasses, or helmets, capture physiological data [13, 14, 22–24, 26, 27, 29–33, 38, 40–44]. In this category, we can also cite phones [21, 30, 39, 40, 43, 44] that represent a rich information source. **b) Sensors embedded in the office** may be computer peripherals, including computer mouse [12, 15, 16], and keyboards [16], used to measure behavioral data. They may also be office chairs [17, 20, 35]. In [20], chairs contain pressure sensors allowing them to measure body movements. **c)** ***Thermal, traditional, and depth-sensing cameras*** *may also be used as information sensors for measuring stress* [11, 18, 22, 36]. *Furthermore,* **d)** ***surveys*** [23, 29] *allow for gathering information about individuals' well-being through a set of questions that were carefully analyzed. Even if they were used in 19 works, they remain a secondary resource of information and are generally combined with other sensors.* We can see that physiological data were the most popular data used in our selected papers, notably collected through sensors worn or manipulated by users, being available and easy to use in daily life.

### 4.3   RQ3: How is Collected Stress-Related Data Analyzed?

**Artificial Intelligence-Based solutions.** We distinguish traditional Machine Learning (ML) and Deep Learning (DL) algorithms.

**ML-Based Solutions**
On the one hand, ***supervised algorithms*** are suitable for labeled data, which allow the model to be trained and then used to make predictions. In several papers selected for the SLR, researchers aim at predicting numerical values, which corresponds to a *regression* problem. For example, authors of [32] propose a predictive model of stress to determine an individual's emotional state based on physiological data from the previous day. To do this, they use the Random Forest (RF) algorithm, which determines an individual's well-being level with a value varying from 0 to 100. With an average of the Mean Absolute Error (MAE) of 15, their solution proved to be quite effective. Then, in [35], Linear Regression (LR) is utilized to directly associate a stress level from the variation in heart rate. Other solutions called *classification* algorithms exist to predict categories rather than numerical values. The number of desired categories must be made carefully beforehand as the data must be labelled according to this parameter. In [19], the problem is seen as a binary classification with 2 categories: "low stress" and "high stress", achieved with a RF algorithm. Similarly, authors of [36] define "normal state" and "stressed state" categories and use a Hidden Markov Model (HMM). Moreover, Support Vector Machine (SVM) is trained in [20] to perform a first classification with "stress" and "no stress" labels, then a second one with "low stress", "medium stress", and "high stress" labels. n the other hand, ***unsupervised algorithms*** are useful to deal with unlabelled data. Among them, K-Means and Gaussian Mixture Modelling (GMM) were used in some papers of the SLR. In [36], the authors highlight the benefits of these methods that do not require labeling the data. However, they are less represented in our corpus of articles.

**DL-Based Solutions.**  When dealing with images or sounds, ML algorithms cannot be efficiently trained, as they mostly require numerical input data. Thus, researchers resort to DL algorithms. In [37], authors leverage Long-Short-Term-Memory networks (LSTM), which is effective in numerous speech-related applications such as emotion recognition in speech. LSTM belongs to Recurrent Neural Networks (RNN) defined as memory neural networks because the output data depends on both input data and previous states of the network. Furthermore, in [30], the bidirectional variant (Bi-LSTM) is used to enhance stress-related emotion detection in speech. Besides, in [11], the Faster R-CNN algorithm is utilized to extract data from video recordings.

Table 3 summarizes AI algorithms used in SLR papers, considering if they belong to Supervised Learning (Classification or Regression) or Unsupervised Learning (Clustering) algorithms. We note that RF and Decision Tree (DT) algorithms figure in both Supervised Learning categories, as in some works, they were used for classification; and in others, they were utilized for Regression.

**Table 3.** AI Algorithms used in SLR Papers.

| Algorithm type | References |
| --- | --- |
| Classification (Supervised Learning) | SVM [11, 14, 16, 17, 19, 20, 24, 29, 31, 42, 43], DT[11, 16, 24, 39, 42, 43], K-Nearest Neighbours (KNN) [11, 16, 17, 24, 42], RF [16, 19, 24, 43, 44], Multilayer Perceptron (LMP) [19, 42, 44], LSTM [30, 37, 44], LR [24, 43], Gradient Boosting (XGB) [19], Ridge Classifier [19], Naïve Bayes [24], Decision Jungle [34], HMM [36], Neural Network (NN) [24], Elastic Net (EN) [44], AdaBoost [43] |
| Regression (Supervised Learning) | Linear Regression (LR) [21, 35], RF [21, 32], Ridge Regression [21], DT [21] |
| Clustering (Unsupervised Learning) | GMM [45], K-Means [25, 45] |

**Other Solutions.** ML algorithms require large datasets to be trained, when considering different usage contexts [41]. Few works in our corpus use other solutions. [12] utilizes the Poisson regression model, a statistical prediction model, to assess stress-related data obtained from a computer mouse. Moreover, [27] creates a correspondence table to correlate heart rate and skin galvanic response to a potential level of stress and danger for firefighters. In addition, mathematical algorithms are applied in [41], to discretize the dataset into values ranging from 1 (completely relaxed) to 5 (very stressed), using the Symbolic Aggregate Approximation algorithm, and to detect changes in it utilizing ADaptive WINdowing algorithm.

By studying the distribution of adopted solutions, we can see that AI was the most popular solution, being used in 77% of papers selected for the SLR, even though it requires resources to be implemented, such as the collection of large datasets.

## 5    RQ4: What Challenges and How Are They Addressed?

We dedicate this section to the challenges researchers have identified in our corpus, and to their suggestions to meet them. Most challenges raised using affective computing to measure stress at work are related to *ethics, confidentiality, and security*:

**Challenge 1: Ethical Data Collection.** Nowadays, international privacy laws exist, which may make data collection challenging [34]. However, it is possible to mitigate ethical and confidentiality issues.

*Using Less Sensitive Data.* In [19], authors distinguish between "low-level" data characterizing sensitive information such as email or text content, and "high-level" data, corresponding to less sensitive information like the number of emails or the duration of a call. Physiological data or user videos belong to the former category, resulting in significant ethical issues and discomfort among participants due to being observed all day [19]. Therefore, various works collect interaction data with work electronic devices (e.g., phones, computers), which appear less intrusive, besides not requiring the addition of extra sensors for data collection [16]. In the same context, authors of [12] use keyboard strokes as input data. In [30], data extraction is performed from audio recordings to store only the data without the raw audio. Similarly, [S30] retains only metadata related to communication such as voice activity or message length, while discarding raw content, notifications, or audio recordings.

*Considering User Preferences.* In [19], authors propose considering participants' preferences regarding the data they would prefer to share. In this context, users prefer to share data collected from mouse activity and keyboard. We then find data collected from computer usage, wearable sensors, microphones and phones. Finally, we find webcams. These preferences depend on concerned users and their activities.

**Challenge 2: Data Anonymization.** Various solutions exist to this end. First, it is possible to remove data which directly identifies an individual, such as names or phone numbers [12]. At this stage, it is nearly impossible to correlate the data with a specific individual afterward, but it is possible to add a random and unique identifier, known only by the individual concerned, to maintain this link [40]. Another possibility is to encode the data, i.e., replace information about an individual with identifiers or numerical codes [21, 26, 37, 41], and ensure that this process is irreversible [37]. Furthermore, authors in [21] proposed to label data in order to classify responses into categories. For example, categories like "employee under X years old" and "employee over X years old" does not require storing the precise age of the individual.

**Challenge 3: Data Secure Storage.** Another crucial issue addressed across various studies concerns data security. Due to the sensitive nature of incoming and outgoing data, it should only be accessible to specific individuals, and intrusions or data breaches should be avoided as much as possible. Two storage methods were proposed in our corpus. *Local storage*, as proposed in [19], conserves data directly on the devices used for data collection and processing. This method significantly decreases the risk of data theft since nothing is transmitted nor made accessible to third parties. Other experiments store data on specialized *remote servers* with considerable processing power. This solution may raise concerns regarding access to data by unauthorized individuals or potential cyberattacks. More secure options exist, where access is only granted to authorized individuals through their local internet connection or a Virtual Private Network (VPN), as seen in [40]. Among these options, [19] investigated users' perceptions seeing their preferred storage method, showing a preference for local storage, as it allows better data management. In conclusion, the choice depends on user preferences, experiments' objectives, and the needs of the collection, among other factors.

**Challenge 4: Results Accessibility.** To communicate data analysis results, two models are commonly used: in the *personal model*, individuals have exclusive access to their results [26]. However, this does not necessarily imply that the model is trained only on data about concerned individuals. Whereas in the *communal model*, the results are visible to a group of individuals, such as a team of colleagues. In this case, the objective is to create a supportive team environment where everyone is aware of the stress experienced by their teammates, allowing them to overcome stress-related issues through discussion or relaxation [23]. In this model, data can still be anonymous, meaning that all results can be accessed without knowing to whom they are associated, but data can easily be attributed to a specific person [35]. Users' feedbacks reflect mixed opinions [23]. While those who are close to their colleagues are comfortable with sharing such data, some show discomfort. This model is much less used than the personal model, as it may even induce more stress among participants, seeing that it invades privacy [23].

The second category of challenges identified in the literature is related to the ***reliability*** of affective computing tools, i.e., their ability to produce rigorous and consistent stress assessments.

**Challenge 5: Reliable Stress Identification.** The proposed models should reliably differentiate between stress and other emotional states. In [23], participants gave their feedback on a stress assessment system, and one of them noted that the system seemed more related to overall mood than to stress specifically.

**Challenge 6: Real-Environment Implementation.** This challenge highlights the importance of conducting experiments in contexts that are as realistic as possible, or directly in workplace environments. In [22], authors criticized scientific articles employing artificially replicated conditions for stress assessment using affective computing, where users perform standard daily tasks. These tasks may induce significantly higher stress levels than those encountered naturally, which may lead to inaccurate conclusions about the model's effectiveness in measuring stress at work. Besides, during data collection, noise may occur in diverse ways. For example, body movements can cause wearable sensors to shift, causing incorrect physiological data collection. Similarly, environmental noises such as conversations or external sounds may impact the quality of the audio recording. Noise can also arise from the quality of the devices used to collect data [42]. It is almost inevitable, whether in a controlled environment or not [23]. It is essential to control it by applying filters or removing excessively noisy data [31].

## 6   Conclusion

We performed a systematic literature review in which we reviewed a large body of literature describing how affective computing can be used to assess stress at work. We identified an appropriate collection of scientific papers according to some inclusion/exclusion criteria. Subsequently, we analyzed them revealing four different aspects: Application domains, Data collection, Data analysis, and Challenges. Using affective computing to measure stress at work is promising. However, the challenges characterizing these technologies remain alarming. Results interpretability is typically difficult when using deep learning methods, which is appropriate for facial expression analysis. In our future work, we aim to conduct experiments in a real context using a deep learning algorithm to detect negative emotions (such as stress, and despair) by analyzing facial expressions, in larger populations, to have concrete and efficient solutions for stressed and desperate workers.

## References

1. Jeanguenat, A.M., Dror, I.E.: Human factors effecting forensic decision making: workplace stress and well-being. J. Forensic Sci. **63**(1), 258–261 (2018)
2. Mittal, S., Mahendra, S., Sanap, V., Churi, P.: How can machine learning be used in stress management: a systematic literature review of applications in workplaces and education. Int. J. Inf. Manage. Data Insights **2**(2), 100110 (2022)
3. Carneiro, D., Novais, P., Augusto, J.C., Payne, N.: New methods for stress assessment and monitoring at the workplace. IEEE Trans. Affect. Comput. **10**(2), 237–254 (2017)
4. Munoz, S., Iglesias, C.Á., Mayora, O., Osmani, V.: Prediction of stress levels in the workplace using surrounding stress. Inf. Process. Manage. **59**(6), 103064 (2022)
5. Richardson, S.: Affective computing in the modern workplace. Bus. Inf. Rev. **37**(2), 78–85 (2020)
6. Kitchenham, B., Charters, S.: Guidelines for performing systematic literature reviews in software engineering (2007)
7. Lopes, L., Rodrigues, A., Cabral, D., Campos, P.: From monitoring to assisting: a systematic review towards healthier workplaces. Int. J. Environ. Res. Public Health **19**(23), 16197 (2022)
8. Can, Y.S., Arnrich, B., Ersoy, C.: Stress detection in daily life scenarios using smart phones and wearable sensors: a survey. J. Biomed. Inform. **92**, 103139 (2019)
9. Khakurel, J., Melkas, H., Porras, J.: Tapping into the wearable device revolution in the work environment: a systematic review. Inf. Technol. People **31**(3), 791–818 (2018)
10. Aiello, J.R., Kolb, K.J.: Electronic performance monitoring and social context: impact on productivity and stress. J. Appl. Psychol. **80**(3), 339 (1995)
11. Lawanot, W., Inoue, M., Yokemura, T., Mongkolnam, P., Nukoolkit, C.: Daily stress and mood recognition system using deep learning and fuzzy clustering for promoting better well-being. In: IEEE International Conference on Consumer Electronics (ICCE), pp. 1–6 (2019)
12. Banholzer, N., Feuerriegel, S., Fleisch, E., Bauer, G.F., Kowatsch, T.: Computer mouse movements as an indicator of work stress: longitudinal observational field study. J. Med. Internet Res. **23**(4), e27121 (2021)
13. Ubilluz, C., Delgado, R., Marcillo, D., Noboa, T.: Brain waves processing, analysis and acquisition to diagnose stress level in the work environment. In: Rocha, Á., Adeli, H., Reis, L.P., Costanzo, S. (eds.) WorldCIST 2018. AISC, vol. 746, pp. 859–866. Springer, Cham (2018). https://doi.org/10.1007/978-3-319-77712-2_81
14. Leone, A., Rescio, G., Siciliano, P., Papetti, A., Brunzini, A., Germani, M.: Multi sensors platform for stress monitoring of workers in smart manufacturing context. In: IEEE International Instrumentation and Measurement Technology Conference (I2MTC), pp. 1–5. IEEE (2020)
15. Androutsou, T., Angelopoulos, S., Kouris, I., Hristoforou, E., Koutsouris, D.: A smart computer mouse with biometric sensors for unobtrusive office work-related stress monitoring. In: 43rd Annual International Conference of the IEEE Engineering in Medicine and Biology Society (EMBC), pp. 7256–7259 (2021)
16. Pepa, L., Sabatelli, A., Ciabattoni, L., Monteriu', A., Lamberti, F., Morra, L.: Stress detection in computer users from keyboard and mouse dynamics. IEEE Trans. Consum. Electron. **67**(1), 12–19 (2020)
17. Yu, B., Zhang, B., An, P., Xu, L., Xue, M., Hu, J.: An unobtrusive stress recognition system for the smart office. In: 41st Annual International Conference of the IEEE Engineering in Medicine and Biology Society (EMBC), pp. 1326–1329 (2019)

18. Rahman, M.F.A., Giovanni, V.C., Warnars, H.L.H.S., Aryono, G.D.P., Megantoro, B.: Facial recognition development to detect corporate employees stress level. In: IEEE International Conference on Engineering, Technology and Education (TALE), pp. 1–6 (2019)

19. Morshed, M.B., et al.: Advancing the understanding and measurement of workplace stress in remote information workers from passive sensors and behavioral data. In: 10th International Conference on Affective Computing and Intelligent Interaction (ACII), pp. 1–8 (2022)

20. Kuroha, M., Ban, Y., Fukui, R., Warisawa, S.I.: Chronic stress level estimation focused on motion pattern changes acquired from seat pressure distribution. In: IEEE International Conference on Cyberworlds (CW), pp. 135–142 (2019)

21. Stefanescu, V.A., Radoi, I.E.: Stress level prediction using data from wearables. In: 18th RoEduNet Conference: Networking in Education and Research (RoEduNet), pp. 1–6 (2019)

22. Akbar, F., Mark, G., Pavlidis, I., Gutierrez-Osuna, R.: An empirical study comparing unobtrusive physiological sensors for stress detection in computer work. Sensors 19(17), 3766 (2019)

23. Xue, M., Liang, R.H., Hu, J., Yu, B., Feijs, L.: Understanding how group workers reflect on organizational stress with a shared, anonymous heart rate variability data visualization. In: CHI Conference on Human Factors in Computing Systems Extended Abstracts, pp. 1–7 (2022)

24. Fauzi, M.A., Yang, B.: Continuous stress detection of hospital staff using smartwatch sensors and classifier ensemble. In: pHealth, pp. 245–250. IOS Press (2021)

25. Makowska-Tlomak, E., Nielek, R., Skorupska, K., Paluch, J., Kopec, W.: Evaluating a sentiment analysis tool to detect digital transformation stress. In: IEEE/WIC/ACM International Conference on Web Intelligence and Intelligent Agent Technology, pp. 103–111 (2021)

26. Chodan, W., et al.: The SEBA system: a novel approach for assessing psychological stress continuously at the workplace. In: Proceedings of the 6th International Workshop on Sensorbased Activity Recognition and Interaction, pp. 1–6 (2019)

27. Raj, J.V., Sarath, T.V.: An IoT based real-time stress detection system for fire-fighters. In: International Conference on Intelligent Computing and Control Systems (ICCS), pp. 354–360. IEEE (2019)

28. Tsuji, S., Sato, N., Ara, K., Yano, K.: Estimating group stress level by measuring body motion. Front. Psychol. 12, 634722 (2021)

29. Gavas, R.D., et al.: A sensor-enabled digital trier social stress test in an enterprise context. In: 41st Annual International Conference of the IEEE Engineering in Medicine and Biology Society (EMBC), pp. 1321–1325. IEEE (2019)

30. Gaballah, A., Tiwari, A., Narayanan, S., Falk, T.H.: Context-aware speech stress detection in hospital workers using Bi-LSTM classifiers. In: IEEE International Conference on Acoustics, Speech and Signal Processing (ICASSP), pp. 8348–8352 (2021)

31. Tiwari, A., Narayanan, S., Falk, T.H.: Stress and anxiety measurement "in-the-wild" using quality-aware multi-scale HRV features. In: 41st Annual International Conference of the IEEE Engineering in Medicine and Biology Society (EMBC), pp. 7056–7059. IEEE (2019)

32. Umematsu, T., Sano, A., Taylor, S., Tsujikawa, M., Picard, R.W.: Forecasting stress, mood, and health from daytime physiology in office workers and students. In: 42nd Annual International Conference of the IEEE Engineering in Medicine and Biology Society, IEEE (EMBC), pp. 5953–5957 (2020)

33. Paletta, L., et al.: Towards Large-scale evaluation of mental stress and biomechanical strain in manufacturing environments using 3D-referenced gaze and wearable-based analytics. Electron. Imaging 33, 1–7 (2021)

34. Nkurikiyeyezu, K., Shoji, K., Yokokubo, A., Lopez, G.: Thermal comfort and stress recognition in office environment. In: HEALTHINF, pp. 256–263 (2019)

35. Hoekstra, M., Lu, P.L., Lyu, T., Zhang, B., Hu, J.: Collective stress visualization enabled by smart cushions for office chairs. In: Streitz, N.A., Konomi, S. (eds.) HCII 2022. LNCS, vol. 13325, pp. 278–290. Springer, Cham (2022). https://doi.org/10.1007/978-3-031-05463-1_20

36. Vildjiounaite, E., Huotari, V., Kallio, J., Kyllönen, V., Mäkelä, S.M., Gimel'farb, G.: Unobtrusive assessment of stress of office workers via analysis of their motion trajectories. Perv. Mobile Comput. **58**, 101028 (2019)

37. Bromuri, S., Henkel, A.P., Iren, D., Urovi, V.: Using AI to predict service agent stress from emotion patterns in service interactions. J. Serv. Manag. **32**(4), 581–611 (2021)

38. Mättig, B., Döltgen, M., Archut, D., Kretschmer, V.: Intelligent work stress monitoring: prevention of work-related stress with the help of physiological data measured by a sensor wristband. In: Arai, K., Kapoor, S., Bhatia, R. (eds.) Intelligent Systems and Applications: 2018 Intelligent Systems Conference (IntelliSys), vol. 2, pp. 1211–1222. Springer, Cham (2019). https://doi.org/10.1007/978-3-030-01057-7

39. Maxhuni, A., Hernandez-Leal, P., Morales, E.F., Sucar, L.E., Osmani, V., Mayora, O.: Unobtrusive stress assessment using smartphones. IEEE Trans. Mob. Comput. **20**(6), 2313–2325 (2020)

40. Bolliger, L., Lukan, J., Luˇstrek, M., De Bacquer, D., Clays, E.: Protocol of the STRess At Work (STRAW) project: how to disentangle day-to-day occupational stress among academics based on EMA, physiological data, and smartphone sensor and usage data. Int. J. Environ. Res. Public Health **17**(23), 8835 (2020)

41. Suni Lopez, F., Condori-Fernandez, N., Catala, A.: Towards real-time automatic stress detection for office workplaces. In: 5th International Conference In Information Management and Big Data, SIMBig, pp. 273–288 (2018)

42. Jebelli, H., Choi, B., Lee, S.: Application of wearable biosensors to construction sites. I: assessing workers' stress. J. Constr. Eng. Manage. **145**(12), 04019079 (2019)

43. Khowaja, S.A., Prabono, A.G., Setiawan, F., Yahya, B.N., Lee, S.L.: Toward soft real-time stress detection using wrist-worn devices for human workspaces. Soft. Comput. **25**, 2793–2820 (2021)

44. Booth, B.M., Vrzakova, H., Mattingly, S.M., Martinez, G.J., Faust, L., D'Mello, S.K.: Toward robust stress prediction in the age of wearables: modeling perceived stress in a longitudinal study with information workers. IEEE Trans. Affect. Comput. **13**(4), 2201–2217 (2022)

45. Mozgovoy, V.: Longitudinal estimation of stress-related states through bio-sensor data. Appl. Comput. Inform. (2021)

# Rough Set Decision Rules for Usage-Based Churn Modeling in Mobile Telecommunications

Małgorzata Przybyła-Kasperek[1]([✉]) [iD] and Piotr Sulikowski[2] [iD]

[1] Institute of Computer Science, University of Silesia in Katowice, Będzińska 39, 41-200 Sosnowiec, Poland
malgorzata.przybyla-kasperek@us.edu.pl
[2] Faculty of Computer Science and Information Technology, West Pomeranian University of Technology in Szczecin, ul. Żołnierska 49, 71-210 Szczecin, Poland
piotr.sulikowski@zut.edu.pl

**Abstract.** The concern of customer churn significantly impacts the telecommunications industry, given the considerable costs associated with acquiring new customers compared to retaining existing ones. To effectively guide anti-churn initiatives, it becomes crucial to point profitable clients with the highest likelihood of churning. However, the data utilized for identifying potential churners often carries inherent imprecision. In this study, we use a rough sets modeling approach for discerning churn intent based on usage data within the mobile telecommunications domain. This approach takes into account the uncertainty in the data and provides concise, easy-to-interpret rules. In the paper we use four approaches: exhaustive algorithm, genetic algorithm, covering algorithm and LEM2 algorithm and attribute discretization method. Finally, it was found that the best results are obtained using the rough set rule-based systems with the LEM2 algorithm and attribute discretization. A very high accuracy of 0.997 was achieved. The results for the analyzed case are better than those generated by a fuzzy rule-based system.

**Keywords:** Rough set theory · decision rules · churn modeling · telecommunication · artificial intelligence

## 1 Introduction

The issue of customer churn poses a significant challenge for various industries, notably within the mobile telecommunications sector, where the cost of acquiring customers is substantial. Recent years have witnessed a rapid expansion of the mobile telecommunications market across numerous countries. The escalating competitiveness in this market, marked by a surge in telecom operators, increased ease of number portability between operators, the introduction of sophisticated products and services, as well as enticing special offers, underscores the growing importance of addressing customer churn for operators [6, 26].

© The Author(s), under exclusive license to Springer Nature Switzerland AG 2024
N.-T. Nguyen et al. (Eds.): ICCCI 2024, CCIS 2165, pp. 57–70, 2024.
https://doi.org/10.1007/978-3-031-70248-8_5

Predictive modeling techniques, such as machine learning algorithms and data mining, have been widely employed to forecast churn in telecom markets. Real-time analysis of customer data enables companies to detect early signs of churn and take proactive measures to retain customers [19]. Different machine learning approaches are being used for this purpose: neural networks [9], decision trees [8] and ensembles of classifiers [22]. Segmentation of telecom customers based on their characteristics, preferences, and behavior is instrumental in devising targeted retention strategies and is often used in churn prediction. Very often the presented churn analyses are region or country specific [20,23]. There are many studies in the literature on the use of rough sets and fuzzy sets in this field of customer churn [1,12]. Decision rules based on rough sets as well as reducts are used in [2,29]. However, in all these works, the research was conducted on benchmark data sets or mainly on categorical attributes. In contrast, in the present work, the research is conducted on a real data set from a major Polish mobile telecommunication operator. In addition, the advanced analysis and selection of attributes performed prior to using the rough set approach is used. Also, clustering algorithms were used in the literature to analyze customer churn data [11,31].

Rough set theory was proposed in the 1980s by Z. Pawlak [15]. This theory provides methods for describing uncertain concepts by using lower and upper approximation, techniques for reducing multidimensionality in data sets, and strategies for generating rules-based classifiers. Rough sets are very popular and widely developed [16,17,27].

The aim of the study is to present and discuss the results of a churn rough set modeling conducted on real-world data obtained from one of the largest mobile operators in Poland. For this purpose, a comprehensive analysis of the importance and relevance of attributes was first conducted, and then four different approaches were used to generate a rule-based system. In addition, systems with and without attribute discretization were analyzed. The quality of the generated systems was analyzed using 10-fold cross validation with measures such as precision, recall, F1 score and accuracy, as well as decision rule coverage. The number and length of decision rules generated were also compared. It was found that the best system in terms of both classification quality, clarity, conciseness and readability of rules was obtained using the LEM2 algorithm with attribute discretization.

In the paper, the rough set rule-based approach is compared with the approach based on fuzzy sets. It was shown that the rough set approach gives better results and provides an improvement in the quality of classification.

The structure of the paper is organised as follows. Section 2 is dedicated to the rough set theory and used algorithms. Section 3 addresses the data sets that are used, the methodology and presents the obtained results and comparisons. The final section provides conclusions and future research plans.

## 2 Methodology and Basic Concepts

The primary objective of this study is to introduce a rough set methodology for churn modeling in the realm of mobile telecommunications. This is aimed at enhancing retention activities by targeting customers with a heightened likelihood of churn.

Our goal is to justify and demonstrate the use of rough sets in creating a churn modeling system that is both highly accurate and explainable. Given the indeterminate nature of the relationship between customer behavior, as manifested in usage data, and propensity to churn, we decided to adopt rough sets theory to model these relationships and express them through a set of understandable rules.

Rough set methodologies have two key features: managing uncertainty by describing knowledge through rough sets and rules, and offering excellent interpretability through simple language rules. In our presented approach to churn modeling, using real telecom data, we will follow the rough set modeling steps in four different approaches.

Professor Z. Pawlak introduced the rough set theory [14] as an approach to handle incomplete data and identify the crucial attributes necessary for maintaining discernibility among objects. The fundamental concept revolves around the indiscernibility relation, defined for an information system $S = (U, A)$ where $U$ represents the universe (a set of objects), $A$ is a set of attributes, and $a : U \to V_a$ maps objects to values in $V_a$ for attribute $a$. The indiscernibility relation is expressed as

$$IND(A) = \{(x, y) \in U \times U : \forall_{a \in A} a(x) = a(y)\}.$$

This relation is an equivalence relation, exhibiting reflexivity, symmetry, and transitivity, thereby forming partitioned classes of objects known as granules (atoms) of information.

The significance of attributes in the system lies in those whose removal from set $A$ induces changes in information granules compared to the original set of attributes. Another key concept in rough set theory is that of a reduct, defined as the smallest subset of attributes that yields the same information granules as the original set of attributes. This concept proves instrumental in significantly reducing the multidimensionality of data without losing knowledge.

When a decision attribute is identified in an information system, it transforms into a decision table. More precisely, $S = (U, A \cup d)$ is recognized as a decision table, where $A$ represents a set of condition attributes, and $d \notin A$ is the decision attribute. Within the decision table, a decision reduct is ascertained. This reduct is the smallest subset of conditional attributes, denoted as $B \subseteq A$, such that objects from distinct decision classes are distinguishable, i.e., $IND(B) \subseteq IND(d)$.

Decision rule is a formula of the form

$$(a_{i_1} = v_1) \wedge \ldots \wedge (a_{i_m} = v_m) \Rightarrow (d = k)$$

where $a_{i_j} \in A$, $v_i \in V_{a_i}$ for $j \in \{1, \dots, m\}$, $i$ is a natural number. Atomic formulas $(a_{i_j} = v_j)$ are called descriptors.

In existing literature, diverse techniques for identifying decision rules are proposed. One approach relies on a brute-force strategy, but its applicability is restricted to decision tables with a small number of attributes. Alternatively, numerous heuristic algorithms have been proposed to efficiently determine decision rules within a reasonable time frame. Some of these methods draw inspiration from an ant colony [10], approximations [13], or employ alternative methodologies [24, 28].

This study employs four rough set rule-based systems using methods based on: exhaustive algorithm [3], genetic algorithm [5], covering algorithm [4] and LEM2 algorithm [7]. The motivation for choosing these four approaches was their diversity and sensitivity to different aspects when building decision rules. The exhaustive search algorithm guarantees that the optimal set of rules for the data set will be found within the search space as the brute-force search is used. The genetic algorithm is effective for exploring large search spaces and handling high-dimensional data. It can also handle non-linear and complex relationships between attributes. The covering algorithm incrementally builds rules by focusing on the most relevant conditions, leading to interpretable and concise rule sets. The LEM2 algorithm handles inconsistency and uncertainty in the data by incorporating multiple examples and exceptions. When we want to explore the capabilities of rough sets approaches to churn data, all these approaches should be tested. They are a good representation of the other methods related to rough sets theory.

The exhaustive algorithm aims to generate possible rules from the given input data by systematically exploring all object oriented reducts (or local reducts). These algorithms ensure a thorough examination of the entire solution space, leaving no possibility unexplored but are time consuming. The primary objective of the genetic algorithm is to identify a specified number of shortest reducts within a reasonable time, even when dealing with multidimensional data. Each individual in the population is encoding a potential reduct subset of attributes. Generation of the population involves classic binary operators, including mutation, crossover, and the roulette wheel selection algorithm. The covering algorithm focuses on identifying minimal (or very close to minimal) set of rules that cover all objects within a data set. The LEM2 algorithm – Learning from Examples Module 2 – is specifically designed for rule induction within the rough set framework. It aims to generate decision rules that accurately capture the discernibility between different classes in the data. It is particularly effective in situations where the data exhibit inherent uncertainty or imprecision, providing a mechanism to discern meaningful patterns despite the presence of noise. LEM2 starts by initializing an empty rule set. Then it iteratively generates rules to cover instances in the data set. For each attribute in the data set, LEM2 explores potential conditions that can be used to partition the data into subsets. It selects the attribute and its corresponding value that maximally reduce the impurity (entropy) of the resulting subsets. Then the rules are evaluated using measures such as support and confidence. LEM2 employs a pruning strategy to refine the generated rules and improve their generalization ability.

In this article we use implementation of all the above mentioned approaches in the Rough Sets Exploration System (RSES) program [4]. In the paper also rough set rule-based systems were examined that were generated after discretizing the attributes present in the churn data set. The algorithm implemented in RSES is based on searching of the optimal consistent set of cuts in the attribute's values. In the discretization algorithm optimization criteria were established based on the number of cuts and consistency being defined as the preservation of the discernibility relation between objects belonging to different decision classes. In the paper, we used a local approach, which means that for each attribute discretization is conducted independently. The entropy-based method for attribute discretization in RSES uses information theory concepts to determine optimal split points for transforming continuous attributes into discrete intervals. More formally for each attribute separately, the following procedure is implemented. Initially, RSES calculates the entropy of the decision attribute. RSES identifies potential split points along the range of the continuous attribute based on the unique attribute values. It calculates the entropy for each resulting interval. RSES chooses the split point where the most significant reduction in entropy occurs. This selected split point becomes the boundary between two intervals for discretization.

In order to verify churn prediction capabilities with rough set rule-based system, first we divided data into a training and testing sets. For performing validation, we incorporated the 10-fold cross validation technique. As the final step the obtained rough set rule-based model was evaluated using measures such as Kappa and F1 score.

## 3    Experiments Results and Comparisons

This section describes three key aspects: the selection of attributes, the classification quality obtained for the rough set rule-based systems and the analysis of the generated decision rules.

### 3.1    Source Data and Features

We use a data set from one of the major Polish mobile telecommunication operators, comprising information on 15,000 randomly selected clients, for our analysis. The data set was initially segmented to focus on a relatively homogeneous group, specifically individual (non-corporate) post-paid customers. Criteria included having only one active SIM card at the beginning of the analyzed period, maintaining a consistent residential address, and allowing for a maximum change of the quantity of active SIM cards once.

Comprising three tables, the source data set included the BASE table with variables describing customers and contracts, the STATUS table containing activity status history, and the USAGE table detailing service usage per month [25].

To construct a rough set rule-based model for churn modeling based on usage data, payment history, and interactions, a set of 148 potential features for churn modeling had been prepared, including 10 directly extracted from the source data set and 138 constructed through domain knowledge and expert consultation. These features underwent preprocessing, including aggregation and relative changes calculations [32].

We decided to remove non-numerical features, resulting in a data set of 138 features divided into three main groups presented in Table 1: time-related, usage-related, and payment-related. Missing values were addressed using a model-based imputer employing a simple decision tree [18, 32].

**Table 1.** Feature types used in the experiment.

| Time-related features |
|---|
| cooperation_days_with_operator |
| number_of_days_between_deactivation_inquiry_and_first_activation |
| number_of_days_between_deactivation_inquiry_and_date_of_last_promotion |
| number_of_days_between_start_of_analysis_and_resignation |
| user_age_at_the_end_of_analyzed_period |
| **Usage-related features** [total in period, and changes between periods of: 1, 3 and 6 months] |
| sms_number_out |
| mms_number_out |
| outgoing_calls_number |
| outgoing_calls_minutes |
| **Payment-related features** [total in period, and changes between periods of: 1, 3 and 6 months] |
| subscription_total_amount |
| invoice_total_amount |

The data set was too extensive for generating explainable rough set rules. Feature selection was performed using the Gini ranking algorithm [30] and the final set consisted of eight features, emphasizing financial aspects, customer seniority, and outgoing calls. These features are as follows: invoice_total_amount_change_quarter_1, invoice_total_amount_month_1, invoice_total_amount_month_0, cooperation_days_with_operator, invoice_total_amount_change_quarter_0, outgoing_calls_minutes_quarter_0, subscription_total_amount_change_quarter_1, subscription_total_amount_quarter_1. The target variable, called "churned", indicated whether a customer churned (value 1) or remained with the operator (value 0).

To ensure comparable periods, clients with a cooperation history shorter than three quarters were removed, resulting in a final data set of 8,577 records, including 278 churners.

## 3.2   Classification Quality Evaluation

Table 2 presents the confusion matrices and the classification quality metrics obtained for four different methods of generating decision rules without attribute discretization. Table 3 shows the results obtained when discretization of all attributes was applied. The results were the averages obtained using 10-fold cross validation. The following parameters were used in the RSES program when the rough set rule-based systems were generated: for the genetic algorithm – number of reducts 200 and High speed; for the covering algorithm and LEM2 algorithm – cover parameter 0.99;

**Table 2.** Confusion matrix with accuracy metrics for rough set rule-based systems without attribute discretization.

| Rough set rule-based systems – Exhaustive algorithm | | | | | | |
|---|---|---|---|---|---|---|
| Churned | TP | FP | TN | FN | Precision | Sensitivity | F1 Score |
| 0 | 817.9 | 10.9 | 23 | 4.7 | 0.99 | 0.99 | 0.99 |
| 1 | 23 | 4.7 | 817.9 | 10.9 | 0.83 | 0.68 | 0.75 |

Overall accuracy (OA) 0.982;      Kappa 0.737;      Total coverage 0.999

| Rough set rule-based systems – Genetic algorithm | | | | | | |
|---|---|---|---|---|---|---|
| Churned | TP | FP | TN | FN | Precision | Sensitivity | F1 Score |
| 0 | 819.2 | 9.7 | 22.9 | 4.7 | 0.99 | 0.99 | 0.99 |
| 1 | 22.9 | 4.7 | 819.2 | 9.7 | 0.83 | 0.70 | 0.76 |

Overall accuracy (OA) 0.983;      Kappa 0.752;      Total coverage 0.999

| Rough set rule-based systems – Covering algorithm | | | | | | |
|---|---|---|---|---|---|---|
| Churned | TP | FP | TN | FN | Precision | Sensitivity | F1 Score |
| 0 | 818.6 | 7.7 | 18.9 | 7.7 | 0.99 | 0.99 | 0.99 |
| 1 | 18.9 | 7.7 | 818.6 | 7.7 | 0.71 | 0.71 | 0.71 |

Overall accuracy (OA) 0.982;      Kappa 0.701;      Total coverage 0.995

| Rough set rule-based systems – LEM2 algorithm | | | | | | |
|---|---|---|---|---|---|---|
| Churned | TP | FP | TN | FN | Precision | Sensitivity | F1 Score |
| 0 | 679 | 0.8 | 15.1 | 1 | 1 | 1 | 1 |
| 1 | 15.1 | 1 | 679 | 0.8 | 0.95 | 0.94 | 0.94 |

Overall accuracy (OA) 0.997;      Kappa 0.942;      Total coverage 0.812

In all tested systems, it is apparent that the model has noticeably better predictions for those who did not churn, which is connected with an unbalanced data set. We will address this issue further in the future research. Nevertheless, the model offers good predictions for real churners as well.

Figure 1 shows a comparison of the three measures: overall accuracy, kappa and total coverage obtained for the four rough set rule-based systems: Exhaustive algorithm, Genetic algorithm, Covering algorithm and LEM2 algorithm with and without attribute discretization. The conclusions that arise from the study are as follows:

**Table 3.** Confusion matrix with accuracy metrics for rough set rule-based systems with attribute discretization.

| Rough set rule-based systems – Exhaustive algorithm | | | | | | |
|---|---|---|---|---|---|---|
| Churned | TP | FP | TN | FN | Precision | Sensitivity | F1 Score |
| 0 | 826.5 | 2.7 | 24.3 | 3.5 | 1 | 1 | 1 |
| 1 | 24.3 | 3.5 | 826.5 | 2.7 | 0.87 | 0.90 | 0.89 |
| Overall accuracy (OA) 0.993;      Kappa 0.883;      Total coverage 1 | | | | | | | |
| Rough set rule-based systems – Genetic algorithm | | | | | | | |
| Churned | TP | FP | TN | FN | Precision | Sensitivity | F1 Score |
| 0 | 826.9 | 2.3 | 24.2 | 3.6 | 1 | 1 | 1 |
| 1 | 24.2 | 3.6 | 826.9 | 2.3 | 0.91 | 0.87 | 0.89 |
| Overall accuracy (OA) 0.993;      Kappa 0.888;      Total coverage 1 | | | | | | | |
| Rough set rule-based systems – Covering algorithm | | | | | | | |
| Churned | TP | FP | TN | FN | Precision | Sensitivity | F1 Score |
| 0 | 225.6 | 0 | 0 | 0.4 | 1 | 1 | 1 |
| 1 | 0 | 0.4 | 225.6 | 0 | 0 | 0 | 0 |
| Overall accuracy (OA) 0.998;      Kappa 0;      Total coverage 0.264 | | | | | | | |
| Rough set rule-based systems – LEM2 algorithm | | | | | | | |
| Churned | TP | FP | TN | FN | Precision | Sensitivity | F1 Score |
| 0 | 808 | 0.5 | 18.8 | 2.2 | 1 | 1 | 1 |
| 1 | 18.8 | 2.2 | 808 | 0.5 | 0.95 | 0.94 | 0.94 |
| Overall accuracy (OA) 0.997;      Kappa 0.931;      Total coverage 0.968 | | | | | | | |

- Attribute discretization definitely improves the quality of classification in terms of all measures for systems constructed using Exhaustive algorithm, Genetic algorithm and LEM2 algorithm. Only in the case of Covering algorithm the results we obtained after discretization are not valuable due to very low coverage and practically no possibility of recognizing the minority class Churned = 1.
- In the situation where we do not discretize the attributes, the best turned out to be the system using LEM2 algorithm here we have an accuracy of 0.997 however we get it with reduced coverage equal to 0.812. This is probably because the LEM2 algorithm achieves generalisation and noise immunity through rule pruning. The rules can then cover fewer cases. Thus, if we care about recognizing all objects, a better approach would be to use Genetic algorithm, which also generates high accuracy equal to 0.983 with almost complete coverage equal to 0.999.
- If we use attribute discretization, the LEM2 algorithm with the same accuracy as without attribute discretization achieves a higher coverage 0.968. Attribute discretization also improves classification accuracy for the approaches Exhaustive algorithm and Genetic algorithm to 0.993 with complete coverage.

In paper [32], the same data set was studied using a fuzzy set approach and two systems the Mamdani model and the Sugeno model. The main results

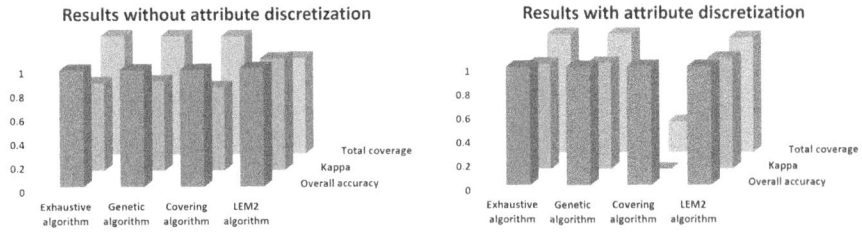

**Fig. 1.** Comparison of results obtained for the rough set rule-based systems without and with attribute discretization.

obtained in paper [32] are quoted in Table 4. As can be seen, the better results in terms of Overall accuracy (OA) and Kappa were obtained using rough set rule-based systems with discretization of attributes and Exhaustive algorithm or Genetic algorithm. For the rough set rule-based systems with LEM2 algorithm, the results are also better than for the fuzzy approach however here the coverage is smaller. Thus, it can be concluded that for a real data set provided by one of the biggest Polish mobile telecommunication operator regarding the risk of churn, the rough set approach gives better results in terms of classification quality than the fuzzy set approach.

**Table 4.** Comparison. Confusion matrix with accuracy metrics for fuzzy rule-based system.

| Fuzzy rule-based systems – Mamdani model | | | | | | | |
|---|---|---|---|---|---|---|---|
| Churned | TP | FP | TN | FN | Precision | Sensitivity | F1 Score |
| 0 | 825.6 | 10.3 | 17.5 | 4.3 | 0.99 | 1 | 0.99 |
| 1 | 17.5 | 4.3 | 825.6 | 10.3 | 0.80 | 0.63 | 0.71 |
| Overall accuracy (OA) 0.983;       Kappa 0.697;       Total coverage 1 | | | | | | | |
| Fuzzy rule-based systems – Sugeno model | | | | | | | |
| Churned | F1 Score | | | | | | |
| 0 | 0.51 | | | | | | |
| 1 | 0.98 | | | | | | |
| Overall accuracy (OA) 0.98;       Kappa 0.502 | | | | | | | |

### 3.3   Analysis of Rough Set Rules Generation

Now we compare all the analyzed rough set rule-based systems in terms of the characteristics of the generated rules. Table 5 presents values such as the number of rules generated (No. rule) for each decision class, support of rules (Supp) and length of rule premises (Len).

**Table 5.** Comparison of rough set rule-based systems in terms of rules characteristics.

| No. rule | No. rule Churned = 0 | No. rule Churned = 1 | Supp Max | Supp Min | Supp Avg | Len Max | Len Min | Len Avg |
|---|---|---|---|---|---|---|---|---|
| Exhaustive algorithm without attribute discretization | | | | | | | | |
| 19021 | 17174 | 1847 | 730 | 1 | 2.4 | 3 | 1 | 1.2 |
| Genetic algorithm without attribute discretization | | | | | | | | |
| 18917 | 17084 | 1833 | 744 | 1 | 2.5 | 3 | 1 | 1.2 |
| Covering algorithm without attribute discretization | | | | | | | | |
| 14485 | 14004 | 481 | 1025 | 1 | 2.5 | 1 | 1 | 1 |
| LEM2 algorithm without attribute discretization | | | | | | | | |
| 1599 | 1456 | 143 | 721 | 1 | 7.7 | 6 | 1 | 4.5 |
| Exhaustive algorithm with attribute discretization | | | | | | | | |
| 2034 | 1435 | 599 | 2172 | 1 | 30.3 | 6 | 1 | 2.9 |
| Genetic algorithm with attribute discretization | | | | | | | | |
| 2063 | 1438 | 625 | 2136 | 1 | 30.2 | 6 | 1 | 2.9 |
| Covering algorithm with attribute discretization | | | | | | | | |
| 3 | 1 | 2 | 1831 | 1 | 611 | 1 | 1 | 1 |
| LEM2 algorithm with attribute discretization | | | | | | | | |
| 256 | 164 | 92 | 1594 | 1 | 61.9 | 8 | 2 | 5.5 |

Naturally, it is quite apparent that after attribute discretization, the number of decision rules obtained in rough set rule-based systems is much smaller than for systems without attribute discretization. This was confirmed in the above results. The ratio of the number of rules without and with attribute discretization reaches almost ten times the reduced number of rules. The exception is the LEM2 algorithm where six times fewer rules were obtained after discretization. Also with using attributes discretization, one can see a much higher rule support, which also seems natural. Based on the results, it can also be concluded that discretization of attributes increased the average length of rule premises. Comparing the different approaches: Exhaustive algorithm, Genetic algorithm, Covering algorithm and LEM2 algorithm, it can be concluded that the smallest number of rules is generated by the Covering algorithm and LEM2 algorithm approaches. Together with the analysis of classification quality, it appears that the best system from both of these perspectives is the LEM2 algorithm with attribute discretization. This system generates the highest classification accuracy with relatively high coverage and the least number of decision rules. Thus, it can be concluded that it is the most efficient and concise system in this case. The top five rules (with the most support) obtained for this system as well as the two decision rules with the most support for the decision class churned = 1 are shown below:

IF (subscription_total_amount_quarter_1 > 122.5) AND (subscription_total_amount_month_0 > 40.945) THEN (churned = 0) with Support 1594

IF (subscription_total_amount_change_quarter_1 ∈ (−0.0435, 1.0405)) AND (subscription_total_amount_change_quarter_0 ∈ (−0.0135, 0.0535)) AND (subscription_total_amount_month_0 > 40.945) THEN (churned = 0) with Support 1344

IF (subscription_total_amount_change_quarter_1 ∈ (−0.0435, 1.0405)) AND (subscription_total_amount_quarter_1 < 122.5)) AND (invoice_total_amount_change_quarter_0 > −0.136) AND (subscription_total_amount_change_quarter_0 ∈ (−0.0135, 0.0535)) AND (outgoing_calls_minutes_quarter_0 ∈ (5422.5, 16967.0)) THEN (churned = 0) with Support 926

IF (subscription_total_amount_change_quarter_1 ∈ (−0.0435, 1.0405)) AND (subscription_total_amount_quarter_1 < 122.5) AND (invoice_total_amount_change_quarter_0 > −0.136) AND (subscription_total_amount_change_quarter_0 ∈ (−0.0135, 0.0535)) AND (outgoing_calls_minutes_quarter_0 ∈ (1140.0, 5422.5)) THEN (churned = 0) with Support 847

IF (subscription_total_amount_quarter_1 < 122.5)) AND (subscription_total_amount_change_quarter_1 ∈ (−0.0435, 1.0405)) AND (subscription_total_amount_change_quarter_0 ∈ (−0.0135, 0.0535)) AND (cooperation_days_with_operator > 1357.5) AND (outgoing_calls_minutes_quarter_0 ∈ (5422.5, 16967.0)) THEN (churned = 0) with Support 495

As can be seen from the above rules, attributes: subscription_total_amount_change_quarter_1 and subscription_total_amount_month_0 appear in most of the rules listed above. In addition, we can also see that the rules with the highest support have quite a large number of premises, in most cases it is 5 premises. We may also notice that features denoting total invoice amount and change in subscription as well as change in total invoice may indicate that the user intends to change operators.

## 4   Conclusions

Churn modeling holds significance in enhancing retention strategies within the mobile telecommunications industry. This study focuses on the creation of eight different rule-based systems using rough sets theory for churn modeling, specifically tailored for one of the prominent Polish mobile operators, to help identify customers likely to churn. The successful implementation of the models incorporates six crucial measures, demonstrating solid prediction capabilities. Notably, this achievement remains noteworthy even when considering the inherent homogeneity within the customer group and the source data set imbalance.

The paper uses four different approaches to generate rough set rule-based systems: exhaustive algorithm, genetic algorithm, covering algorithm and LEM2 algorithm. Results were obtained with and without attribute discretization. It was found that definitely the use of attribute discretization improves the quality of the system both in terms of classification quality, increased accuracy and coverage, as well as in terms of model readability – significantly fewer numbers of rules are generated if attribute discretization is used. The best results both in terms of classification quality and the number of decision rules (only 344 rules) were generated for the system using the LEM2 algorithm and applying attribute discretization. For this system accuracy is equal to 0.997 and the number of rules generated is equal to 256. The results obtained were compared with the results that had been generated using fuzzy sets. It was shown that with the use of rough set theory the initial results obtained for the analyzed data set are better.

In future research, we are planning to investigate the effects of prior utilization of methods dedicated to imbalanced data to equalize the number of objects in the decision classes.

# References

1. Amin, A., et al.: Customer churn prediction in the telecommunication sector using a rough set approach. Neurocomputing **237**, 242–254 (2017)
2. Amin, A., Shehzad, S., Khan, C., Ali, I., Anwar, S.: Churn prediction in telecommunication industry using rough set approach. In: Camacho, D., Kim, S.-W., Trawiński, B. (eds.) New Trends in Computational Collective Intelligence. SCI, vol. 572, pp. 83–95. Springer, Cham (2015). https://doi.org/10.1007/978-3-319-10774-5_8
3. Bazan, J.G., Szczuka, M.: The rough set exploration system. In: Peters, J.F., Skowron, A. (eds.) Transactions on Rough Sets III. LNCS, vol. 3400, pp. 37–56. Springer, Heidelberg (2005). https://doi.org/10.1007/11427834_2
4. Bazan, J.G., Szczuka, M.: RSES and RSESlib - a collection of tools for rough set computations. In: Ziarko, W., Yao, Y. (eds.) RSCTC 2000. LNCS (LNAI), vol. 2005, pp. 106–113. Springer, Heidelberg (2001). https://doi.org/10.1007/3-540-45554-X_12
5. Bazan, J.G., Nguyen, H.S., Nguyen, S.H., Synak, P., Wróblewski, J.: Rough set algorithms in classification problem. In: Polkowski, L., Tsumoto, S., Lin, T.Y. (eds.) Rough Set Methods and Applications, pp. 49–88. Physica-Verlag, Heidelberg (2000). https://doi.org/10.1007/978-3-7908-1840-6_3
6. Freeland, J.: The Ultimate CRM Handbook: Strategies and Concepts for Building Enduring Customer Loyalty and Profitability. McGraw-Hill, New York (2002)
7. Grzymała-Busse, J.: A new version of the rule induction system LERS. Fund. Inform. **31**(1), 27–39 (1997)
8. Höppner, S., Stripling, E., Baesens, B., vanden Broucke, S., Verdonck, T.: Profit driven decision trees for churn prediction. Eur. J. Oper. Res. **284**(3), 920–933 (2020)
9. Jafari-Marandi, R., Denton, J., Idris, A., Smith, B.K., Keramati, A.: Optimum profit-driven churn decision making: innovative artificial neural networks in telecom industry. Neural Comput. Appl. **32**, 14929–14962 (2020)
10. Jensen, R., Shen, Q.: Finding rough set reducts with ant colony optimization. In: Proceedings of the 2003 UK Workshop on Computational Intelligence, vol. 1, no. 2, pp. 15–22 (2003)
11. Khan, M.A., Khan, M.A.I., Aref, M., Khan, S.F.: Cluster & rough set theory based approach to find the reason for customer churn. Int. J. Appl. Bus. Econ. Res. **14**(1), 439–455 (2016)
12. Muhammad, A., Usman, M., Cheuk Fong, A.: A churn prediction model for prepaid customers in telecommunication using fuzzy classifiers. Telecommun. Syst. **66**, 603–14 (2017)

13. Nguyen, H.S., Ślęzak, D.: Approximate reducts and association rules. In: Zhong, N., Skowron, A., Ohsuga, S. (eds.) RSFDGrC 1999. LNCS (LNAI), vol. 1711, pp. 137–145. Springer, Heidelberg (1999). https://doi.org/10.1007/978-3-540-48061-7_18

14. Pawlak, Z.: Rough sets. Int. J. Comput. Inf. Sci. **11**, 341–356 (1982)

15. Pawlak, Z.: Rough sets. Int. J. Parallel Program. **11**(5), 341–356 (1982)

16. Peters, J.F., Skowron, A., Stepaniuk, J.: Rough sets: foundations and perspectives. In: Lin, T.Y., Liau, C.J., Kacprzyk, J. (eds.) Granular, Fuzzy, and Soft Computing, pp. 877–889. Springer, New York (2023). https://doi.org/10.1007/978-1-0716-2628-3_461

17. Przybyła-Kasperek, M.: Study of selected methods for balancing independent data sets in k-nearest neighbors classifiers with Pawlak conflict analysis. Appl. Soft Comput. **129**, 109612 (2022)

18. Rahman, G., Islam, Z.: A decision tree-based missing value imputation technique for data pre-processing. In: Proceedings of the Ninth Australasian Data Mining Conference, vol. 121, pp. 41–50 (2011)

19. Rahman, S.U., Amjad, T., Hussain, S.: Customer churn prediction in telecommunication industry using machine learning. IEEE Access **8**, 101719–101730 (2020)

20. Saleh, S., Saha, S.: Customer retention and churn prediction in the telecommunication industry: a case study on a Danish university. SN Appl. Sci. **5**(7), 173 (2023)

21. Sharaf Addin, E.H., Admodisastro, N., Mohd Ashri, S.N.S., Kamaruddin, A., Chong, Y.C.: Customer mobile behavioral segmentation and analysis in telecom using machine learning. Appl. Artif. Intell. **36**(1), 2009223 (2022)

22. Sharma, T., Gupta, P., Nigam, V., Goel, M.: Customer churn prediction in telecommunications using gradient boosted trees. In: Khanna, A., Gupta, D., Bhattacharyya, S., Snasel, V., Platos, J., Hassanien, A.E. (eds.) International Conference on Innovative Computing and Communications. AISC, vol. 1059, pp. 235–246. Springer, Singapore (2020). https://doi.org/10.1007/978-981-15-0324-5_20

23. Shrestha, S.M., Shakya, A.: A customer churn prediction model using XGBoost for the telecommunication industry in Nepal. Procedia Comput. Sci. **215**, 652–661 (2022)

24. Son, N. H., Ślęzak, D.: Approximate reducts and association rules-correspondence and complexity results. In: Proceedings of the 7th RSFDGrC 1996, Yamaguchi, Japan, pp. 107–115 (1999)

25. Sulikowski, P., Zdziebko, T.: Churn factors identification from real-world data in the telecommunications industry: case study. Procedia Comput. Sci. **192**, 4800–4809 (2021). https://doi.org/10.1016/j.procs.2021.09.258

26. Sulikowski, P., Mobile operator customer classification in churn analysis. In: Proceedings of the SAS®Global Forum 2008 Conference, San Antonio, Texas, 16–19 March 2008. SAS Institute Inc., Cary (2008). Paper 344-2008

27. Suraj, Z., Grochowalski, P.: The RSDS-Bibliographic Database for Rough Sets and Related Fields. In: Peters, J.F., Skowron, A., Bhaumik, R.N., Ramanna, S. (eds.) Transactions on Rough Sets XXIII. LNCS, vol. 13610, pp. 99–117. Springer, Heidelberg (2023). https://doi.org/10.1007/978-3-662-66544-2_7

28. Susmaga, R., Słowiński, R.: Generation of rough sets reducts and constructs based on inter-class and intra-class information. Fuzzy Sets Syst. **274**, 124–142 (2015)

29. Vijaya, J., Sivasankar, E.: Computing efficient features using rough set theory combined with ensemble classification techniques to improve the customer churn prediction in telecommunication sector. Computing **100**, 839–860 (2018)

30. Vora, S., Yang, H.: A comprehensive study of eleven feature selection algorithms and their impact on text classification. In: Computing Conference, pp. 440–449. IEEE (2017)
31. Yuhang, Q., Chen, P., Lin, Z., Yang, Y., Zeng, L., Fan, Y.: Clustering analysis for silent telecommunication customers based on k-means plus. Paper presented at 4th IEEE Information Technology, Networking, Electronic and Automation Control Conference (ITNEC), Chongqing, China, 12–14 June 2020, pp. 1023–1027 (2020)
32. Zdziebko, T., Sulikowski, P., Sałabun, W., Przybyła-Kasperek, M., Bąk, I.: Optimizing customer retention in the telecom industry: a fuzzy-based churn modeling with usage data. Electronics **13**(3), 469 (2024)

# Deep Learning Techniques

# CNN Classifier for Helicobacter Pylori Detection in Immunohistochemically Stained Gastric WSI

Pere Lloret[1,2], Pau Cano[1,2(✉)], Eva Musulen[3,4], and Debora Gil[1,2]

[1] Department of Computer Science, Universitat Autònoma de Barcelona,
Bellaterra, Spain
[2] Computer Vision Center, Campus UAB, Barcelona, Spain
{plloret,pcano,debora}@cvc.uab.cat
[3] Pathology Department, Hospital Universitari General de Catalunya-Grupo
QuironSalud, Sant Cugat del Vallès, 08915 Barcelona, Spain
eva.musulen@quironsalud.es, emusulen@carrerasresearch.org
[4] Institut de Recerca contra la Leucèmia Josep Carreras (IJC), 08916 Badalona,
Barcelona, Spain

**Abstract.** This work addresses the detection of Helicobacter pylori, a bacterium classified since 1994 as class 1 carcinogen to humans. Due to its high specificity and sensitivity, the preferred diagnosis technique is the analysis of histological images with immunohistochemical staining, a process in which certain stained antibodies bind to antigens of the biological element of interest. This analysis is a time consuming task, which is currently done by an expert pathologist that visually inspects the digitized images.

We propose the use of a Convolutional Neural Network (CNN) at sample level, in conjunction with a simple classifier to determine the final diagnosis at patient level. The designed CNN architecture is able to discern intricate patterns at different levels, particularly focusing on the distinctive color and structure that *H. pylori* displays in the samples. We have tested our model using a cross-validation on a set of patients with annotated patches, at sample and diagnosis level, and assessed reproducibility of diagnosis prediction on an independent set of patients. In the 5-fold cross-validation, our CNN model has an overall $92\% \pm 3\%$ of accuracy and $91\% \pm 6\%$ sensitivity in classification of *Helicobacter pylori* at sample level, and an accuracy of $92\% \pm 2\%$, with sensitivity of $92\% \pm 5\%$ at patient diagnosis level. In the independent set an accuracy of $84\%$, with a sensitivity of $84\%$ at patient diagnosis level validates the reproducibility of results.

**Keywords:** Helicobacter pylori · WSI · CNN

## 1 Introduction

*Helicobacter pylori* (H. pylori), a bacterium, stands as the primary culprit behind gastritis-an inflammation of the gastric mucosa that can escalate to more severe

© The Author(s), under exclusive license to Springer Nature Switzerland AG 2024
N.-T. Nguyen et al. (Eds.): ICCCI 2024, CCIS 2165, pp. 73–82, 2024.
https://doi.org/10.1007/978-3-031-70248-8_6

conditions, like gastric ulcers and, in extreme cases, cancer. This pervasive bacterium has infected over 50% of the global population, reaching a prevalence exceeding 80% in adults over fifty [6]. The quick and precise identification of this bacterium is crucial for efficient diagnosis, infection treatment, and the prevention of secondary pathologies.

The identification of *H. pylori* is commonly accomplished through the examination of gastric mucosa samples. While conventional H&E stains can reveal the bacterium, enhanced visualization is achieved with histochemical techniques like Giemsa or Warthin-Starry (WS). Notably, immunohistochemical detection stands out as the most precise diagnostic method [1], enabling the visualization of *H. pylori* through specific protein staining on its membrane. In this technique, the bacterium takes on a distinct color, avoiding false detections caused by other gram-negative bacteria in the sample. Immunohistochemical staining imparts a reddish hue to the specific protein of *H. pylori*, while the surrounding tissue retains the bluish hue of the hematoxylin counterstain. Despite facilitating visual identification, pathologists must meticulously inspect entire immunohistochemistry images, particularly focusing on regions at the borders where *H. pylori* is predominantly located. Given the vast image dimensions ($120,000 \times 16,000$ pixels) and the presence of multiple tissue sections in one digitized image, this manual inspection becomes a time-consuming task, especially at low concentrations of *H. pylori*.

Figure 1 illustrates a gastric mucosa sample with positive immunohistochemical staining for *H. pylori*, with close-ups of regions at the border of the tissue with varying bacterial densities (None, Low density, and High density) in window images on the right side. While the window with a high concentration is easily discernible, careful inspection is required for the window with low density to detect the reddish spots of *H. pylori* and avoid confusion with other artifacts.

The recent digitization of histopathological images has created a void in artificial intelligence methods for their analysis. In this study, we propose an automated method for analyzing immunohistochemically stained histological images of gastric tissue to detect H. pylori.

## 1.1   State-of-the-Art

While Deep Learning (DL) models have exhibited commendable efficacy in numerous histopathological tasks [5], there remains a notable scarcity of studies addressing the detection of *H. pylori*. Existing works [3,4,7] predominantly utilize DL methods based on convolutional neural networks, focusing on classifying cropped images extracted from tissue samples into *H. pylori* positive and negative categories.

In the study by Klein(2020) [3], a compact VGG-style architecture was trained on both Giemsa and H&E slides. The network, employed as a decision support system for pathologists, demonstrated the ability to identify regions with *H. pylori* presence, and exhibited a sensitivity of 1 but a low specificity of 0.66 for Giemsa stained samples. In a similar vein, [4] utilized a model akin to [3], trained on silver staining samples, achieving a sensitivity and specificity of

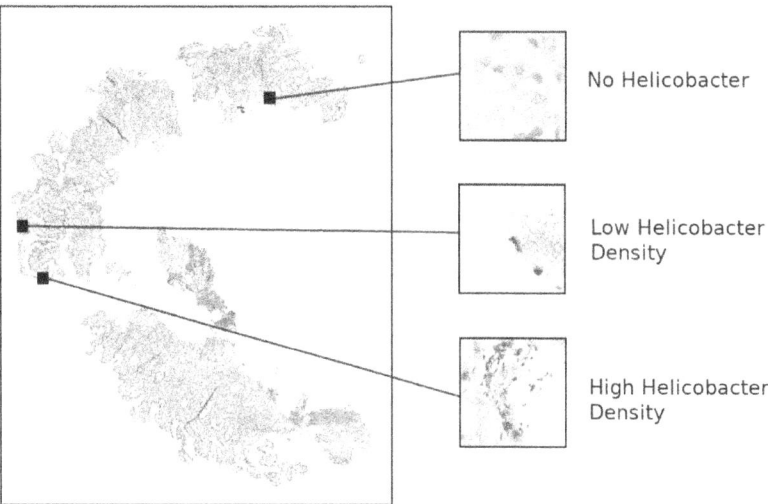

**Fig. 1.** Left: Histology sample with immunohistochemical staining. Right: 3 windows of the same histological sample showing different levels of *Helicobacter pylori* density

0.89 and 0.87, respectively, at the expense of a notable number of false positives, resulting in a precision of 77% for *H. pylori* patch detection.

Zhou(2020) [7] proposed an ensemble model integrating the output probabilities of three ResNet-18 and three DenseNet-21 models trained on patches from H&E-stained Whole-Slide Images (WSI). This ensemble, when applied to WSI-level diagnosis, demonstrated a sensitivity of 0.87, specificity of 0.92, and an F1-score of 0.89. The model, serving as DL support for pathologists, exhibited enhanced accuracy in diagnosing *H. pylori* positive samples but introduced higher uncertainty in identifying *H. pylori* negative samples, ultimately not significantly impacting diagnostic confidence or performance compared to a pathologist without DL support.

As far as we know there are no works addressing the diagnosis of immunohistochemically stained WSI using CNNs. However, Cano (2023) [2] uses an autoencoder architecture trained on samples without *H. pylori* to detect it on immunohistochemically stained WSI, obtaining 91% accuracy, 86% sensitivity and 96% specificity at cropped patch level.

In this paper, we present a method for the detection of *H. pylori* in immunohistochemically stained WSI based on a classification CNN trained on an augmented annotated set of patches extracted from an own collected database. In order to avoid overfiting, augmentation is based on a sampling of patches across tissue borders in a neighbourhood of patches annotated by experts. In order to assess reproducibility of results, the set of patients with annotated patches are used to train and validate the model in a cross-validation scheme, while the remaining cases are used as an independent validation set. Validation is conducted at patch sample and patient diagnosis level.

## 2   Detection of *H. Pylori* Using CNN

Our method has the following steps (sketched in Fig. 2): detection of areas of interest in the image, detection of anomalous stained elements in each region of interest, and aggregation of each region of interest in the image for the diagnosis of the sample. Since *H. pylori* is located along the border, first a series of contour detections around an automatically detected mask are used to detect the borders of the tissue sample. Patches are defined by sliding windows of $256 \times 256$ pixels cropped along the contour of such borders. This set of windows are the input to the CNN for their classification into positive (*H. pylori* presence in the window) or negative (no *H. pylori* in the window) case. Finally, the total number and percentage of positive windows determine the final classification of the sample.

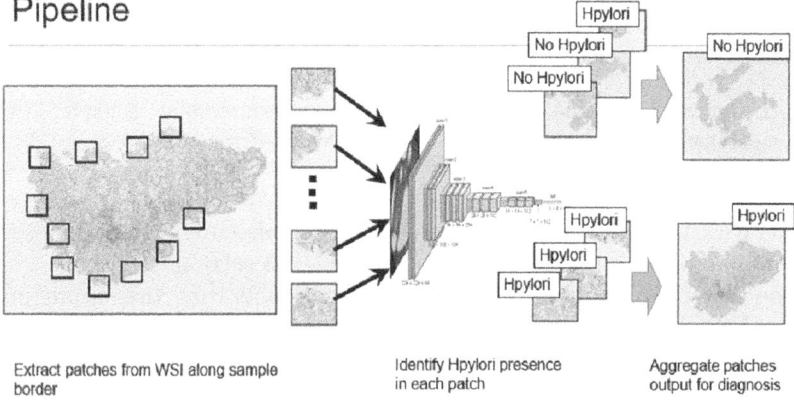

**Fig. 2.** Schema of the Pipeline

The first task is to segment the laboratory images containing two Whole-Slide Images (WSI) of the patient stained with immunohistochemistry, along with a control sample. Next, one of the two Whole-Slide Images is selected, and using image processing functions, the extraction of multiple cropped patches is performed. These patches are 3-channel images of size $256 \times 256$, each representing a fragment of the outline of the sample contained in the selected Whole-Slide Image. The analysis is focused solely on the sample's outline since that is where the bacteria are located.

Classification problems require a good set of annotated data; we have a set of annotated cropped patches provided by the doctor, which is insufficient in size to train the classifier. Therefore, we have performed data augmentation on the annotated patches with the presence of the bacterium. From each patch with the presence of the bacterium, we generated 8 new patches at specific distances from the original patch, directed towards neighboring patches (4 patches generated per direction), as illustrated in the schematic diagram in Fig. 3,the continuous arrows indicate the direction over which we will select the centers of the new patches. Using this method instead of flipping the patches we avoid working with synthetic data. To have a completely balanced dataset, we have selected random patches from diagnostically negative patients, as their patches should always be negative.

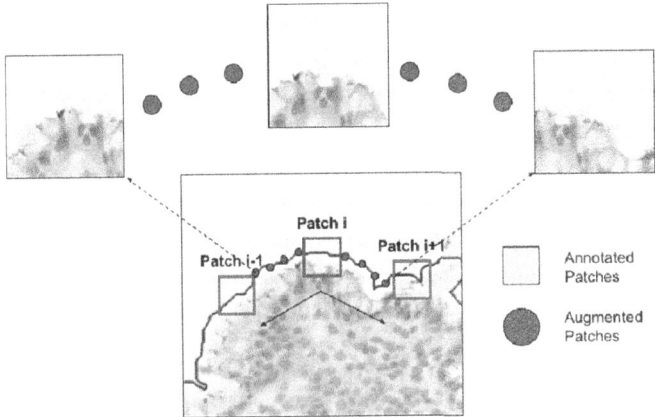

**Fig. 3.** Schema of the Data Augmentation

The model employed to address the classification of patches is a Convolutional Neural Network (CNN). It consists of three convolutional layers with a max-pooling layer and a ReLU activation layer after each convolutional layer. The network is designed to autonomously discern intricate patterns within images, allowing it to identify and categorize objects with a high degree of accuracy. Through the convolutional layers, the model learns hierarchical features, recognizing simple shapes in early layers and progressively more complex structures in deeper layers. The fully connected layers then consolidate this information for final classification.

## 3   Experiments

In this project, we conducted two experiments. In the initial experiment, we performed cross-validation to assess the classifier's performance at both the window level and in delivering accurate patient diagnoses. This dual evaluation allowed us to obtain detailed metrics for window classification, providing insights into the model's fine-grained proficiency. Simultaneously, we scrutinized its capacity to provide precise and reliable patient diagnoses, highlighting the broader clinical implications of the classifier's outcomes. Conducting cross-validation provides us with an estimation of the model's performance with new data, allowing us to verify its generalization.

In the second experiment, we broadened the model's training scope to include data from all annotated patients, facilitating a thorough evaluation of its diagnostic capabilities on patients beyond the initial training set. This allowed us to assess the model's ability to generalize insights to new patient data, providing valuable insights into its adaptability and reliability across diverse cases. Exposure to patients not encountered during the initial training phase deepened our understanding of the model's performance and behavior in the face of unseen medical scenarios.

Figure 4 represents a conceptual map illustrating how we have partitioned the data for the execution of the two experiments. We have both annotated and non-annotated data, and as mentioned earlier, data augmentation has been performed on the annotated data for both positive and negative windows. In the first experiment, we employed GroupKFold, where the groups correspond to patients, utilizing 5 folds. For each fold, a new classifier was trained with the data from the patients in that fold and subsequently tested. The subsequent step involved saving the percentage of windows diagnosed as positive by the classifier in each fold, relative to the total number of windows. We then trained the threshold for diagnosing a patient as positive for *H. pylori* based on the training patients in that fold. This threshold was later tested on the test patients, and finally, metrics were extracted for each fold.

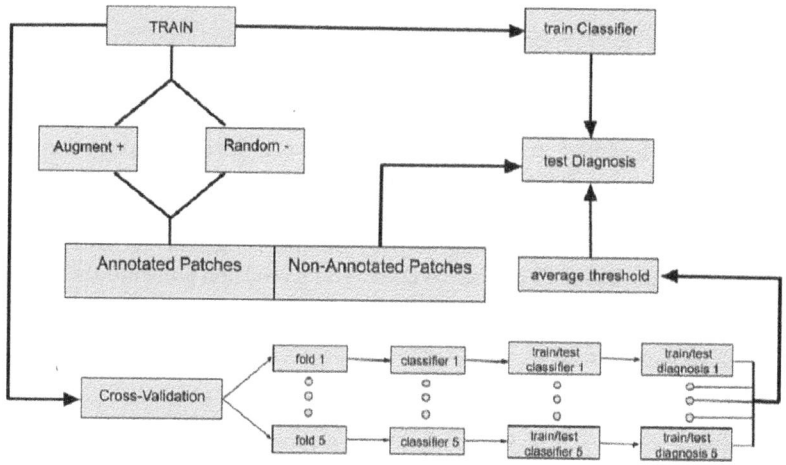

**Fig. 4.** Schema of the Data Split

For both experiments training took place over 50 epochs, employing the cross-entropy loss criterion from the torch.nn library and utilizing the Adam optimizer from the torch.optim library. A learning rate of 1e-3 was chosen to fine-tune the model parameters during the training process

## 3.1   Results

**Experiment 1:** The performance metrics we have considered are the area under the curve (AUC), accuracy, and recall for each diagnostic class (positive *H. pylori* or negative *H. pylori*). Table 1 reports a statistical summary of the quality metrics of the patch classifier for each fold, including the mean and standard deviation for each metric. The proposed system achieves an optimal average specificity of 0.95 along with an excellent average sensitivity of 0.91.

**Table 1.** Statistical Summary of the 5-fold Validation for the Classifier

| Folds | AUC | Accuracy | Recall Class 0 | Recall Class 1 |
|-------|-----|----------|----------------|----------------|
| fold 1 | 0.97 | 0.96 | 0.96 | 0.98 |
| fold 2 | 0.89 | 0.89 | 0.94 | 0.84 |
| fold 3 | 0.92 | 0.94 | 0.89 | 0.96 |
| fold 4 | 0.92 | 0.91 | 0.98 | 0.85 |
| fold 5 | 0.93 | 0.92 | 0.96 | 0.90 |
| average | $0.93 \pm 0.03$ | $0.92 \pm 0.03$ | $0.95 \pm 0.03$ | $0.91 \pm 0.06$ |

Table 2 reports a statistical summary of the quality metrics of the patient diagnosis for each fold, including the mean and standard deviation for each metric and the threshold for diagnosing a patient as positive for each fold. The proposed system achieves an excellent average specificity of 0.92 along with an excellent average sensitivity of 0.92. The mean threshold of 0.082 is subsequently used to determine the percentage at which a patient will be diagnosed as positive in the second experiment. Table 3 shows the confusion matrix of the annotated patients diagnosis. Of 183 patients, only 14 have been incorrectly classified.

**Table 2.** Statistical Summary of the 5-fold Validation for Patient Diagnosis

| Folds | AUC | Accuracy | Recall Class 0 | Recall Class 1 | Threshold |
|-------|-----|----------|----------------|----------------|-----------|
| fold 1 | 0.99 | 0.94 | 0.96 | 0.90 | 0.086 |
| fold 2 | 0.99 | 0.94 | 0.90 | 1 | 0.072 |
| fold 3 | 0.98 | 0.89 | 0.83 | 0.95 | 0.095 |
| fold 4 | 0.98 | 0.89 | 0.91 | 0.87 | 0.086 |
| fold 5 | 0.96 | 0.94 | 1 | 0.89 | 0.076 |
| average | $0.98 \pm 0.01$ | $0.92 \pm 0.02$ | $0.92 \pm 0.06$ | $0.92 \pm 0.05$ | $0.083 \pm 0.01$ |

**Table 3.** Confusion matrix of the annotated patients diagnosis

| Predicted | Ground Truth | |
|---|---|---|
| | H. pylori | No H. pylori |
| H. pylori | 72 (TP) | 8 (FP) |
| No H. pylori | 6 (FN) | 97 (TN) |

**Experiment 2:** In the second experiment, the classifier has been trained with patient data annotated post data augmentation, as observed in the top part of Fig. 4. All available data has been allocated for training, as the performance of the patch classifier has already been validated in the cross-validation (Table 1). The subsequent step involved validating the model's performance in diagnosing patients. To achieve this, we selected all patients with non-annotated data (holdout), those the classifier has never encountered. These patients were then processed through the classifier trained with all patients having annotated data. We obtained the percentage of positive patches per patient and determined their diagnosis using the threshold extracted from the average of the thresholds from each fold of the cross-validation. Table 4 provides a statistical summary with the same metrics used in experiment 1, offering insights into the performance of the classifier with new patients. This allows us to assess its applicability to real-world scenarios where the model encounters unseen information. The proposed system achieves a good specificity and sensitivity of 0.86 and 0.82, respectively. Table 5 shows the confusion matrix of the samples' diagnosis. Of 86 patients, 14 have been incorrectly classified.

**Table 4.** Statistical Summary of the Validation for Patient Diagnosis

| | AUC | Accuracy | Recall Class 0 | Recall Class 1 | Threshold |
|---|---|---|---|---|---|
| Global Model | 0.92 | 0.84 | 0.86 | 0.82 | 0.083 |

**Table 5.** Confusion matrix of the holdout patients diagnosis

| Predicted | Ground Truth | |
|---|---|---|
| | H. pylori | No H. pylori |
| H. pylori | 47 (TP) | 4 (FP) |
| No H. pylori | 10 (FN) | 25 (TN) |

## 4   Conclusions

We have presented a first DL system for the diagnosis of *H. pylori* on immunohistochemically stained gastric mucosa based on a CNN classifier, trained to learn patterns within patches containing H. pylori, the model aims to accurately classify the windows. One of the main challenges for the use of classification approaches for the identification of *H. pylori* in histological images is the collection of enough annotated data, since this implies a time consuming visual inspection and identification of patches containing the bacteria. Common data augmentation techniques base on transformations of annotated data which are prone to introduced dependencies across training samples and, thus, favour model overfitting. In this work we have presented an augmentation that bases on an oversampling across WSI tissue border of the annotated patches in order to introduce a realistic variability in samples.

Results in a cross-validation scheme and in the independent set of patients show the higher reproducibility of a model trained using our data augmentation. Comparing to SoA methods, our patch classification has higher sensitivity and specificity (92% of both versus 89% of sensitivity and 87% specificity achieved in [4]). This results in a higher diagnostic prediction with average 92% of sensitivity and specificity in a 5-fold validation and 82% sensitivity and 86% specificity in the independent hold out test. These numbers are higher to the 86% sensitivity achieved by the 10-fold validation of the auto-encoder for detection in immunohistochemical staining of [2] with comparable specificity ranges (92%±6% versus 96%±9%). Comparing to methods using a different staining, our method also achieves better performance in cross-validation averages, and similar performance in the independent hold out test (87% sensitivity with 92% specificity in [7]).

The analysis of the failing cases, shows that there is room for improvements. Up to now, the patient diagnosis has been considered in relation to the Whole Slide Image (WSI). However, there are WSI's with multiple tissue samples, and it's possible that only one sample shows the presence of *H. pylori*. In such cases, the prediction might conclude that the percentage of positive patches does not surpass the threshold. To address this situation, patients could be analyzed based on individual tissue samples instead of WSI's. Additionally many of the failed cases are patients with a very low density of H. pylori. This could be solved by annotating more patches with low density, providing the classifier with increased exposure to this type of cases during training.

# References

1. Batts, K., Ketover, S., Kakar, S., et al: Gastrointestinal pathology society. appropriate use of special stains for identifying helicobacter pylori: recommendations from the rodger c. haggitt gastrointestinal pathology society. Am. J. Surg. Pathol. **37**(11), e12–22 (2013)
2. Cano, P., Caravaca, A., Gil, D., Musulen, E.: Diagnosis of helicobacter pylori using autoencoders for the detection of anomalous staining patterns in immunohistochemistry images (2023). https://arxiv.org/abs/2309.16053
3. Klein, S., Gildenblat, J., Ihle, M.A.: Deep learning for sensitive detection of helicobacter pylori in gastric biopsies. BMC Gastroenterol. **20**(1), 1–11 (2020)

4. Liscia, D.S., D'Andrea, M., Biletta, E., et al.: Use of digital pathology and artificial intelligence for the diagnosis of helicobacter pylori in gastric biopsies. Pathologica **114**(4), 295 (2022)
5. Salto-Tellez, M., Maxwell, P., Hamilton, P.: Artificial intelligence-the third revolution in pathology. Histopathology **74**(3), 372–376 (2019)
6. Yang, J.: Treatment of helicobacter pylori infection: current status and future concepts. World J. Gastroenterol. **20**(18), 5283–93 (2014)
7. Zhou, S., Marklund, H., Blaha, O., et al.: Deep learning assistance for the histopathologic diagnosis of helicobacter pylori. Intell.-Based Med. **1**, 100004 (2020)

# Complete Convolutional Neural Networks Environment for Computer Vision Problems With Nvidia Drive AGX Xavier

Sorin Valcan[1,2]([✉]) [iD] and Mihail Gaianu[1,2] [iD]

[1] West University of Timisoara, Timisoara, Romania
sorin.valcan96@e-uvt.ro
[2] Continental Automotive Romania, UX Department, Timisoara, Romania

**Abstract.** Development of Convolutional Neural Networks (CNN) models with supervised learning in computer vision tasks can be a demanding process. Multiple components lead to creation of the final model including data acquisition, image labeling, data preprocessing, model training, testing and key performance indicators computation. In general these tasks require significant hardware resources from both storage and computational power points of view. In this paper we propose an alternative lighter environment for the entire process based on the Nvidia Drive AGX Xavier development platform. Using methods for automatic image labeling and fully automated environments of CNN models training and testing, this method is much easier to setup and use. Additional advantages of portability to use it in real-life scenarios makes it a very flexible development environment. We present experiments performed for driver monitoring computer vision tasks with similar applicability in other detection problems.

**Keywords:** Convolutional neural networks · Object detection · Driver monitoring · Safety features · Ground truth data · Infrared sensors

## 1 Introduction

Convolutional Neural Networks (CNN) usage in computer vision tasks require multiple fields of expertise that used together lead to a final model capable of performing the desired detection or classification. Because of different tasks involved, multiple hardware and human resources are necessary.

In real-life scenarios and in specific fields like automotive or medical imaging there are additional restrictions that constrain the entire process even more. The usage of synthetic generated data for training and testing is forbidden because of the risks involved of missing the necessary details of real samples. Labeling has to be performed manually because of the desire to have full human control of the data used for training the model. CNNs are not yet used in final decision making systems because of the huge importance of the outputs correctness with life threatening consequences.

N.-T. Nguyen et al. (Eds.): ICCCI 2024, CCIS 2165, pp. 83–93, 2024.
https://doi.org/10.1007/978-3-031-70248-8_7

Even with these factors involved in the real situation of Machine Learning (ML) usage in automation of such important tasks, development of models capable to perform better and better detections is ongoing. Depending on the technical capability to control and understand the ML outputs and on the legal evolution of acceptance for such solutions, development of models may become common and in high demand.

Foreseeing this future, we propose in this paper a light and flexible ML development environment based on the Nvidia Drive AGX Xavier development platform with experiments performed in automotive area with the main focus on driver monitoring. Other existing software components with the purpose of removing the manual working effort are used on this hardware to generate Ground Truth (GT) data and train CNN models.

## 2  Related Work

This paper is focused on creating a portable environment that reduces the human manual work required in multiple ML tasks. This is a time consuming process regardless of the task being performed. It is defined by a predictable task with known output and an incapacity of reducing the amount of repetitive work necessary for producing the output. It also assumes a required quality of the output which is very questionable in practice.

Manual work generates fatigue, boredom and stress even after the actual work has been performed. Studies on the psychological effects of repetitive work are presented in [3,4,6]. These effects directly influence the capacity and motivation of each individual to perform a quality work. The effect on cardiovascular reactivity and other observational methods during repetitive work are presented in [1,7]

To atomize a series of tasks in CNN development for driver monitoring systems, different components require different software solutions. For image labeling, a series of facial features detection solutions may be used as GT data generator based on existing images [2,8]. In [11] we presented a GT data generation for eyes with focus on accuracy instead of consistency. There are also semi-automatic methods developed for easing the necessary manual effort based on interactive labeling tools [5,9].

For data preprocessing, an automated system of data sets preparation for training and testing can be used to avoid the manual effort and improve the amount of experiments with ML models [10]. With the agreed CNN architectures, an automatic environment for model training can be created to be able to focus on results and improvements.

## 3  Problem Description

The entire process of creating a CNN model with supervised learning capable to perform detections involve multiple working areas.

First the data acquisition process requires at least one trained user to operate the recording systems and various subjects to be recorded if it is the case(ex. humans, animals, objects). In the current context of automotive computer vision tasks we eliminate the posibility of using synthetic generated data.

Once the data is available partial or completely the labeling process can start. This usually involves teams of people marking the images with the required labels, task that is complicated and tiring. To perform it, the hardware infrastructure must be provided for the labeling team. The data movement from the recording scene to the labeling infrastructure and the data transfer time must also be taken into consideration for the required effort.

With the labeled data available, a series of data preprocessing tasks must be performed to obtain the final sets of training and testing data. Depending on the problem the split can be straightforward or more complicated. If not performed in an automatic way, the preprocessing can produce lot of human repetitive work and limit the capacity to perform multiple ML tests with the available data.

With the prepared data sets ready, the training and testing of CNN models can be performed. Depending on the dimension of the problem and the used architectures, this can require significant processing power, reason why it is usually performed in data centers with powerful processing units installed and with significant cooling required.

With the CNN model created, an additional Key Performance Indicators (KPI) process can be performed to compare the model outputs and the manual labeling work. This requires the usage of the model to process the existing data that was not used in training phase.

We consider multiple tasks presented above to be repetitive and automatable. Also the data centers with very high computing capabilities may not be any longer required if the software modules and CNN architectures are light enough and used at correct times alongside the human manual work. In the following chapter we will present a light hardware architecture with efficient software modules capable to perform the entire chain previously described in a more automated way.

## 4   Methods

The proposed environment contains relative limited hardware components with efficient software modules and neural networks architectures that reduce the processing times using Graphics Processing Unit (GPU) accelerated functionalities. Based on it we created a general CNN development environment to be used with a relative reduced number of necessary users. The environment is used for generation of authentic data sets for driver monitoring in real-life scenarios and conditions and for creation of CNN models including the steps of image labeling, data preprocessing, training/tesing and KPI generation.

### 4.1  Environment Concept

The general concept uses multiple automations that eliminate most of the human work. In Fig. 1 is presented an architecture containing the necessary human presence to operate the systems and the automated sections where the system performs ML computations outside the human working times.

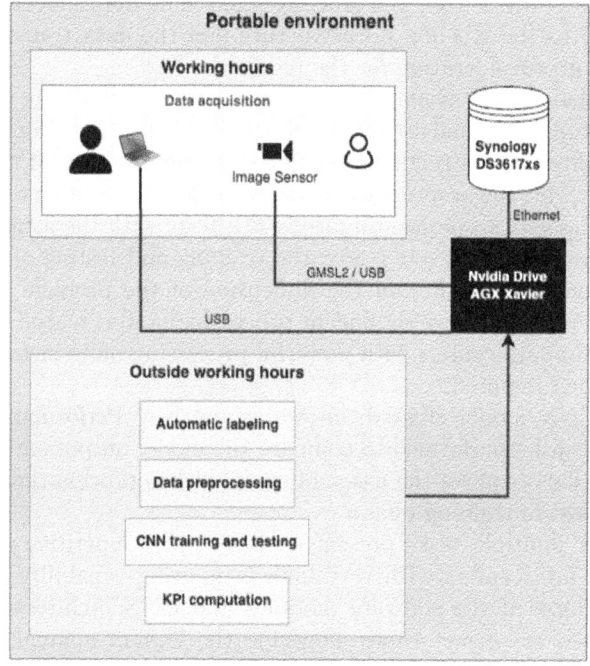

**Fig. 1.** Concept of data acquisition and automated processing

The hardware architecture used for our computations is also presented in Fig. 1. The main component is the Nvidia Drive AGX Xavier development platform created specially for acceleration of ML automotive solutions. Multiple connections used in automotive are available: GMSL2, ethernet, USB, CAN, Flexray or GPIO. The platform also contains the Xavier GPU unit with 8 multiprocessors, 64 CUDA cores per multiprocessor, and 28 GB RAM. It should be operated at ambient temperatures between 0 to 45 °C. The storage accepted temperature is between −40 to 65 °C.

The second component is represented by a Synology DS3617xs storage. The storage can operate at temperature between 5 to 40 °C and stored between -20 and 60 °C. This makes it usable in tougher conditions inside the car for real time processing. The available space is 108 terabytes and it is connected to the Xavier platform via ethernet in a local network.

The environment is portable enough to be used either in a simulator, real car or simply in the office. This opens the flexibility to generate and process the data independent of the location.

## 4.2  Working Hours

In the presented concept, a human work day is filled by the data acquisition process which must be done with real persons. An image sensor is connected to the Xavier platform via GMSL2 or USB connection. A trained user with knowledge of the system is connecting to the Xavier platform and controls the software recording process. The recordings are stored locally on the Xavier device and copied to the storage immediately after the process is stopped.

## 4.3  Outside Working Hours

After a working day of approximately 8 h of recording data, normally the system will remain off. In our architecture the other 16 h are the times when the automation comes alive with a huge possibility of performing work that otherwise would require other days of effort.

First, the automatic labeling must be executed on the newly acquired data to be usable in the CNN training process. For driver monitoring systems such automatic labeling modules exist. In [11] we presented an automatic labeling module with the main focus on detection accuracy. The method uses multiple GPU accelerated functions that fit very well the Xavier GPU device [12]. This makes the presented algorithm and the processing platform to be an ideal match for image processing.

The data preprocessing step represents the preparation of existing frames and labels for the training and testing process. Existing subjects must be separated between the two phases. CNNs have a very high capacity to learn the subjects used in training and perform much better on them in comparison to the ones never seen before. In training it is important to use a relative equal number of frames per subject, otherwise the model might be very biased by the features of a very few persons and unable to generalize to perform detections on different faces.

The CNN training and testing phase can be computationally expensive depending on the chosen architecture and the input size. However, there are CNN architectures used for facial features location that obtain similar accuracy with reduced dimensionality of the original input [10]. With relative light architectures and a correct selection of the important frames generated for each individual subject the training and testing of the CNN can be a matter of minutes rather than hours or days.

The KPI computations represents a computationally light process to compare different sets of labels. This process is brought to the presented environment next to the existing data sets and results generated by the previous computations.

The usage of this process brings multiple advantages in comparison to a classic approach as presented in Sect. 3. First of all, a classic method requires a more complicated hardware architecture: recording setup in simulator or real cars, temporary storage in recording setup, permanent storage at a fixed location, working stations or virtual machines for the labeling team and high transfer speed for manual data preprocessing steps. For each different task previously described, different people would be required to perform manual work: a team required for labeling, dedicated persons for data preprocessing every time new data arrives, users to manually perform multiple CNN experiments with different subjects in the training and testing phase and people dedicated for data transfer from recording scene to other tasks that require it's usage.

In our presented architecture these tasks are automated, executed in offline hours and capable to be performed without important location restrictions. The number of hardware resources and locations necessary for the entire chain is drastically reduced. Labeling and CNN experiments are performed automatically on the same portable setup.

In architecture from Fig. 1 the most inefficient task is represented by the data acquisition process that is limited by a variety of factors:

- Recordings scenarios might be required and a trained user needs to coordinate the subject
- Subjects might be difficult to find and the data acquisition process is limited by individual personal schedule
- Systems must be controlled by an experienced user
- A faster data acquisition process would require the task parallelization, meaning multiple setups to perform more recordings
- Data need to be authentic and generated from the real world. Automation does not depend on software modules but on real world conditions and resources.

Other persons involved on the technical side are represented by software developers required for development of an automatic labeling algorithm, automatic data preprocessing environment and CNN architectures creation for specific problems. These represent development tasks that require knowledge and innovation, which is the opposite of the tiring and error prone repetitive manual work.

## 5   Experiment

In this section we will show an experiment performed with the presented environment for a driver monitoring system CNN development with the efficiency of the utilized modules.

The data acquisition part was only simulated because of the preexistence of a driver monitoring data set. This process is unpredictable and irrelevant from the automation point of view and execution times.

A data set with 113 subjects and approximately 12.400 frames per subject was used. The frame dimension was $1280 \times 800$ pixels. An example of frame with

**Fig. 2.** Example of frame with eye labels from driver monitoring CNN development experiments

eye labels generated is presented in Fig. 2. The time necessary to perform the recordings for one subject is approximately 1 h and 15 min.

Every day for 11 consecutive days the recordings of 10 different subjects were added to the storage. This represents a more than ideal case of data acquisition. It is very unlikely to be able to record such high number of subjects in one working day.

At the end of the day at approximately 6:00 PM an automatic toolchain started to perform the tasks described in Fig. 1 for outside working hours.

First, the toolchain will check the new recordings added in the storage and will start the process of GT data generation on the new images using the algorithm previously presented in [11]. The necessary time for execution on approximately 124.000 frames requires between 4 and 7 h depending on the number of generated eye labels. The algorithm requires less time if more labels are generated because of the tracking system used to make it more efficient. In case a subject is rarely detected the execution time grows. Around midnight the newly generated GT data is ready to be used for CNN training.

The architecture used for eye detection is based on two CNNs: first network is used to extract two bounding boxes of approximate eyes location and the second one is used to process individual patches and search around the initial location to fix the bounding box at the correct coordinates.

The architecture of the first CNN is presented in Fig. 3. In the preprocessing part the image dimension is reduced using 4 average pooling layers. Two pairs of convolutional-ReLU-max pooling layers are applied for feature extraction before a flatten layer is used to initiate the section of a classic neural network. One fully connected layer with 500 neurons is applied before the output layer containing eight neurons representing two bounding boxes defined as $(x, y, w, h)$.

The second CNN is described in Fig. 4 and is used for patch classification. The input is represented by a patch generated with the first CNN of the system. The preprocessing part contains two pairs of convolutional-ReLU-max pooling

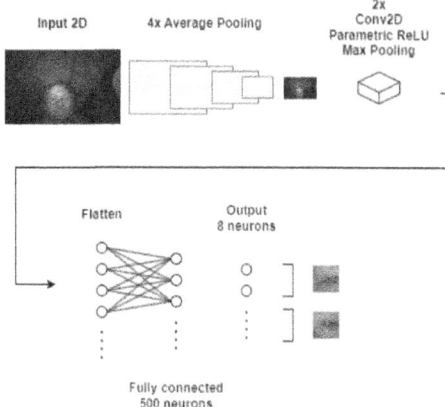

**Fig. 3.** First CNN of the system used for extracting an initial location for both eyes

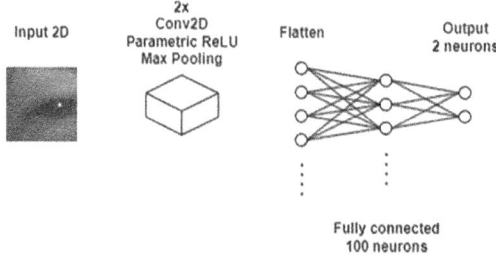

**Fig. 4.** Second CNN of the system used for confirmation of a patch if the desired feature is present

layers followed by a flatten layer. One fully connected layer with 100 neurons is used for learning and the output layer contains two neurons representing values for true or false.

For the first CNN of the system an automatic subsets generation process is executed multiple times and used for separating subjects frames with GT labels between the training and testing phase to avoid redundant data by imposing distance restrictions between selected frames. The time required for one subset generation grows with the number of subjects available: with 10 subjects the required time approximately 2 min and with 113 subjects it takes approximately 8 min.

The second CNN uses a similar subsets generation process but this time the focus is on the patches selection. The ones obtained with the GT data generation algorithm are directly used as correct samples containing an eye and the other half of incorrect ones are obtained by moving a correct bounding box in a random direction to prevent it from fully containing the eye. This required time also depends on the number of recorded subjects: with 10 subjects the generation process takes 6 min while for 113 subjects it takes up to 17 min.

To train the first CNN of the system it takes approximately 6 min with 10 epochs. The second one requires 5 min with 3 epochs.

From midnight until 9:00 AM the time allows for generation of approximately 15 CNN models trained on various subjects in different split configurations if we consider the biggest processing times. In the morning these results are ready for analysis by the CNN development responsible while the data acquisition process can continue during the day.

The process of KPI generation is very light with computations performed between two sets of labels. Obtaining the data using the generated models can be timely expensive because it requires the CNN system to be executed on multiple subjects compared to a few frames for testing purposes. The execution of the CNN on 16 subjects representing approximately 15% of subjects not used in the training phase would require approximately 6 h. This process should be performed in general on models with good testing accuracy in order to avoid unnecessary processing effort on the hardware.

## 6 Discussion

In [11] the automatic GT data generation algorithm was meant to move the focus from the manual work effort of labeling images to a more efficient algorithm development effort that would facilitate the generation of training and testing GT data.

With the software modules and process architecture described in this paper we observe that the main manual work problem remains the data acquisition processes. As already discussed, specific fields like automotive or medical imaging can not rely on synthetic generated data because the level of details in the generated images is limited.

Performing the recordings with real persons can not be accelerated by software modules. Alternative methods for data acquisition exist in theory with millions of cars on the roads constantly capable to generate data. However, the legal side of this story makes it unfeasible because people would not like to know they are permanently monitored in their own cars by various companies. In medical imaging data acquisition is even more complicated. A real image can be obtained from a patient with a real existing tumor. No one can predict or organize a number of images generated on a period of time with the purpose of training ML models.

Three main tasks remain in the focus: manual data acquisition, GT data generation software development and neural networks architectures development for specific problems.

## 7 Future Work

The future work for this paper is represented by algorithms efficiency improvements and experiments performed in other computer vision problems.

The GT data generation process is going in the direction of being applied in various use cases: traffic signs recognition, cabin monitoring or medical imaging tumor detection. Ideally, a more general set of rules must be developed to be able to avoid very specific software development that is not reusable for other areas.

The times required for GT data generation can be improved with smarter GPU implementations for various specific functions. Also in the CNN architecture presented in this paper the searching process used with the second CNN takes a very long time and should be replaced by a one time execution network to output a better final location.

Future developments can lead to connection of such system to a cloud infrastructure especially for big data storage purposes, even though an impressive capacity can be mounted on such a portable setup.

## 8    Conclusions

This paper presented a hardware architecture and working schedule for human and hardware machines that together create a minimalist environment for CNN models development.

The purpose of this work is to bring smart solutions for automation in areas where excessive human work is required for repetitive tasks. Using software modules for GT data generation, data preprocessing and CNN training, testing and execution we are capable to reduce the necessary effort of human manual work.

The concept presents a combination between human effort and hardware effort working together and using the available 24 h in a smart manner. The environment brings advantages of portability and flexibility regarding the location where computations have to be performed in comparison to classic approaches of fixed data centers.

With automotive being one of the most demanding area in terms of temperature conditions where a system has to run, this minimal hardware setup presents a viable solution to be used in complicated conditions for a high variety of tasks.

## References

1. Carvalho, D., et al.: Cardiovascular reactivity (CVR) during repetitive work in the presence of fatigue, February 2023. https://doi.org/10.54941/ahfe1002833
2. Gu, H., Cheng, G.S.: Feature points extraction from faces (2003)
3. Häusser, J.A., Schulz-Hardt, S., Schultze, T., Tomaschek, A., Mojzisch, A.: Experimental evidence for the effects of task repetitiveness on mental strain and objective work performance. J. Organ. Behav. 35(5), 705–721 (2014). https://doi.org/10.1002/job.1920, https://onlinelibrary.wiley.com/doi/abs/10.1002/job.1920
4. Illankoon, P., Manathunge, Y.: Psychological effects of short, repetitive and easy tasks on production lines, June 2008
5. Le, V., Brandt, J., Lin, Z., Bourdev, L.: Interactive facial feature localization, October 2012. https://doi.org/10.1007/978-3-642-33712-3_49

6. Mixter, S., et al.: Stress-related responses to alternations between repetitive physical work and cognitive tasks of different difficulties. Int. J. Environ. Res. Public Health **17**(22) (2020). https://doi.org/10.3390/ijerph17228509, https://www.mdpi.com/1660-4601/17/22/8509

7. Nyman, T., et al.: Reliability and validity of six selected observational methods for risk assessment of hand intensive and repetitive work. Int. J. Environ. Res. Public Health **20**(8), 5505 (2023)

8. Paul, S., Uddin, M.S., Bouakaz, S.: Face recognition using eyes, nostrils and mouth features, March 2004. https://doi.org/10.13140/2.1.4070.5605

9. Tian, Y., Liu, W., Xiao, R., Wen, F., Tang, X.: A face annotation framework with partial clustering and interactive labeling. In: 2007 IEEE Conference on Computer Vision and Pattern Recognition, pp. 1–8 (2007). https://doi.org/10.1109/CVPR.2007.383282

10. Valcan, S., Găianu, M.: Nostrils and mouth detection for drivers using convolutional neural networks with automatically generated ground truth data. In: 2022 International Conference on Computational Science and Computational Intelligence (CSCI), pp. 1497–1503. IEEE Computer Society, Los Alamitos, CA, USA, December 2022. https://doi.org/10.1109/CSCI58124.2022.00265, https://doi.ieeecomputersociety.org/10.1109/CSCI58124.2022.00265

11. Valcan, S., Gaianu, M.: Ground truth data generator for eye location on infrared driver recordings. J. Imaging **7**(9) (2021). https://doi.org/10.3390/jimaging7090162, https://www.mdpi.com/2313-433X/7/9/162

12. Valcan, S., Găianu, M.: Cuda implementation for eye location on infrared images. Scalable Computing: Pract. Experience **23** (2022). https://doi.org/10.3390/jimaging7090162, https://www.scpe.org/index.php/scpe/issue/view/156

# The Development of an Application-Specific Instruction Set Processor Specialized on a Convolutional Neural Network Trained on MNIST

Dávid Nevezi-Strango[1,3]($\boxtimes$) (ID), Daniel Grosu[2,3] (ID), and Sorin Valcan[1,3] (ID)

[1] West University of Timişoara, Timişoara, Romania
david.nevezi00@e-uvt.ro
[2] Polytechnic University of Timişoara, Timişoara, Romania
[3] Continental Automotive Romania, UX Department, Timişoara, Romania

**Abstract.** This paper will present the development of an ASIP design and it's toolchain for integrating a convolutional neural network in the ONNX format that works on the MNIST dataset. Additionally, it will present the results of the design's deployment onto a Xilinx Zynq Ultra-Scale+ MPSoC ZCU102 FPGA. As a toolchain, the LLVM umbrella project is being used. LLVM started as a compiler project, aiming for modularity but later became an umbrella project due to the large number of contributions and subprojects. The three main components are the ONNX-MLIR project, the LLVM main project (mainly the clangd, the optimizer and a specific RISC-V backend developed by us) and an HLS Simulator project called Comet, used to design the architecture. After a series of modifications, Comet was renamed to AIDA.

**Keywords:** RISC-V · LLVM · ONNX · Comet · HLS · C++ · ASIP

## 1 Introduction

The project has the intention of creating a hardware design that is specialized in executing machine learning (ML) models, the main motivation being to have a proof of concept. According to paper [11], neural networks showed their success in 2017, they have been accustomed for a wide range of problems besides image classification and computer vision, like for example, natural language processing, text classification, speech recognition and so on, we have chose a Convolutional Neural Network (CNN) trained on MNIST [8].

So the first step for us was to create a compiler environment and find a simulator that we can test our compiler with. We have chosen the Low Level Virtual Machine (LLVM) project due to it's modularity. There is another scientific paper that presents the work in progress for the Maxpool operator that is to be published [15]. In short, the LLVM is a compiler that is divided into 3 modules, the frontend, the optimizer and the backend [12]. It's philosophy is that instead of

© The Author(s), under exclusive license to Springer Nature Switzerland AG 2024
N.-T. Nguyen et al. (Eds.): ICCCI 2024, CCIS 2165, pp. 94–105, 2024.
https://doi.org/10.1007/978-3-031-70248-8_8

creating a new compiler for each processor architecture every single time there is a new programming language, one could just simply develop a frontend module, attach it to the LLVM environment and it is ready to be used, thus, reducing the time and effort to be put in creating a new compiler and understanding the architecture of every existing processor. This is also true the other way around, in the case of creating a new processor. More details regarding the LLVM project can be found in [15].

LLVM's main language is C++ along with a domain specific language called TableGen [14].

Our plan was to expand the RISC-V instruction set to be able to work with ML operators, create a compiler that is able to compile a model in ONNX format for our expanded RISC-V instruction set architecture (ISA) and execute the model using a High-Level Synthesis (HLS) Simulator, called Comet [16]. According to Coussy and Morawiec in [7], HLS has a substantial potential due to the fact that it may be one solution to the demand for productivity beyond state-of-the-art methods and flows. Another reason would be the possibility to maintain a design capability for a growing demand for wide-range of functionalities and increase chip manufacturing capacities. This possibility would be an answer for concerns within the industry. The product of a synthesis would be an Application-Specific Instruction-set Processors (ASIP).

According to the paper [9], ASIPs have appeared due to the requirements for an expanding application area. At that time, new application ideas have appeared with conflicting requirements like for example low power consumption and performance. To offer a solution, people have come up with the idea of customizing a processor so that it can solve a certain problem class. When taking ASIPs into consideration, the pros and cons would be the following: lower system cost, satisfying performance, while the trade-offs being that it would do poorly for other types of problem. Nevertheless, they make the argument that when designing an ASIP, it is actually a hardware/software co-design, since issues might appear not only due to software implementation but due to hardware design as well, such as design size, power consumption and achievable processor frequency [9].

The general method of creating an ASIP is by taking an already existing processor and it's ISA and modifying or extending it with new functionalities.

As Gschwind mentions in [9], the problem of creating or extending the instruction sets has been treated as a pipeline scheduling or a module selection problem which are usually delegated to a compiler to solve, but this inherently means that no logic capabilities will be generated, defeating the purpose of ASIPs. The approach which can actually generate new logic capabilities would be the one that uses Field-Programmable Gate Arrays (FPGA). This approach consists of using a special purpose compiler that "extracts functionality from a high-level languages description and implements it" [9]. The downside of this solution is that there is an idiom limitation of the system and that there is a high communication overhead between the processor and the attached FPGA.

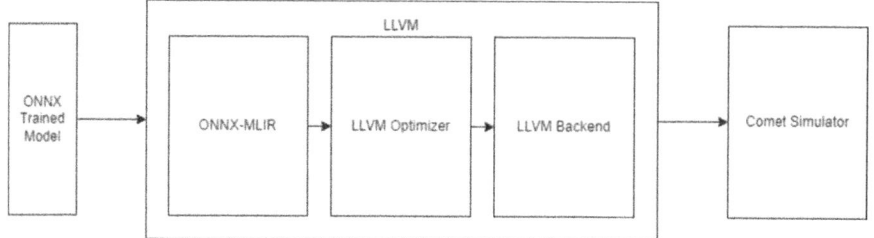

**Fig. 1.** Architecture diagram

In summary, as visible in Fig. 1, this was the architecture of our toolchain, a pretty standard one for an LLVM project. (This diagram can also be observed in [15]).

For our development, we have chosen a simple convolutional neural network in the ONNX format, trained on the MNIST dataset.

The Open Neural Network Exchange (ONNX), "is an open-source machine-independent format and widely used for exchanging neural network models" [10]. It is an active project with contributors from the open-source communities. This format gives the possibility to define models as extensible computational acyclic graph model with it's own built-in operators and standard data types. This graph is usually represented as a list of nodes. These nodes are actually calls to their corresponding operators and are not restricted to one single input and one single output. These operators are implemented externally for any applicable data type but are portable across frameworks [3].

The famous MNIST dataset [8] is a good one that offers the possibility of comparing one ML technique to other ones with a high level of simplicity which is greatly appreciated among researchers and students alike. It is said, that in 2012 there were up to 68 such techniques listed on a website [13] which also provides references and their test error rates. All these 68 ML classifiers can be categorized into 6:

- linear classifiers
- k-nearest neighbors
- boosted stumps
- nonlinear classifiers
- support vector machines
- non-convolutional neural networks
- convolutional neural networks

Also, in paper [8] is pointed out an interesting fact, that convolutional neural networks offer better performance and an even better one if they are stacked. Supposedly, a whole "commitee" of convolutional neural networks trained on an augmented dataset using elastic distortion gives 0.27% error rate (meaning 27 errors in 10.000 test set) meanwhile one single and large convolutional neural network gives 0.35% error rate. Also, another comparison which points out the benefits of using a deep neural network, is that it gives 0.35% error rate meanwhile a shallow one gives somewhere around 0.40%-0.60%. Without pre-processing and convolutional structure, a deep neural network achieves 0.83% error rate and a shallow one 1.10% [8].

Overall, this dataset is really good as a common ground for students and researchers to learn and invent new techniques.

## 2    Related Work

To the best of our knowledge, there is no published work where an HLS-based approach is used for a RISC-V ASIP core design, specialized on ML algorithms.

A similar work that somewhat resembles our intentions would be the Open Neural Network Compiler (ONNC) project [1], which is accordingly, a retargetable compilation framework that is meant for proprietary NVDLA-based deep learning accelerators. Since our accelerator would not be NVDLA-based, and the project can also be deemed as obsolete having the last release supposedly in 2020, we have chosen the approach mentioned earlier (ONNX-MLIR coupled with LLVM). Another similar project is Apache TVM which according to paper [6], is an "end-to-end optimizing compiler stack to lower and fine-tune deep learning workloads". The developers of TVM emphasize on the rate of rapid development of ASIPs specializing on deep learning algorithms. They have adopted to follow a design which separated algorithm description, schedule and hardware interface and has the unique approach of having two optimization layer, one addressing computation graph related improvements, the other one aiming to optimize tensor related aspects. We have considered using TVM but have decided to use ONNX-MLIR, the reasons being that it uses similar technologies as LLVM, more comprehensive documentation and experience with LLVM.

According to paper [9], examples of ASIPs would be the "MIPS-1" core by "Lexra", the "ARC" core or the "Carmel" architecture by Siemens, but a more recent paper [17], which presents an experiment of a synthesized RISC-V core deployed onto a Xilinx FPGA. Unfortunately, no performance results were presented in the mentioned paper.

## 3    Methods

In this section, the whole toolchain and it's development will be discussed, starting from the instruction set architecture design, going through the toolchain and until the synthesis of the proposed ASIP architecture is being deployed onto the FPGA, that being a Xilinx Zynq UltraScale+ MPSoC ZCU102. Our toolchain has grown to a complexity where it needed to follow a certain architecture regarding the dataflow. Our current environment tries to follow the Von-Neumann processor architecture's sequential flow, having the instruction encoding done at the beginning, after which would follow the implementation regarding the compiler toolchain and in the end the execution of the model on our simulator.

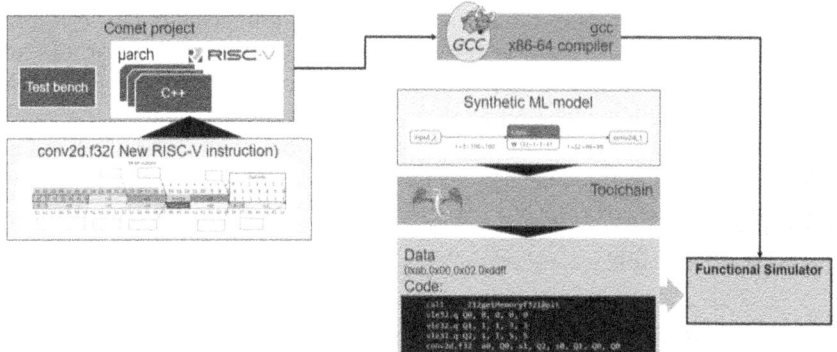

**Fig. 2.** Toolchain Architecture

The compile and runtime environment (when testing) however, fully respects this architecture.

ISA is a concept that represents certain instructions in values expressed in bits which are details regarding the memory locations, used data types and the identifier of the instruction. RISC-V defines multiple types of instructions, of which 6 are the most common [19].

According to Fig. 3, taken from the RISC-V literature [19], an R-type instruction is for operation using registers, an I-type is for immediate operands and memory loading operations, an S-type is used for storage operations, a B-type is for conditional instructions with branching, an U-type is for immediate operands of large size, and lastly, a J-type is for unconditional jump instruction [5]. It is also essential to mention the endianness of such instructions. Luckily, the RISC-V standard is flexible, allowing to extend the length of the instructions if additional space is required for an encoding. It is recommended to extend these instructions with a multiple of 32 bits.

| 31 | 30 | 25 24 | 21 | 20 | 19 | 15 14 | 12 11 | 8 | 7 | 6 | 0 | |
|---|---|---|---|---|---|---|---|---|---|---|---|---|
| | funct7 | | rs2 | | rs1 | funct3 | | rd | | opcode | | R-type |
| | imm[11:0] | | | | rs1 | funct3 | | rd | | opcode | | I-type |
| | imm[11:5] | | rs2 | | rs1 | funct3 | | imm[4:0] | | opcode | | S-type |
| imm[12] | imm[10:5] | | rs2 | | rs1 | funct3 | imm[4:1] | imm[11] | | opcode | | B-type |
| | imm[31:12] | | | | | | | rd | | opcode | | U-type |
| imm[20] | imm[10:1] | | imm[11] | | imm[19:12] | | | rd | | opcode | | J-type |

**Fig. 3.** RISC-V ISA format diagram

So as a first step, we had to design the new instruction's encoding for ML operators such as convolution or maxpooling. As specification, we used the one given by Tensor Operator Set Architecture (TOSA) [4] and have encoded the operators parameters as register addresses. Besides creating new R-type instructions, we have also worked on I-type and S-type, since we have devised a new data type called "QWord", having the structure of a sequence of four numeric values (usually integers). The need for a new data type was to be able to memorize the shape of the tensors, thus limiting the maximum tensor rank to 4. There was another reason for choosing this data type, that being the handling of cases where operators had too many parameters to be encoded onto 64 bit. In such cases, we have decided to group single or paired numeric values into an array of four values.

Next, we had to work on our compiler that was capable of taking a CNN model in the ONNX format and compile it into machine code that later had to be executed by our simulator.

This meant that first we had to create a connection between LLVM and the ONNX format. Thankfully, there is an already existing LLVM subproject called Open Neural Network Exchange - Multi-Level Intermediate Representation (ONNX-MLIR) [2]. To sum things up, this project works on multiple dialects and converts (or as the authors call it, lowers) instructions from one dialect to another, eventually reaching the point of having the said instructions in LLVM Intermediate Representation (LLVM IR). Designing an ASIP architecture meant that we will be mainly working with intrinsics, since the implementations of the operators would be done at the hardware level.

For this reason and because of using our new data type, we had to create new dialects that matched our intrinsic definitions and conversion codes that handled lowering between each dialect. Figure 4 is an illustrative representation of the lowering stages of a convolution operator, using our implementation. It is necessary to mention that these lowering stages are produced by the ONNX-MLIR module using the dialects and conversions that are available at it's disposal, and that there is not meant to be any intervention or modification done by a developer.

For a faster development process, we have concluded that we would need a code generator. Our problem was that whenever we were developing a new operator, we had to take the base codeline and modify it to our operator. This meant that we could make the base codeline into a template and create a code generator which would make the necessary modification and provide a separate implementation. All these modifications are noted inside a JSON file. We have concluded that the best programming language for this task would be Python, since it offers a high degree of flexibility regarding string manipulation. The end result of the same convolution that was lowered in ONNX-MLIR, would be similar to the source code presented below.

```
define ptr @main_graph(ptr %0) {
  %2 = call ptr @_Z12getMemoryf32l(i64 22)
  %3 = call <4 x i32> @llvm.conti.onnx.conv2df32(ptr %2, ptr %0,
       <4 x i32> <i32 1, i32 3, i32 100, i32 100>,
       ptr @constant_0, <4 x i32> <i32 32, i32 3, i32 3, i32 3>,
       <4 x i32> zeroinitializer, <4 x i32> zeroinitializer)
  ret ptr %2
}
```

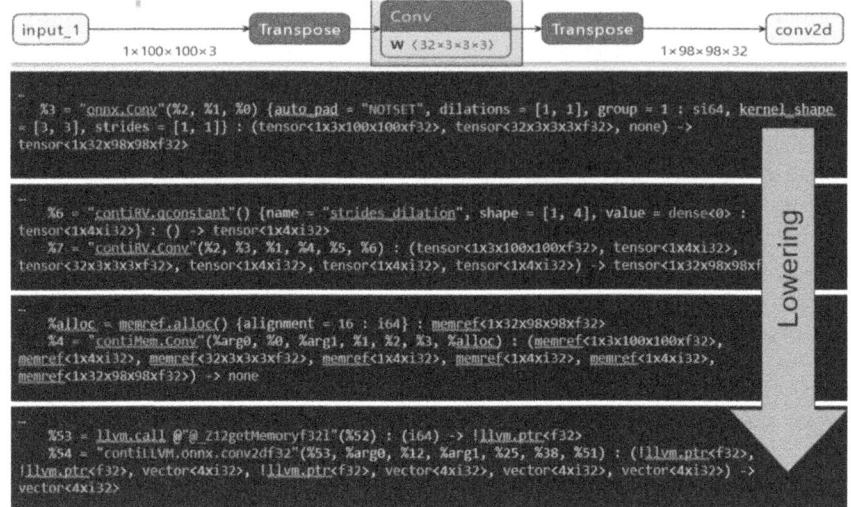

**Fig. 4.** Lowering instructions in ONNX-MLIR

After getting the instruction in LLVM IR syntax, we had to start developing the intrinsic for it in LLVM. Such process normally means defining the intrinsic, creating the mapping between the parameters and register locations and instantiating all these for different base data types (integer and float values). At this point, the importance of having at hand the ISA encoding can be clearly observed, since without it, the mapping cannot be done.

More details can be found in [15].

Compiling the code in LLVM IR syntax would result in an Assembly code, example of a convolution can be seen below.

```
main_graph:                      # @main_graph
...
    call    _Z12getMemoryf32l@plt
    lui     a1, %hi(constant_0)
    addi    a1, a1, %lo(constant_0)
    vle32.q Q0, [0, 0, 0, 0]
    vle32.q Q1, [32, 3, 3, 3]
    vle32.q Q2, [1, 3, 100, 100]
    conv2d.f32      a0, Q0, s0, Q2, a1, Q1, Q0, Q0
    lw      ra, 12(sp)
    lw      s0, 8(sp)
    addi    sp, sp, 16
...
```

Finally, the last step is the operator's implementation and integrating it into our functional simulator. As Gschwind mentions in [9], this means adding functionality to the arithmetic logic unit (ALU) of the processor which is essentially the extension of the Instruction Fetch (IF) and Execution (EX) pipeline stages. Besides taking this approach, we also expanded the DC (Decode) stage in an unconventional way. Whenever we were implementing a new operator, we have extended all these stages for our custom OpCode. We have decided to have one custom OpCode for all of our implementation, making the differentiation between instructions using the func7 parameter (excluding the QWord load and store instructions which are differentiated using the func3 parameter). However, at the EX stage, we designed the runtime in an unorthodox way: we start the execution stage by decoding the instruction parameters, loading them into variables and only after this we start executing the operator. Handling the QWord data type meant extending the load and store instruction types with our's by verifying the func7. After a number of modifications made to the Comet project, we have deemed that we should rename our "fork", thus becoming AIDA, the used abbreviation for Application-specific RISC-V Hardware Accelerator. We would like to mention that our guideline for implementation was the TOSA Specification [4], including tensor handling as well. Thus, we treat tensors as a continuous memory space and use offset techniques based on this specification to access values within tensors. After the first trials, we have since optimized these access techniques because otherwise one operator would have had very poor time performance.

Due to the embedded nature of the synthesis to an FPGA, we were required to "tailor" our implementation, more exactly, including or excluding certain operators.

For testing our implementation, we usually created a test application in C++ programming language. During compilation, we also save the input shapes of each layer into a file in YAML format, which is then used by another code generator that translates it into C++ memory definitions. This is necessary for the purpose of storing the intermediary results within the test application.

Arriving to a point where the majority of compilation is done automatically either by the compiler or by the helper code generators, we have concluded that we succeeded in creating an environment based upon the Von-Neumann architecture style.

## 4   Obtained Results

During our planning, we have decided that we would rather divide our development process into smaller milestones than leaving it as a lengthy one with a huge end-goal. Thus, our results are a bit unusual to what one may expect. So far, we had the target of verifying if the implementation is right, meaning that we did not made any accuracy tests on our classifier but compared the execution on an FPGA to the execution on a traditional x86 processor, using well-known frameworks. This comparison was made using the outputs of several or all layers (depending on the test case), with a relative tolerance of $10^{-4}$ and with an absolute tolerance of $10^{-6}$ (otherwise said, 4 decimal digits relative and 6 decimal digits absolute tolerance). Fortunately enough, our test comparisons have all passed.

However, to be able to show it's performance, we have also made measurements related to execution time, power consumption and speed defined as FLOPS (Floating Point Operations Per Seconds).

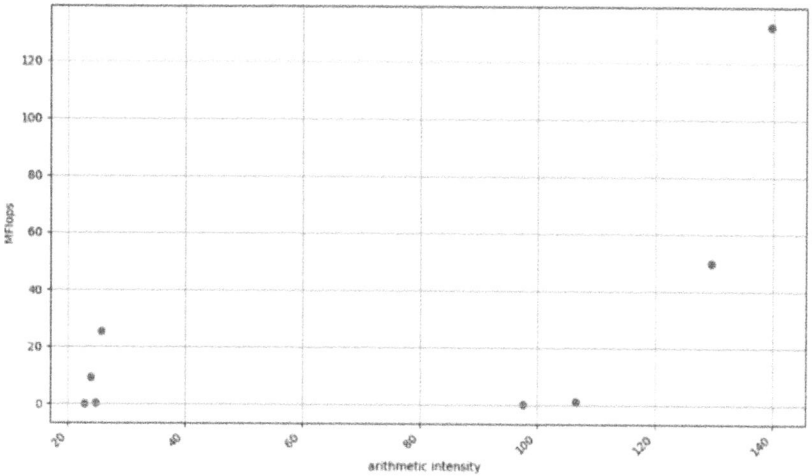

**Fig. 5.** Flops VS arithmetic intensity

In Fig. 5 is shown the performance of a convolution, shown as number of floating point operations per second (also known as Flops, expressed as MFlops in Fig. 5) with regard to number of such operations per memory usage (otherwise known as arithmetic intensity, expressed in MFlops/MB in our case). The peak performance that we have recorded was 132.73 MFlops. We also have recorded that our total on-chip power was 4.936 W, the FPGA working on a frequency of 75 MHz. The performance is a good starting point for us but indicates that more work has to be done in order to achieve satisfiable results. Power consumption-wise, we are satisfied with this value, as we intend to keep it as low as possible.

In Fig. 6 can be observed the execution time per each cycle. As we expected, there are considerable time spikes on some cycles, those cycles actually being the ones where our operators take place, which are much more complex instructions than a simple addition or load.

The recorded execution times for the operators were the following:

– Convolution2D: 8.65 milliseconds
– ReLU: 5.85 milliseconds
– Maxpool2D: 2.76 milliseconds
– Transpose: 1.44 milliseconds
– Reshape: 1.44 milliseconds
– GEMM: 9.11 milliseconds
– Softmax: 0.02 milliseconds

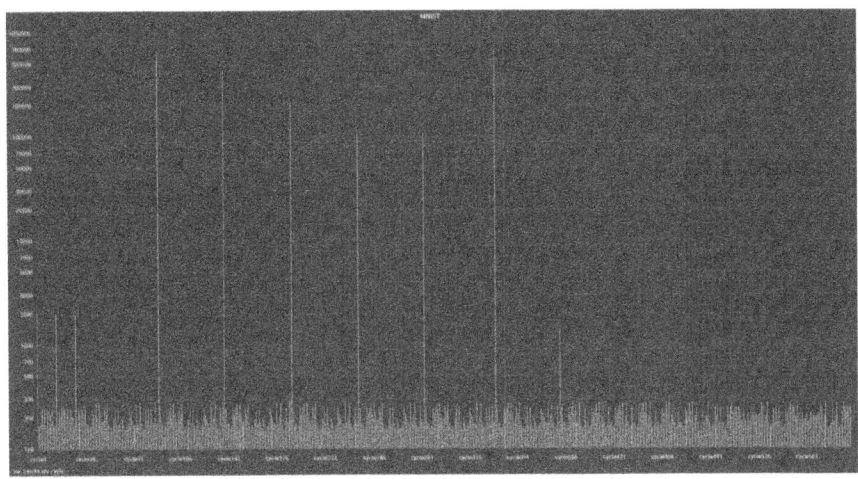

**Fig. 6.** Execution Times per Cycles

After analyzing these execution times, Conv2D and General Matrix Multiplication (GEMM) was expected to have a big execution time. However, to our surprise, ReLU had almost double times bigger execution time compared to Maxpool2D. Further investigation indicated that our memory access operations (load and store) to tensors slowed dramatically our operators. These access operations recalculate memory addresses every time from scratch, without using an already existing offset value to which the core would increment it to the address of the next value from a certain tensor. This approach was mandatory for our kernel-based operations (Conv2D, Maxpool2D, GEMM). Thus, Maxpool2D executed more quickly than ReLU, which was applying the operation element-wise and storing to a separate output tensor.

## 5    Future Work

The next stage that would happen in the near future is a benchmark of execution performance of our ASIP in comparison to an x86 or other cores of different architectures. Additionally, there is of course, room for more improvements we would like to work on.

First of them all, would be replacing our initial neural network with the one that is being developed in [18]. The neural network mentioned in that paper is actually composed of two networks, thus actually being a whole system. This network system has the aim of doing automatic data annotation of eyes, nose and mouth on recordings. Since it is not using any framework, like for example PyTorch or TensorFlow, we are in the need of changing the toolchain's frontend module. Hopefully, it should not be that big of a problem.

Other improvements would concern our implementation of the hardware architecture. After some time, we have noticed that the simulator we were using has problems in it's pipeline implementation. At some point there were quite a number of memory leaks and even after we have tried multiple solutions, there are still a small number of them present. Besides this, some data hazards have also persisted. Thus, we propose re-implementing the pipeline from scratch with the hope that our approach will be a better one and will be easier to fix in the case of issues appearing. Another improvement regarding our implementation would be related to our operators. In the first iteration, our operator's execution time was not up to the modern day's standards. Even though, as it can be seen in the previous section, we would like to optimize further our code so it would be competitive against other processors.

There are some automatization related aspects that would require some work to be done, for example, regarding the automatic inclusion or exclusion of code.

We would also like to adapt our compile environment to one that would follow the Harvard architecture, by parallelizing as much as possible.

## 6  Conclusions

To sumarize, this paper discussed about the implementation process of a toolchain that is meant for an ASIP design by the authors. This toolchain consists of the LLVM compiler (composed of the base project, the ONNX-MLIR frontend and our own backend), our RISC-V simulator, AIDA, and the helper utilities such as the Python scripts and code generators, implemented with the intention of gaining a higher degree of automation for the toolchain. The presented results included our current implementation and the comparison with a traditional x86 processor using the output tensor of each layer. These are important and satisfying steps towards the end-goal of developing an ASIP with low power consumption and high performance ML models execution.

**Acknowledgement.** This paper is supported by XReco project that is a Horizon Europe Innovation Project co-financed by the EC under Grant Agreement ID: 101070250.

## References

1. ONNC. https://onnc.ai/, Accessed January 2024
2. ONNX-MLIR. https://onnx.ai/onnx-mlir/, Accessed March 2023
3. Open Neural Network Exchange. https://onnx.ai, Accessed March 2023
4. TOSA specification. https://www.mlplatform.org/tosa/tosa_spec.html, Accessed July 2023
5. RISC-V Instruction Set Explanation. Technical Report, Fraser Innovation (2020). https://fraserinnovations.com/wp-content/uploads/2020/12/RISCV-Instruction_Set_introduction_20201130.pdf, Accessed May 2023
6. Chen, T., et al.: Tvm: end-to-end optimization stack for deep learning. arXiv preprint arXiv:1802.04799 **11**(2018), 20 (2018)
7. Coussy, P., Morawiec, A.: High-level synthesis, vol. 1. Springer (2010)

8. Deng, L.: The mnist database of handwritten digit images for machine learning research [best of the web]. IEEE Signal Process. Mag. **29**(6), 141–142 (2012)
9. Gschwind, M.: Instruction set selection for asip design. In: Proceedings of the Seventh International Workshop on Hardware/Software Codesign, pp. 7–11 (1999)
10. Jin, T., et al.: Compiling onnx neural network models using mlir. arXiv preprint arXiv:2008.08272 (2020)
11. Kadam, S.S., Adamuthe, A.C., Patil, A.B.: CNN model for image classification on MNIST and fashion-MNIST dataset. J. Sci. Res. **64**(2), 374–384 (2020)
12. Lattner, C.: The architecture of open source applications (Volume 1) LLVM. http://www.aosabook.org/en/llvm.html, Accessed May 2023
13. LeCun, Y., Cortes, C., Burges, C.J.: The MNIST database of handwritten digits. https://yann.lecun.com/exdb/mnist/, Accessed January 2024
14. LLVM Project: LLVM Documentation. https://llvm.org/docs/, Accessed May 2023
15. Nevezi-Strango, D., Rotar, D., Valcan, S., Găianu, M.: Maxpool operator for risc-v processor (2023), accepted for publication at 25th International Symposium on Symbolic and Numeric Algorithms for Scientific Computing, DIPMAI Workshop
16. Rokicki, S., Pala, D., Paturel, J., Sentieys, O.: What you simulate is what you synthesize: design of a risc-v core from c++ specifications. In: RISC-V Workshop 2019, pp. 1–2 (2019)
17. Toker, O.: A high-level synthesis approach for a risc-v rv32i-based system on chip and its fpga implementation. Eng. Proc. **58**(1) (2023). https://doi.org/10.3390/ecsa-10-16212, https://www.mdpi.com/2673-4591/58/1/72
18. Valcan, S., Găianu, M.: Eye detection for drivers using convolutional neural networks with automatically generated ground truth data. In: 2022 24th International Symposium on Symbolic and Numeric Algorithms for Scientific Computing (SYNASC), pp. 239–244 (2022). https://doi.org/10.1109/SYNASC57785.2022.00045
19. Waterman, A., Asanović K.: The RISC-V instruction set manual volume i: user-level ISA. Technical Report, RISC-V International (2017). https://riscv.org/wp-content/uploads/2017/05/riscv-spec-v2.2.pdf, Accessed May 2023

# Detection and Localization of Covid-19 on Chest Radiographs by Deep Learning Algorithms

Ahmed Balaazi[1](✉), Najeh Nafti[2,3](✉), Asma Ben Abdallah[2],
and Mohamed Hedi Bedoui[2]

[1] Faculty of Medicine of Tunis, University of Tunis El Manar, 1006 Tunis, Tunisia
ahmed.balaazi@etudiant-fmt.utm.tn
[2] Faculty of Medicine of Monastir, University of Monastir Technology and Medical Imaging
Research Laboratory - LTIM - LR12ES06, 5000 Monastir, Tunisia
naftinajeh94@gmail.com
[3] University of Sfax, National School of engineering of Sfax, Sfax, Tunisia

**Abstract.** The COVID-19 pandemic has caused one of the most severe global health crises in human history. Despite the progress made, the risk of contamination still exists due to new variants and insufficient booster vaccinations. Deep learning algorithms could be beneficial for the early diagnosis of Covid-19. This paper presents an application of deep learning techniques for detecting and localizing Covid-19 lung lesions on chest radiographs. It is a multi-class classification approach using three deep learning architectures: DensNet169, VGG16 and a non-pretrained sequential architecture. Indeed, transfer learning and ensemble learning techniques were used to classify the radiographs into three classes: "COVID", "Pneumonia" and "Normal". The database used consisted of 3225 chest radiographs selected by the team radiologist from the COVIDx - CXR version 8 database, one of the largest public databases. State of the art results were obtained for the three architectures and their different combinations, with "Accuracy" values above 83% for all architectures and reaching 96% for ensembles, which is consistent with results found in the literature. Class activation mapping (CAM) techniques were employed to localize and visualize the COVID lesions detected on chest radiographs.

**Keywords:** COVID-19 detection and localization · Transfer Learning · Chest X-ray · CAM

## 1 Introduction

COVID-19 pneumonia is a respiratory infection secondary to the SARS-COV2 virus, which caused one of the most severe pandemics and global health crises in human history [1]. Initial symptoms such as fever, cough and asthenia may not be alarming, but can quickly progress to severe respiratory failure requiring hospitalization in an intensive care unit. Mortality is high, particularly in people with a weakened immune system [2]. Despite the progress made in diagnosis, treatment, and prevention since the virus first appeared in Wuhan, China in 2019, over 600 million positive cases and

N.-T. Nguyen et al. (Eds.): ICCCI 2024, CCIS 2165, pp. 106–118, 2024.
https://doi.org/10.1007/978-3-031-70248-8_9

6.5 million deaths have been recorded to date [3]. The risk of contamination remains present due to new variants of the virus and insufficient booster vaccinations [4]. The most common diagnostic technique for COVID-19 is RT-PCR. This method detects the virus by collecting respiratory secretions from nasopharyngeal or oropharyngeal swabs [5]. CT scans are also highly sensitive as a diagnostic and prognostic technique for COVID [6]. However, both techniques are expensive and may lead to delays in disease management. Chest radiography is an inexpensive and easily accessible imaging technique but lacks sensitivity and specificity [7].

This paper contributes to the growing body of research on the application of Deep Learning (DL) techniques in the context of COVID-19 diagnosis [8–11]. Specifically, our focus is on the detection and localization of COVID-19 lung lesions on chest radiographs. In pursuit of robust results, we employ a multi-class classification approach, utilizing three distinct DL architectures: DensNet169, VGG16, and a non-pretrained sequential architecture. Regarding the significance of transfer learning and ensemble learning in enhancing model performance, our methodology incorporates these techniques to classify chest radiographs into three pivotal classes: "COVID," "Pneumonia," and "Normal."

More precisely, this study proposes the implementation of a comprehensive workflow for the 3-class classification of chest radiographs, categorizing them into 'COVID,' 'Normal,' and 'Pneumonia.' The workflow encompasses the following key steps: (i) selection and collection of images from public databases by the team radiologist, (ii) database structuring and image pre-processing, (iii) multi-class classification of radiographs using two pre-trained architectures: VGG16, DensNet169, and a linear non-pretrained sequential architecture. Ensemble Learning techniques were used to apply these architectures individually and in different combinations. (iv) The different architectures were evaluated using the usual metrics. Finally, (v) The anomalies on radiographs classified as 'COVID' and 'Pneumonia' were localized and visualized using the class activation mapping (CAM) techniques.

Our main contributions include:

1. **Balanced Database Construction:** We created a balanced chest X-ray database to ensure fair representation across classes.
2. **Annotation Accuracy Verification:** Our team, led by a radiologist, rigorously verified annotations, and selected relevant images for improved classification.
3. **Transfer Learning and Ensemble Techniques**: We demonstrated the effectiveness of transfer learning and ensemble techniques for multiclass classification of thoracic X-rays.
4. **Region of Interest Localization:** We employed advanced techniques to locate and visualize key regions in the lung, enhancing result interpretation and disease severity estimation.

The rest of this paper is structured as follows. Section 2 reports the related work, Sect. 3 describes the methodology in detail. Section 4 presents the experimental setup and the obtained results including a brief discussion, and finally the paper is concluded in Sect. 5.

## 2  Related Word

Numerous studies have applied Convolutional Neural Networks (CNNs) for the diagnosis of COVID-19 on chest X-rays, demonstrating their high sensitivity and specificity. In this paper we present the different architectures found in the literature, the size and composition of the database used, and the best results obtained:

In their study, Nigam et al. (2021) [9] employed VGG16, DenseNet121, Xception, NASNet, and EfficientNet to detect COVID on chest radiographs. The dataset used in the study contained 16,634 images, which unfortunately have not been made publicly available. The highest achieved accuracy was 93.48%, which was obtained using EfficientNetB7. Ismael and Sengur (2021) [10] proposed a binary classification network for COVID detection. They used ResNet18, ResNet50, ResNet101, VGG16, and VGG19 for feature extraction and SVM (support vector machine) for image classification. The highest accuracy value achieved was 94.7% with ResNet50. The database contained only 380 images. Abbas et al. (2021) [11] validated a CNN network named DeTraC (Decompose, Transfer, and Compose) for classifying COVID-19 images on chest radiographs. The network achieved an accuracy of 93.1% using a combination of two small datasets, totaling 196 images. Pavlova et al. (2021) [12] introduced the COVIDx8B database, which is the largest publicly available chest X-ray database of COVID-19, containing over 19,000 images. They also developed the COVID-Net CXR 2 model, a CNN specifically designed for COVID-19 detection, which achieved an accuracy of 95.5%. Zhao et al. (2021) [13] used ResNet50V2 to classify the COVIDx8B dataset with a maximum accuracy of 96.5%.

Ensemble Learning has been the subject of several studies, including the following: Gianchandani et al. (2020) [14] combined four different architectures: DenseNet201, InceptionResNetV2, ResNet152V2, and VGG16, to classify chest X-ray radiographs into three classes: COVID, normal, and pneumonia. The database consisted of 1203 images, and the accuracy ranged from 96% to 99%. Khin Yadanar Win et al. (2021) [15] proposed a method for detecting COVID in chest X-rays by segmenting both lungs and classifying regions of interest using different models. The database consisted of 15,000 radiographs with an unbalanced distribution of classes. To address this issue, several methods were employed, including weight adaptation and data augmentation. The highest achieved accuracy was 99%. Hemdan et al. (2020) [16] developed the COVIDX-Net using seven CNNs to distinguish between COVID-19 positive and negative radiographs. Their dataset consisted of 25 normal radiographs and 25 COVID-positive radiographs, and they obtained an F-score of 89% for the 'normal' class and 91% for the 'COVID' class. However, the small size of their database may limit the reliability of their deep learning model. In their study, Ben Jabra et al. [17] compared the performance of 16 pretrained CNN architectures and used ensemble learning to combine the best performing ones. The highest achieved accuracy was 99%. The database used in the study consisted of 237 COVID radiographs, 1338 normal radiographs, and 1336 viral pneumonia radiographs.

These studies have limitations, such as imbalanced data between classes and imprecise annotations in some databases, which may result in poor algorithm generalization. Additionally, the lack of precise lesion localization can be problematic for physicians who need to assess the extent of lung affected areas to adapt therapeutic management.

# 3 Materials and Methods

## 3.1 Dataset

The first step of our work was to build a database of frontal chest radiographs including COVID, non-COVID pneumonia and normal images (Fig. 1). To accomplish this, we selected the most relevant images from the COVIDx-CXR version 8 database. This database is one of the largest public databases developed as part of COVID-Net, a global open-source initiative dedicated to accelerating advances in machine learning to assist frontline healthcare workers in the fight against the pandemic [18]. The chest radiographs were selected and labelled by a radiologist in our team to improve the selection criteria and database quality. The radiographs represent all stages of the disease, from moderate to critical lesions, and include patients of different ages. This was done to create a diverse database that accurately represents the general population and improves the model's capability for generalization.

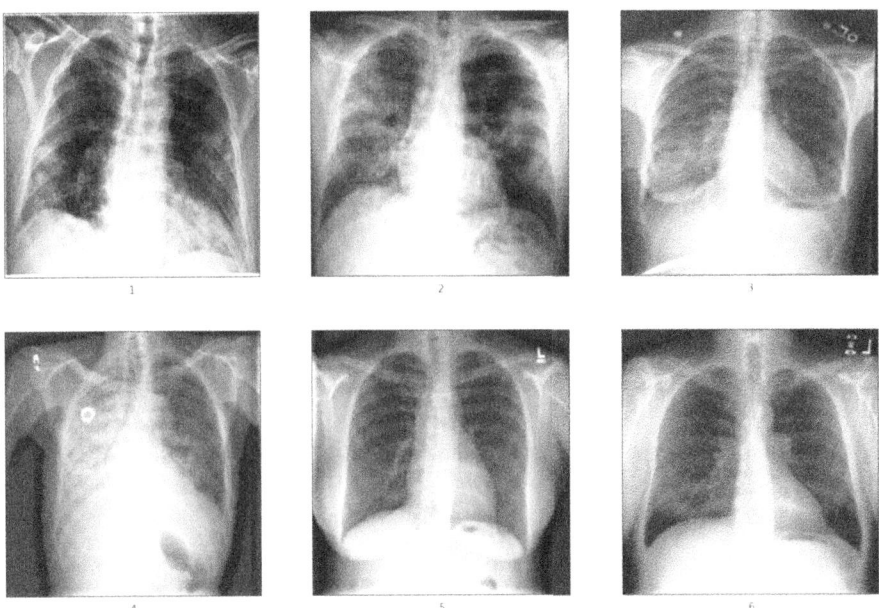

**Fig. 1.** A batch of radiographs from the dataset (1–2: Covid, 3–4: Pneumonia, 5–6: Normal)

The database contained 3,225 frontal chest radiographs with a balanced distribution of the 3 classes. Data were divided into training and test subsets, with 2,884 (89%) radiographs assigned to the training phase and 341 (11%) to the test phase (Table 1).

## 3.2 Proposed Method

The proposed approach included four stages, (i) pre-processing, (ii) classification of the radiographs using transfer and ensemble learning, (iii) localization and visualization of

**Table 1.** Characteristics of the database

|  | Technique | Train | Test | Total |
|---|---|---|---|---|
| «COVID» | Frontal radiographs | 962 | 116 | 1078 |
| «Pneumonia» | | 959 | 113 | 1072 |
| «Normal» | | 963 | 112 | 1075 |
| Total | | 2884 | 341 | 3225 |

anomalies on the radiographs classified as 'COVID' and 'Pneumonia', (iv) evaluation of the results obtained (Fig. 2).

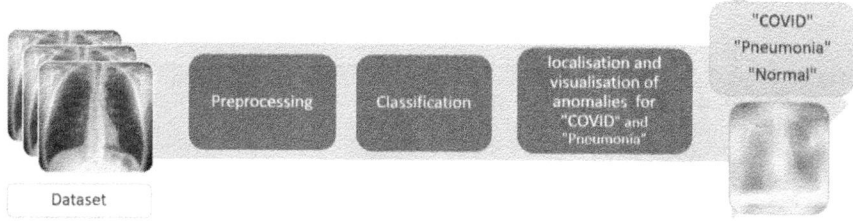

**Fig. 2.** Workflow of the proposed method

**Preprocessing.** The first step was to normalize the input data, which consisted of setting the pixel values to between 0 and 1 by dividing them by 255, and the second step was to apply standardization, which consisted of setting the mean $\mu$ to zero and the standard deviation $\sigma$ to one. Overall, the two processes have the same objective: to adjust the numerical variables so that they are comparable on a common scale, thus making the loss function more symmetrical. Before being processed by the model, all images were cropped and resized simultaneously to $224 \times 224 \times 3$. This process removed irrelevant regions and ensured consistency in the input data fed into the CNN models.

**Classification and Adopted Architectures.** Three convolutional neural network (CNN) architectures were employed to classify radiographs into three classes: "COVID", "Pneumonia" and "Normal". The approach was first to train from scratch an architecture organized in sequential mode. This architecture consisted of two identical blocks. The first block was formed by two Conv2D convolution layers with 32 kernels, each sized 3 $\times$ 3, using the ReLU activation function. The second block was formed by two Conv2D convolution layers with 64 kernels, each sized 3 $\times$ 3, also using the ReLU activation function. Each block was followed by a BatchNormalization layer, a MaxPooling2D layer, and a Dropout layer with a 40% dropout rate. At the top of the architecture, we added two fully connected layers, preceded by a 'Flatten' layer (see Table 2). Then, transfer Learning techniques [19] were utilized to apply pre-trained DenseNet169 [20] and VGG16 [21] architectures as feature extractors. Both models were pre-trained on the ImageNet database [22], which contains over 14 million labelled images divided into more than 1,000 classes. On top of these two architectures, five layers were used in sequential mode, for the final classification: 'Flatten' layer, a first 'Dense' layer, a 'Dropout 40%' layer, a second 'Dense' layer and a final 'Dense' layer (Fig. 3). Only the parameters of these final layers were trainable, while the parameters of the pre-trained models were frozen to maintain the ImageNet weights.

**Table 2.** Architecture of the sequential non pretrained model

| Layer | Type | Parameter size | Activation function | Output shape |
|---|---|---|---|---|
| 1 | Conv2D | 32 (3 × 3) | relu | 224 × 224 x 32 |
| 2 | Conv2D | 32 (3 × 3) | relu | 224 × 224 x 32 |
| 3 | BatchNormalization | – | – | 224 × 224 x 32 |
| 4 | MaxPool2D | 2 × 2 | – | 112 × 112 x 32 |
| 5 | Dropout | 0.4 | – | 112 × 112 × 32 |
| 6 | Conv2D | 64 (3 × 3) | relu | 112 × 112 × 64 |
| 7 | Conv2D | 64 (3 x 3) | relu | 112 × 112 × 64 |
| 8 | BatchNormalization | – | – | 112 × 112 × 64 |
| 9 | MaxPool2D | 2 × 2 | – | 56 × 56 × 64 |
| 10 | Dropout | 0.4 | – | 56 × 56 × 64 |
| 11 | Flatten | – | – | 200704 |
| 12 | Dense | 512 | relu | 512 |
| 13 | Dense | 2 | softmax | 2 |

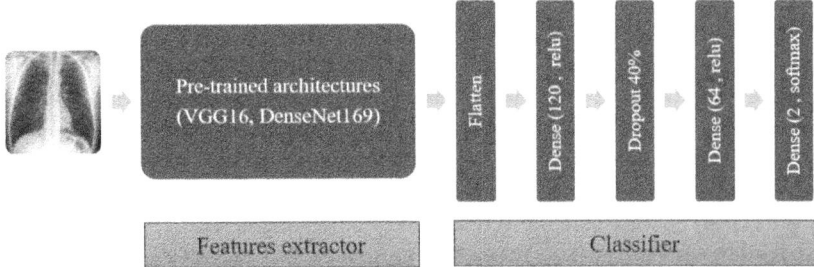

**Fig. 3.** Pre-trained models

**Ensemble Learning** The three architectures were applied in different combinations using ensemble learning techniques [23]. The first method used was the aggregation of the final predictions of the different models which were trained separately on the same database. Stacking was also employed [24], whereby feature vectors were extracted from the final layers of various architectures, concatenated and then used as input for a new model, known as the meta-learner. The meta-learner model comprised of the following layers: Flatten, Dense, Dropout (40%), and Dense, organized in sequential mode, which served as the final classifier (Fig. 4).

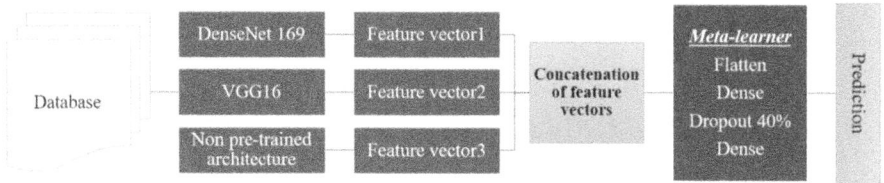

**Fig. 4.** Ensemble learning architecture using Stacking.

**Localization and Visualization of Radiographic Abnormalities for the Classes "COVID" and "Pneumonia" using Class Activation Mapping (CAM).** To enhance the practicality and comprehensibility of deep learning, several studies have been conducted. Zhou et al. [25] introduced the Class Activation Mapping (CAM) technique, which generates a heat map highlighting the significant regions of an image that influenced the prediction, based on the weights of the final convolutional layer. This technique is valuable in comprehending how a convolutional neural network arrived at a classification decision. We applied Gradient-weighted Class Activation Mapping (Grad-CAM), a specialized type of Class Activation Mapping (CAM) technique, to the VGG16 network. Grad-CAM generates a localization map by leveraging the gradients flowing into the final convolutional layer. This map highlights key regions in the image that are important for predicting a specific concept. Notably, this approach does not necessitate any modifications to the network architecture or additional training [26].

**Experiment Setup and Performance Metrics.** The loss function employed is 'categorical_crossentropy' with ADAM (Adaptive Momentum estimation) as the optimizer, which uses the gradient descent method with a learning rate set to 0.0001. The number of epochs is set to 50, and the batch size is 32. To prevent overfitting and save the best performances, we utilized the 'Modelcheckpoint' and 'Earlystopping' criteria as callbacks. All experiments were conducted on the Google Collaboratory platform using Python and the Tensorflow framework. The metrics computed on the test set include Accuracy, F1-score, sensitivity, specificity, positive predictive value (PPV), and negative predictive value (NPV), all of which are derived from the confusion matrix.

## 4  Results

We conducted several experiments using the chosen architectures. First, we applied the method to a binary classification of X-rays into COVID and non-COVID (Sect. 4.1). Next, we performed three-class classification using different models separately (Sect. 4.2). Then, we applied ensemble learning techniques (Sect. 4.3). Finally, we visualized the detected lesions using the CAM technique (Sect. 4.5).

## 4.1  Experimentation 1: Binary Classification

We started by experimenting with binary classification of radiographs into "COVID" and "NON COVID", comparing the results of the different architectures applied to a test dataset of 240 chest radiographs representing the "normal" and "COVID" classes. Table 3 summarizes the performance of different models for binary classification. All models achieved significant classification results. It was observed that the pre-trained models outperformed the non-pre-trained model in terms of execution time and classification results. The accuracy of the non-pre-trained architecture was 92.50%, while the pretrained architecture achieved over 96% accuracy. Best accuracy was 96.6% achieved by DensNet169.

**Table 3.** Binary classification: Performance of the different architectures in %

|  | Accuracy | Sensibility | Specificity | VPP | F1 score |
|---|---|---|---|---|---|
| Sequential model | 92.50 | 98.07 | 88.23 | 86.44 | 92 |
| VGG16 | 96.25 | 98.07 | 94.85 | 93.57 | 96 |
| DenseNet169 | 96.66 | 99.03 | 94.85 | 93.63 | 96 |

It is important to note that all the architectures had good sensitivity (Linear Architecture = 98.07%, VGG16 = 98.07, DensNet169 = 99.03%). These results are supported by the confusion matrices, which show a low number of false negatives across all architectures (no more than 2). However, the evaluated models exhibited lower specificity values, with the non-pre-trained model achieving 88.23% and the pre-trained models achieving 93%. The confusion matrices revealed a relatively high number of false positives: 16 for the non-pre-trained model and 7 for the pre-trained models (see Fig. 5).

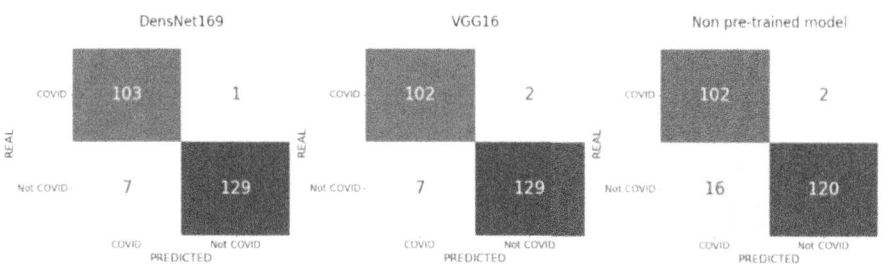

**Fig. 5.** Confusion matrices for binary classification.

## 4.2  Experimentation 2: Three-Class Classification

Table 3 shows the performance of the different architectures for three-class classification. All models achieved accuracy above 80%, with superiority for the pre-trained architectures. VGG16 outperformed DensNet169 with an accuracy of 94.5% (versus 92%). VGG16 and DenseNet169 obtained a sensitivity of 98% and 100% for the "Normal" class, with a number of False Negatives (FN) equal to 0 for VGG16 (Fig. 6). For the "COVID" and "Pneumonia" classes, VGG16 and DenseNet169 have a better specificity (99% for "COVID" with VGG16) (Table 4).

**Table 4.** Three-class classification: Performance of the different architectures in %.

|  | Label | Sensitivity | Specificity | F1-Score | Accuracy |
|---|---|---|---|---|---|
| Sequential model | COVID | 87% | 96% | 90% | 83.2% |
|  | Normal | 89% | 86% | 82% |  |
|  | Pneumonia | 73% | 92% | 77% |  |
| VGG16 | COVID | 95% | 99% | 97% | 94.4% |
|  | Normal | 100% | 95.3% | 95% |  |
|  | Pneumonia | 88% | 97.5% | 91% |  |
| DensNet169 | COVID | 95% | 94.7% | 93% | 92% |
|  | Normal | 98% | 96% | 95% |  |
|  | Pneumonia | 82% | 97.5% | 88% |  |

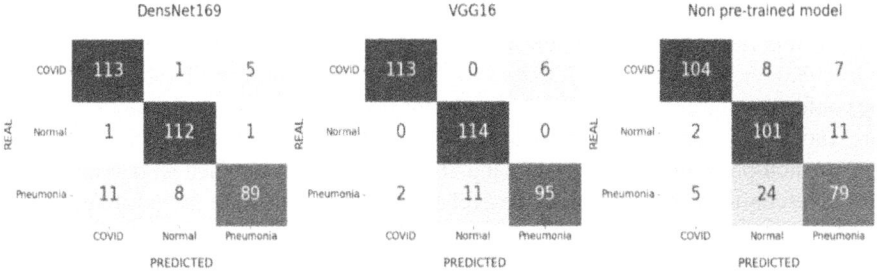

**Fig. 6.** Confusion matrices of different models for three-class classification.

### 4.3  Experimentation 3: Ensemble Learning

Table 5 shows that combining the 3 architectures, by aggregating the final predictions or using the stacking method yielded better results, with accuracies of 95% and 96% respectively. The ensemble learning techniques also improved all other metrics of the three-class classification, resulting in fewer false negatives and positives (Fig. 7).

### 4.4  Localization and Visualization of Lesions for Radiographs Classified as COVID and Pneumonia

Gradient-weighted Class Activation Mapping (Grad-CAM) techniques have been used to visualize and locate the areas of the lung affected by both the 'COVID' and 'Pneumonia' classes. This will assist clinicians in quickly locating lesions and assessing the severity of abnormalities, enabling them to tailor treatment (see Figs. 8 and 9).

**Table 5.** Three-Class classification: Performance of Ensemble learning models

|  | Label | Sensitivity | Specificity | F1-Score | Accuracy |
|---|---|---|---|---|---|
| Aggregation | COVID | 95% | 99% | 97% | 95% |
|  | Normal | 100% | 95% | 95% |  |
|  | Pneumonia | 89% | 98% | 92% |  |
| Stacking | COVID | 96% | 99% | 97% | 96% |
|  | Normal | 100% | 95.4% | 97% |  |
|  | Pneumonia | 92% | 98% | 94% |  |

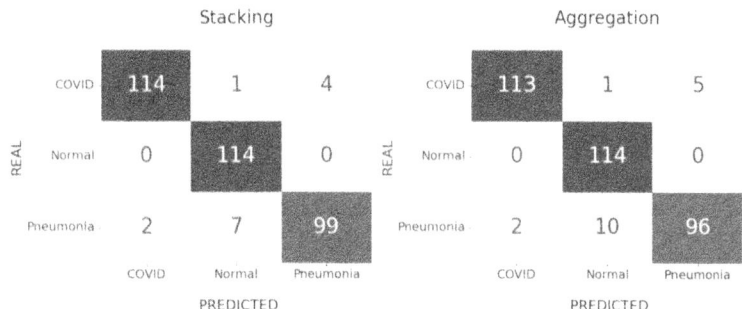

**Fig. 7.** Confusion matrices for ensembles

**Fig. 8.** An example of a Pneumonia X-ray of the test set, the heat map generated, and the superposition of the X-rays and the heat map.

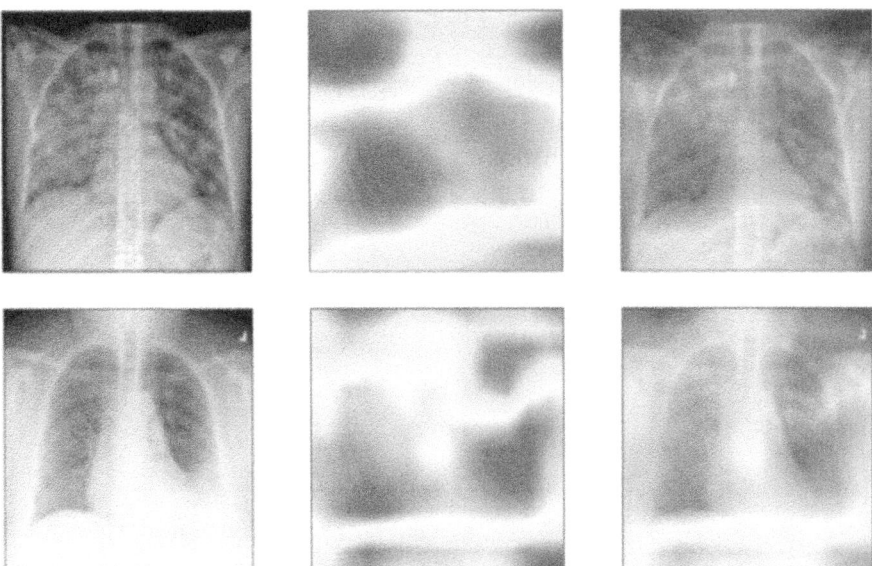

**Fig. 9.** Two examples of COVID X-rays of the test set, the heat maps generated, and the superposition of the X-rays and their heat maps.

## 5   Conclusions

In this paper, we present a multi-class classification approach based on three deep learning architectures: DensNet169, VGG16 and a non-pretrained sequential architecture, to detect COVID and differentiate it from "Normal" chest x-rays and "Pneumonia". We were able to show the effectiveness of Transfer Learning techniques to detect "COVID" on chest radiographs and overcome the huge amount of data required by Deep Learning algorithms. The best performances were achieved by ensemble learning models, so we confirmed that this technique is a solid approach to obtain optimal results with minimum errors. With Class activation mapping (CAM) we generated heat maps for radiographs classified as "COVID" or "Pneumonia" which will be useful for physicians in locating lesions and estimating the severity of the disease. However, it is known that the most efficient deep learning models require a large amount of data, which is difficult to achieve in the biomedical field due to the limited access and the scarcity of labelled databases. A multidisciplinary collaboration between radiologists, computer scientists and data scientists are therefore necessary to overcome these difficulties.

**Data and Code Availability**
The dataset used in this study is publicly available on Figshare [27] and the code implementation can be found on GitHub [28].

**Acknowledgments.** We, the authors, would like to express our collective appreciation for each other's dedicated contributions to this research endeavor. Our collaborative efforts and shared expertise have enriched the content of this paper. Special thanks to our advisors for their guidance and support throughout the process.

**Disclosure of Interests.** The authors have no competing interests to declare that are relevant to the content of this article.

# References

1. A novel coronavirus genome identified in a cluster of pneumonia cases — Wuhan, China 2019–2020 - PMC n.d. https://www.ncbi.nlm.nih.gov/pmc/articles/PMC8393069/. Accessed 14 Jan 2024
2. Wang, D., Hu, B., Hu, C., Zhu, F., Liu, X., Zhang, J., et al.: Clinical characteristics of 138 hospitalized patients with 2019 novel coronavirus-infected pneumonia in Wuhan. China. JAMA **323**, 1061–1069 (2020). https://doi.org/10.1001/jama.2020.1585
3. Coronavirus including COVID-19: symptoms, Outbreaks, Transmission, Treatment & Prevention n.d. https://www.webmd.com/covid/features/coronavirus-disease-2019-covid-19. Accessed 19 Feb 2023. Recherche Google n.d
4. Aleem, A., Akbar Samad, A.B., Vaqar, S.: Emerging variants of SARS-CoV-2 and novel therapeutics against coronavirus (COVID-19). StatPearls, StatPearls Publishing, Treasure Island (FL) (2023)
5. Detection of SARS-CoV-2 in different types of clinical specimens - PubMed n.d. https://pub med.ncbi.nlm.nih.gov/32159775/. Accessed 14 Jan 2024
6. Sensitivity of chest CT for COVID-19: comparison to RT-PCR I Radiology n.d. https://pubs. rsna.org/doi/10.1148/radiol.2020200432. Accessed 14 Jan 2024
7. Imaging profile of the COVID-19 infection: radiologic findings and literature review - PubMed n.d. https://pubmed.ncbi.nlm.nih.gov/33778547/. Accessed 14 Jan 2024
8. Deep learning and its applications in biomedicine - ScienceDirect n.d. https://www.sciencedi rect.com/science/article/pii/S1672022918300020. Accessed 14 Jan 2024
9. Nigam, B., Nigam, A., Jain, R., Dodia, S., Arora, N., Annappa, B.: COVID-19: automatic detection from x-ray images by utilizing deep learning methods. Expert Syst. Appl. **176**, 114883 (2021). https://doi.org/10.1016/j.eswa.2021.114883
10. Ismael, A.M., Şengür, A.: Deep learning approaches for COVID-19 detection based on chest X-ray images. Expert Syst. Appl. **164**, 114054 (2021). https://doi.org/10.1016/j.eswa.2020. 114054
11. Abbas, A., Abdelsamea, M.M., Gaber, M.M.: Classification of COVID-19 in chest X-ray images using DeTraC deep convolutional neural network. Appl. Intell. **51**, 854–864 (2021). https://doi.org/10.1007/s10489-020-01829-7
12. Frontiers I COVID-Net CXR-2: an enhanced deep convolutional neural network design for detection of COVID-19 cases from chest x-ray images n.d. https://www.frontiersin.org/art icles/10.3389/fmed.2022.861680/full. Accessed 14 Jan 2024
13. Diagnostics I Free Full-Text I fine-tuning convolutional neural networks for COVID-19 detection from chest x-ray images n.d. https://www.mdpi.com/2075-4418/11/10/1887. Accessed 14 Jan 2024
14. Gianchandani, N., Jaiswal, A., Singh, D., Kumar, V., Kaur, M.: Rapid COVID-19 diagnosis using ensemble deep transfer learning models from chest radiographic images. J. Ambient. Intell. Humaniz. Comput. **14**, 5541–5553 (2023). https://doi.org/10.1007/s12652-020-026 69-6
15. Applied Sciences I Free Full-Text I Ensemble deep learning for the detection of COVID-19 in unbalanced chest x-ray dataset n.d. https://www.mdpi.com/2076-3417/11/22/10528. Accessed 14 Jan 2024
16. Hemdan, E.E.-D., Shouman, M.A., Karar, M.E.L COVIDX-Net: a framework of deep learning classifiers to diagnose COVID-19 in X-Ray images. arXivOrg 2020. https://arxiv.org/abs/ 2003.11055v1. Accessed 14 Jan 2024

17. Applied Sciences | Free Full-Text | COVID-19 diagnosis in chest x-rays using deep learning and majority voting n.d. https://www.mdpi.com/2076-3417/11/6/2884. Accessed 14 Jan 2024

18. Wang, L., Lin, Z.Q., Wong, A.: COVID-Net: a tailored deep convolutional neural network design for detection of COVID-19 cases from chest X-ray images. Sci. Rep. **10**, 19549 (2020). https://doi.org/10.1038/s41598-020-76550-z

19. Pan, S.J., Yang, Q.: A survey on transfer learning. IEEE Trans. Knowl. Data Eng. **22**, 1345–1359 (2010). https://doi.org/10.1109/TKDE.2009.191

20. [1608.06993] Densely connected convolutional networks n.d. https://arxiv.org/abs/1608.06993. Accessed 16 Jan 2024

21. Simonyan, K., Zisserman, A.: Very deep convolutional networks for large-scale image recognition (2015). https://doi.org/10.48550/arXiv.1409.1556

22. Deng, J., Dong, W., Socher, R., Li, L.-J., Li, K., Fei-Fei, L.: ImageNet: a large-scale hierarchical image database. In: 2009 IEEE Conference on Computer Vision and Pattern Recognition, pp. 248–55 (2009). https://doi.org/10.1109/CVPR.2009.5206848

23. Mohammed, A., Kora, R.: A comprehensive review on ensemble deep learning: opportunities and challenges. J. King Saud Univ. Comput. Inf. Sci. **35**, 757–774 (2023). https://doi.org/10.1016/j.jksuci.2023.01.014

24. Pavlyshenko, B.: Using stacking approaches for machine learning models. In: 2018 IEEE Second Int. Conf. Data Stream Min. Process. DSMP, pp. 255–258 (2018). https://doi.org/10.1109/DSMP.2018.8478522

25. Zhou B, Khosla A, Lapedriza A, Oliva A, Torralba A. Learning Deep Features for Discriminative Localization 2015. https://doi.org/10.48550/arXiv.1512.04150

26. Selvaraju, R.R., Cogswell, M., Das, A., Vedantam, R., Parikh, D., Batra, D.: Grad-CAM: visual explanations from deep networks via gradient-based localization. Int. J. Comput. Vis. **128**, 336–359 (2020). https://doi.org/10.1007/s11263-019-01228-7

27. Balaazi, A.: COVID-pneumonia detection: expert-selected x-ray dataset (COVIDx-CXR) (2024). https://figshare.com/articles/dataset/COVID-Pneumonia_Detection_Expert-Selected_X-ray_Dataset_COVIDx-CXR_/25917340/1, https://doi.org/10.6084/m9.figshare.25917340.v1

28. AhmedBALAAZI.          AhmedBALAAZI/Detection-and-localization-of-COVID-19-and-Pneumonia-on-chest-radiographs 2024

# Big Textual Data Analytics Using Transformer-Based Deep Learning for Decision Making

Omar Haddad(✉)📧 and Mohamed Nazih Omri📧

MARS Research Lab LR17ES05, University of Sousse, Sousse, Tunisia
omarhaddad@fsegs.u-sfax.tn, mohamednazih.omri@eniso.u-sousse.tn

**Abstract.** With the remarkable emergence of significant results for various deep learning-based transformation techniques, which are pre-training models such as BERT and its branches, which have the potential to deal with big data analysis frameworks. This represents a qualitative leap in understanding large-sized textual data of the opinions of Web users in order to classify it into several poles, and it is an important motivation for valuing it and benefiting from it in Decision-Making by managers of various companies such as marketing, health care insurance, finance, protection, etc. This paper explains how to provide high performance in analyzing huge text data by building a solid model suitable for analyzing huge text data, then classifying it by improving the contextual information in the sentence using the BERT technique with mechanism CNN. Extensive experiments on large-scale text data have demonstrated the remarkable efficiency of our model, an estimated percentage 92% compared to new and recent research studies.

**Keywords:** Big Data Analytics · Hadoop · DistilBERT · Convolutional Neural Network

## 1 Introduction

### 1.1 Context and Issues

The amazing developments in technology that the world is witnessing today, especially cloud computing [6,18] and the Internet of Things [11,13], have paved the way to facilitate the ability to produce large amounts of digital data from several different sources, most notably social media networks [2,7]. This requires new, powerful methods that can accommodate the high-precision computing of massive amounts of data. One of the basic aspects of these data is represented by text, which is the largest means of communication between Web users in terms of opinions and expressions. This plays a vital role in pushing for its treatment and raising the challenges facing the efficiency of the state-of-the-art Machine Learning-based models that aim to classify the textual content

© The Author(s), under exclusive license to Springer Nature Switzerland AG 2024
N.-T. Nguyen et al. (Eds.): ICCCI 2024, CCIS 2165, pp. 119–131, 2024.
https://doi.org/10.1007/978-3-031-70248-8_10

according to its polarity: positive or negative. Generally, this classification is performed through integrating several pieces of knowledge, such as the meanings, frequency, and contexts of words [17,22]. The ability of the scientific community to transform this data into valuable information can make a difference in helping corporate decision-makers make appropriate decisions in an Large-Scale decision-making (LSDM) event [1] regarding products or services. It constitutes, in turn, an important factor in the process of improving companies' ability to produce. Although LSDM is a complex and challenging category, achieving a collective decision and taking into account the relationships of agreement or conflict between participants becomes necessary to build a unified society, a sound, and effective decision. In this context, researchers have contributed to presenting many machine learning and deep learning models that work on mining structured and unstructured textual data that will be classified into positive and negative poles. The classification process is made according to multiple and diverse mechanisms using natural language processing tools [14]. However, most of the proposed models suffer from low accuracy in determining the correct data polarity. Likewise, these models remain unable to analyze big data with respect to streaming and batch text data at the same time. Therefore, to achieve the construction of a solid model with high quality in classification accuracy that can accommodate large batch text data [3] and streaming [10,21], we developed an appropriate strategy based on the big data analysis frameworks indicated by the survey [4]. Then the huge textual data is analyzed through multiple means, most notably deep learning and selecting the appropriate feature, which is beneficial in the effectiveness of making effective predictive decisions for companies and organizations in several areas such as sports [20], healthcare [12], agriculture [15], smart cities [16], marketing [9], etc.

## 1.2    Contribution

Following an extensive literature review on the primary approaches suggested for the analysis of big data, we have pinpointed the inherent disadvantages and constraints of the methodologies examined in this study. Despite numerous efforts to tackle the challenges linked to big data analysis, various issues have curtailed the efficacy of these endeavors. The subpar precision levels of these methodologies result in outcomes that frequently contain inaccuracies. Given the emphasis of this paper on concise texts, the ensuing section outlines several pivotal advancements that set our approach apart.

– Instead of employing deep learning methods for review classification, a robust and efficient approach has been devised for analyzing large datasets to assist decision-makers. We suggest utilizing this approach to create preprocessing data analytics that focus on data cleaning, optimizing information retrieval time, and minimizing storage requirements. This methodology involves leveraging the Hadoop big data analysis framework and modifying the structure of the analysis component using Map-Reduce.

– In the preceding stage, the text embedding process incorporates the DistilBERT approach, known for its ability to efficiently capture sentence contexts with lower energy consumption compared to other transformation models. This approach is seen as a forerunner to convolutional neural networks (CNNs), which excel at accurately classifying brief texts across various categories within large datasets.
– To improve the accuracy of the results of the proposed model, we fixed the problems caused by the neural network by adding channels to the CNN technology.

### 1.3   Paper Organization

The remainder of this paper is structured as follows: An overview of some earlier research papers on big data analysis using machine learning approaches is presented in Sect. 2. In Sect. 3, we provide a sophisticated method designed to polarize big data to make predictions, which aids in business managers' decision-making. To verify the effectiveness of our proposed model compared to the primary research study models in the context of this paper, Sect. 4 empirically investigates it. We wrap up our work and offer some suggestions for the direction of our next work in Sect. 5.

## 2   Releated Work

In this section, we will present previous literature studies that have been conducted in the field of big text data analysis [19], to classify the text into different poles. In reviewing them, we understand their limitations and use them to improve our results.

The authors of [9] present the main challenges associated with identifying consumer opinions in short texts on social networks. This is used for evaluating the best brands and products. As a result, the authors developed a false review detection framework based on natural language processing, which shows 85.5% accuracy in the detection of false reviews.

In the DeepEmotionNet model, the authors [23] achieved effective results over recent models with an estimated range of 4.9% to 29. 8% in the F score. To build this architecture, three key components are required: 12-layer Contextual Encoders, Message2Vec, and Emotion2Vec. This model contributes to the analysis of the different feelings of senior managers associated with the financial performance of companies. Therefore, this analysis contributes to improving the predictive power of companies' financial performance.

In [8], the goal of this model is to support the major manufacturers and designers in making the best and most appropriate decisions when engineering electric vehicles. We do this by extracting valuable opinions from automobile manufacturers, which are then used to determine what elements should be updated and improved in electric vehicles. This model consists of a Hadoop big data analysis framework and some deep learning tools such as CNN. Also, the experimental results of this model provide an accuracy of 85% in classifying textual data into different poles.

The authors of the study suggested in [5] introduce PABIDDL, a novel prediction method based on deep learning and big data analysis. The first step in this three-stage technique is to use MapReduce in the Hadoop framework to reduce Big Data. The authors used the GloVe approach to initialize these data in the second step. Ultimately, the text data was classified into benefits and disadvantages poles using CNN deep learning. The findings of the empirical study indicated that the accuracy rate of the PABIDDL approach was 0.93%.

In terms of processing accuracy, the scientific community strives to achieve advanced results by using machine learning techniques, including deep learning, to classify short-texts. Despite the great progress that researchers have made in building multiple and modern models, these efforts remain limited. We can observe this in the level of accuracy of the classification of data for studies related to our work conducted on different models. Moreover, all the proposed methods do not work in a large-scale data analysis system in terms of combining batch and streaming data.

## 3    Proposed Decision-Making Approach

We provide a high-level overview of our proposed architecture in this section, followed by a detailed description of how the data can be preprocessed using the Hadoop big data analytics framework. The second step is to classify big data into multiple and different poles by combining algorithms with deep learning techniques so that companies can extract predictions about their products, services, and more. In this section, a modern and advanced model is designed based primarily on big data analysis methods using deep learning, which has removed some obstacles and challenges associated with the processing of large data sets. The model clarifies some future predictions for corporate decision-makers regarding services and products based on large volumes of textual data circulating on websites, which represent web users' opinions and feelings (Fig. 1).

### 3.1    Pre-Processing : Big Data Analytics

In this stage, large-scale data is cleaned and then verified using standard features that contribute to accurately identifying the data through the characteristics of the concept of big data and natural language processing (stop words, punctuation, URL, etc.). This process is accomplished in MapReduce through the Feature Selection component of both the Hadoop framework. This leads to reducing the amount of useless data and identifying huge and useful data, which contributes to its accurate classification.

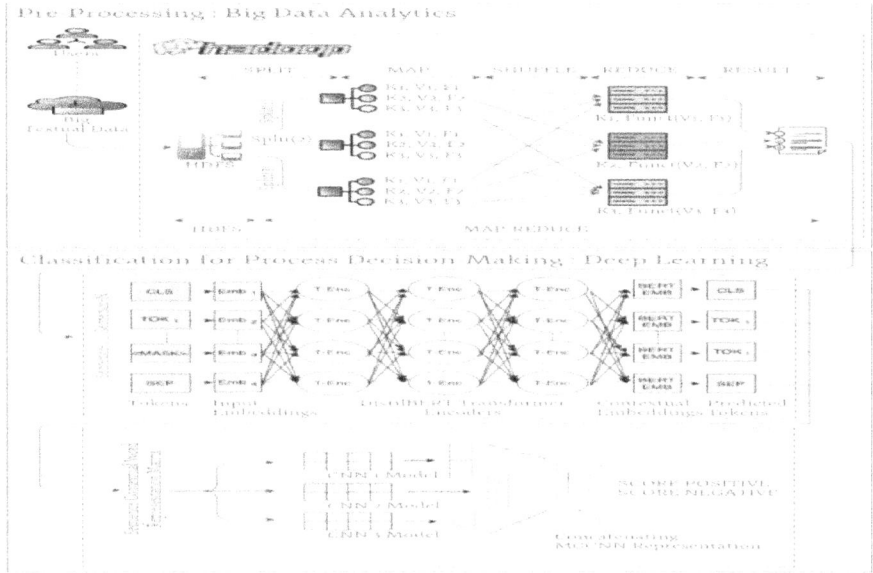

**Fig. 1.** Main stages of proposed model

**Proposed Algorithm for Big Data Pre-Processing.** The purpose of this phase is to develop a robust algorithm capable of preprocessing large-scale data. Following the MapReduce architecture, the first stage of the algorithm contributed to cleaning the data, and the second stage contributed to reducing the data.

## 3.2   Classification : Deep Learning

In the next step, the DistilBERT method is used, which provides each Short-Text word with a large-dimensional vector that represents contexts and connotations. These vectors are then included in the CNN input. This technology devotes its high ability to classifying large-sized data, but during the process, training takes place in the network.

**Proposed Algorithm.** In this algorithm, large-scale data are classified into multiple poles using deep learning methods. As input for this algorithm, we used the results of the first algorithm, which is data preprocessed by the Hadoop framework based on a variety of features. The DistilBERT method is used to extract features from these data using mathematical operations. CNN are run on the data processed from the previous stages.

---

**Algorithm 1:** MAP-REDUCE Task

---

1 **Input :** $HDFS_{Files}$ for a three-node cluster
2 **Ouput :** Generate new parallel parts
3 **Begin**
4 $MAP_{Part} \leftarrow \emptyset$
5 **for** *each* $PART_{Node}$ *in* $HDFS_{Files}$ **do**
6     $MAP_{KeyPart} \leftarrow$ Creation($PART_{Node}$, KeyPart)
7     $MAP_{Value} \leftarrow$ Creation($PART_{Node}$, Value)
8     $MAP_{SelectedFeature} \leftarrow$ Creation($PART_{Node}$, SlectedFeature)
9     $MAP_{Part} \leftarrow MAP_{Part} \cap MAP$
10 **end**
11 **for** *each* $PART$ *in* $MAP_{Part}$ **do**
12     $Collection_{KeyPart} \leftarrow$ UniqueKeyExtraction($PART_{KeyPart}$)
13     $Collection_{Value} \leftarrow Value(PART_{Value})$
14     $Collection_{SelectedFeature} \leftarrow Reducing(PART_{SelectedFeature})$
15 **end**
16 **for** *each* $REDUCE_i$ *in* $Collection$ **do**
17     emit($REDUCE_i$)
18 **end**
19 **end**

---

**Algorithm 2:** Prediction Task

---

1 **Input :** $HDFS_{DataSet}$
2 **Ouput :** $SCORE_{Prediction}$
3 **Begin**
4 **for** $ST_k$ *in* $HDFS_{DataSet}$ **do**
5     $TE \leftarrow TokenEmbeddings(ST_k, Parameter DistilBERT)$
6     $SE \leftarrow SegmentEmbeddings(ST_k)$
7     $PE \leftarrow PositionEmbeddings(ST_k)$
8     $TokenEmbeddings \leftarrow TE \oplus SE \oplus PE$
9     $RM \leftarrow TokenEmbeddings$
10     ConcatenateChannels $\leftarrow \emptyset$
11     **for** $CNN_i$ *in* $MultiChannel - CNN$ **do**
12        Flattening $\leftarrow \emptyset$
13        **for** *each* $CL_Y$ *of* $MST_k$ **do**
14           $CL_Y \leftarrow \sigma(Sum(MST_j \oplus Feature) + b)$
15           $CLM_F \leftarrow CL_Y$
16           Flattening $\leftarrow$ Flattening $\cap$ Max-Pooling($CLM_F$)
17        **end**
18        ConcatenateChannels $\leftarrow$ ReLU(Flattening)
19     **end**
20     print($SCORE_{Prediction}(ConcatenateChannels)$)
21 **end**

# 4    Experimental Study of the Proposed Approach

This section provides an overview of the experimental results of our model on several types of data sets. This process helps confirm the obtained results and consolidate them. In a second stage, it is compared with results from previous studies on deep learning models on large-scala data.

## 4.1    Evaluation Environment

**Simulation Setup.** To run the proposed model, the Google Colab Pro+ platform was used on the Hadoop 3.3 big data analysis framework. This model was simulated on a cluster consisting of three devices :

– RYZEN 5, 5500 SERIES, 24 GB RAM, Ubuntu 23.04 (Professional),
– RYZEN 5, 5500 SERIES, 24 GB RAM, Ubuntu 23.04 (Professional),
– RYZEN 5, 5700 SERIES, 24 GB RAM, Ubuntu 23.04 (Professional).

**Dataset Description.**    Yelp  https://s3.amazonaws.com/fast-ai-nlp/yelp_ review_polarity_csv.tgz is a standard data set that is used by many research studies, as it represents a group of reviews with positive polarity, symbolized by Category 2 The negative symbolizes category 1. For each polarity, 280,000 training samples and 19,000 test samples are randomly taken, which amounts to a total of 560,000 training samples and 38,000 test samples, as shown in the attached Table 1.

**Table 1.** .

| DataSet | Positive Reviews | | Negative Reviews | | Statistic of reviews |
|---------|-------|------------|-------|------------|----------------------|
|         | Train | Validation | Train | Validation |                      |
| YELP    | 280000 | 19000     | 280000 | 19000     | 598000               |

**DataSet for Training and Testing Process.** The data set contributes to the evaluation of models through simulations of algorithms to consider their performance according to the use of part of the data for training, estimated at 93.214% and the other part for testing, estimated at 6.786%. In this aspect, the evaluation process is devoted to deep learning models using standards approved by the scientific and research community, including accuracy, recall, and F1-Score.

This paragraph includes the two figures ?? that show how to train a neural network on large short-text data. Let:

– **TP (True Positives):** If the "short text" category has a positive value and positive ranking results, then the category is considered positive.,

- **TN (true negatives):** In this case, if there is a negative value for the "short text" category, and the ranking after classification is negative,
- **FP (false positives):** The listing after classification is "positive", while the category value for "short text" is "negative", and
- **FN (false negatives):** When the value of the "short text" category is "positive" and the ranking after classification is "negative", the category is considered positive.

The following standard equations *Precision* (Eq. 1), *Recall* (Eq. 2), *Accuracy* (Eq. 3), $F1 - Score$ (Eq. 4), and $G - mean$ (eq. ??) provide an evaluation of the efficiency of our MapReduce+DistilBERTMCCNN approach, these equations are defined as follows :

**Precision:** the subset of true positive predicted instances against the set of all positive predicted instances.

$$Precision = \frac{TP}{TP + FP} \tag{1}$$

**Recall:** the subset of the true positive predicted instances against the set of all actual positive instances.

$$Recall = \frac{TP}{TP + FN} \tag{2}$$

**Accuracy:** the subset of predicted real instances compared to the set of predicted instances.

$$Accuracy = \frac{TP + TN}{TP + TN + FP + FN} \tag{3}$$

**F1-Score:** represents the harmonic mean between the precision measurement and that of the recall.

$$F1 - Score = \frac{2 \times Precision \times Recall}{Precision + Recall} \tag{4}$$

### 4.2    Experimental Study and Results Study

In this part, we conducted many experiments on the large YELP dataset using deep learning mechanisms that are highly efficient in providing impressive results according to the literature. In this context, we highlight our model, which provided superior results to these deep learning mechanisms in terms of accuracy in classifying data, as shown in the Figs. 2, 3, 4. Our model proved very close to the accuracy of training, and our model also has the smallest area of errors. All of this is shown in the Fig. 6. (Fig. 5, Tables 2, 3, 4 and 5)

**Table 2.** Polarity of positive and negative data through precision results

| Approach | YELP DataSet | | |
|---|---|---|---|
| | Positive | Negative | Macro avg |
| DistilBERT+CNN | 86 | 85 | 86 |
| DistilBERT+MultiChannel CNN | 87 | 87 | 87 |
| DistilBERT+GRU | 85 | 78 | 82 |
| DistilBERT+LSTM | 86 | 85 | 85 |
| DistilBERT+Bidirectional LSTM | 87 | 84 | 86 |
| **MapReduce+DistilBERTMCCNN** | **94** | **96** | **95** |

**Table 3.** Polarity of positive and negative data through recall results

| Approach | YELP DataSet | | |
|---|---|---|---|
| | Positive | Negative | Macro avg |
| DistilBERT+CNN | 83 | 88 | 85 |
| DistilBERT+MultiChannel CNN | 86 | 87 | 87 |
| DistilBERT+GRU | 76 | 86 | 81 |
| DistilBERT+LSTM | 85 | 85 | 85 |
| DistilBERT+Bidirectional LSTM | 83 | 88 | 85 |
| **MapReduce+DistilBERTMCCNN** | **96** | **94** | **95** |

**Table 4.** Polarity of positive and negative data through f1-Score results

| Approach | YELP DataSet | | |
|---|---|---|---|
| | Positive | Negative | Macro avg |
| DistilBERT+CNN | 84 | 87 | 86 |
| DistilBERT+MultiChannel CNN | 87 | 87 | 87 |
| DistilBERT+GRU | 80 | 82 | 81 |
| DistilBERT+LSTM | 85 | 85 | 85 |
| DistilBERT+Bidirectional LSTM | 85 | 86 | 85 |
| **MapReduce+DistilBERTMCCNN** | **95** | **95** | **95** |

**Table 5.** Polarity of positive and negative data through accuracy results

| Approach | YELP DataSet | | |
|---|---|---|---|
| | Positive | Negative | Macro avg |
| DistilBERT+CNN | 86 | 86 | 86 |
| DistilBERT+MultiChannel CNN | 87 | 87 | 87 |
| DistilBERT+GRU | 81 | 81 | 81 |
| DistilBERT+LSTM | 85 | 85 | 85 |
| DistilBERT+Bidirectional LSTM | 85 | 85 | 85 |
| **MapReduce+DistilBERTMCCNN** | **95** | **95** | **95** |

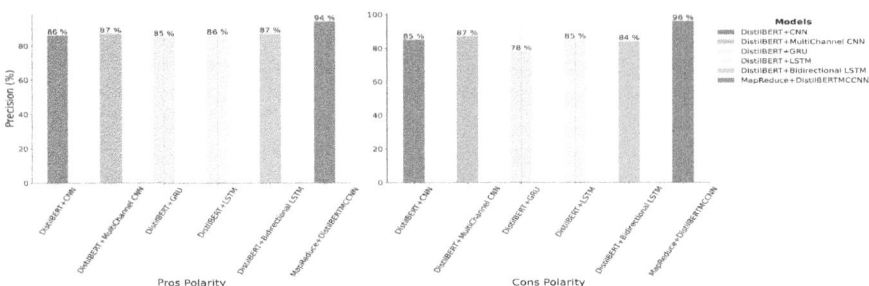

**Fig. 2.** Polarity of positive and negative data through precision results

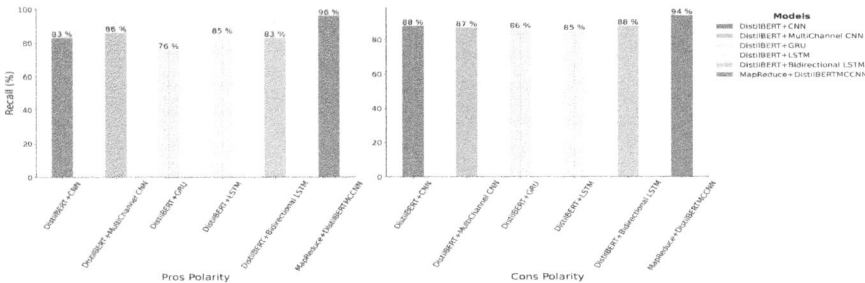

**Fig. 3.** Polarity of positive and negative data through recall results

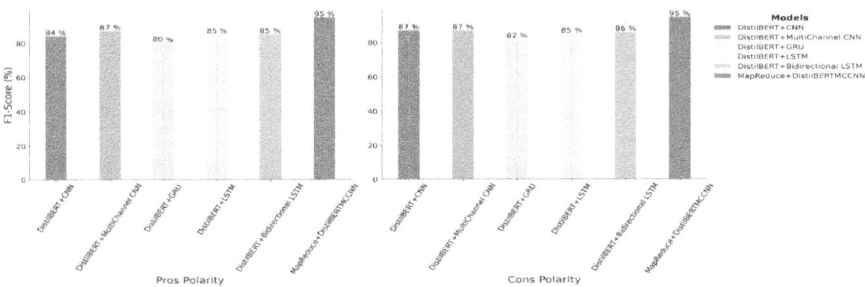

**Fig. 4.** Polarity of positive and negative data through f1-Score results

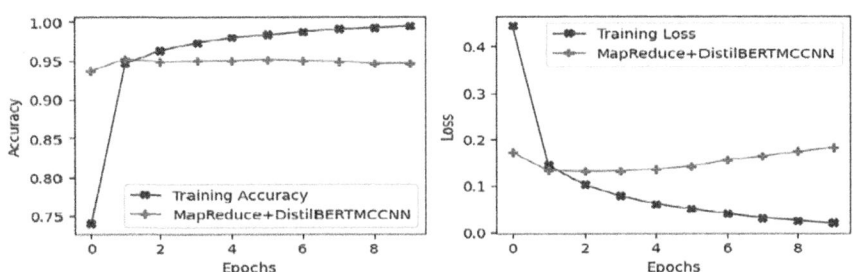

**Fig. 5.** Polarity of positive and negative data through accuracy results

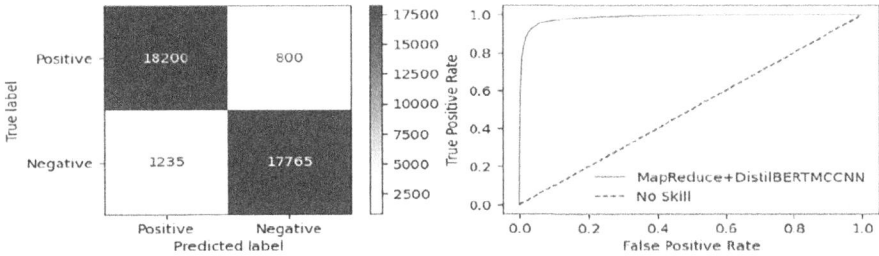

**Fig. 6.** Confusion matrix on detection of polarity positive and negative

# 5  Conclusions

## 5.1  Summary

In this study, we present our suggestion for analyzing large volumes of data and using DistilBERT, which improves the accuracy of text data processing. Based on this, we constructed an integrated model that outperforms the findings found in the literature in terms of accuracy when it comes to categorizing the data into distinct columns in terms of f1-score, accuracy, retrieval accuracy, and classification accuracy. The large data analysis framework with (HDFS and MapReduce) in Hadoop served as the foundation for this approach. The data are inputted and gathered in the HDFS. The MapReduce model analyzes the gathered data, and we suggest using features in this process. The key elements of the suggested strategy are big data engineering, which makes use of big data analysis frameworks and designs the data preparation phases in accordance with the Hadoop framework. In order to make the categorization work more flexible and accurate, this first architecture is centered on setting up the right environment for cleaning and digesting massive data. Then, using deep learning algorithms, especially DistilBERT and MultiChannel CNN, the huge data is categorized in a way that validates the findings of the completed trials proving the advantage of the suggested model over literary studies paradigms.

## 5.2  Prospects

The fundamental principle of the relativity of completed works is consistently seen to imply their various degrees of incompleteness and imperfection. We may discuss suitable prospects for the suggested method based on this. MapReduce+DistilBERTMCCNN Several of the most innovative paths for the suggested model's engineering development. The first tendency prompts us to thoroughly review the numerous current big data analysis research and evaluate the performance value of each with the suggested methodology. Alternatively, we are attempting to fuse concepts from semantics with deep learning methods to enhance the effectiveness of brief text analysis in prediction procedures.

# References

1. Ding, R.X., et al.: Large-scale decision-making: characterization, taxonomy, challenges and future directions from an artificial intelligence and applications perspective. Inf. Fusion **59**, 84–102 (2020). https://doi.org/10.1016/j.inffus.2020.01.006, https://www.sciencedirect.com/science/article/pii/S1566253520300117

2. Fadhli, I., Hlaoua, L., Omri, M.N.: Survey-credible conversation and sentiment analysis. Soc. Netw. Anal. Min. **14**(1), 13 (2023) https://doi.org/10.1007/s13278-023-01176-8

3. Fowler, J.W., Mönch, L.: A survey of scheduling with parallel batch (p-batch) processing. Eur. J. Oper. Res. **298**(1), 1–24 (2022). https://doi.org/10.1016/j.ejor.2021.06.012, https://www.sciencedirect.com/science/article/pii/S037722172172100518X

4. Haddad, O., Fkih, F., Omri, M.N.: A survey on distributed frameworks for machine learning based big data analysis. In: New Trends in Intelligent Software Methodologies, Tools and Techniques, vol. 355, pp. 702–714, September 2022. https://doi.org/10.3233/FAIA220299

5. Haddad, O., Fkih, F., Omri, M.N.: Toward a prediction approach based on deep learning in big data analytics. Neural Comput. Appl. **35**(8), 6043–6063 (2023). https://doi.org/10.1007/s00521-022-07986-9

6. Helali, L., Omri, M.N.: A survey of data center consolidation in cloud computing systems. Comput. Sci. Rev. **39**, 100366 (2021). https://doi.org/10.1016/j.cosrev.2021.100366, https://www.sciencedirect.com/science/article/pii/S157401372100006X

7. Jassim, M.A., Abd, D.H., Omri, M.N.: A survey of sentiment analysis from film critics based on machine learning, lexicon and hybridization. Neural Comput. Appl. **35**(13), 9437–9461 (2023). https://doi.org/10.1007/s00521-023-08359-6

8. Jena, R.: An empirical case study on indian consumers' sentiment towards electric vehicles: a big data analytics approach. Ind. Mark. Manage. **90**, 605–616 (2020). https://doi.org/10.1016/j.indmarman.2019.12.012, https://www.sciencedirect.com/science/article/pii/S0019850118307557

9. Kauffmann, E., Peral, J., Gil, D., Ferrández, A., Sellers, R., Mora, H.: A framework for big data analytics in commercial social networks: a case study on sentiment analysis and fake review detection for marketing decision-making. Ind. Mark. Manage. **90**, 523–537 (2020). https://doi.org/10.1016/j.indmarman.2019.08.003, https://www.sciencedirect.com/science/article/pii/S0019850118307612

10. Kolajo, T., Daramola, O., Adebiyi, A.: Big data stream analysis: a systematic literature review. J. Big Data **6**(1), 47 (2019). https://doi.org/10.1186/s40537-019-0210-7

11. Lv, Z., Lou, R., Li, J., Singh, A.K., Song, H.: Big data analytics for 6g-enabled massive internet of things. IEEE Internet Things J. **8**(7), 5350–5359 (2021). https://doi.org/10.1109/JIOT.2021.3056128

12. Lv, Z., Qiao, L.: Analysis of healthcare big data. Futur. Gener. Comput. Syst. **109**, 103–110 (2020). https://doi.org/10.1016/j.future.2020.03.039, https://www.sciencedirect.com/science/article/pii/S0167739X20304829

13. Mahmoud, R., Belgacem, S., Omri, M.N.: Towards an end-to-end isolated and continuous deep gesture recognition process. Neural Comput. Appl. **34**(16), 13713–13732 (2022). https://doi.org/10.1007/s00521-022-07165-w

14. Min, B., et al.: Recent advances in natural language processing via large pre-trained language models: A survey. ACM Comput. Surv. **56**(2), 1–40 (2023)

15. Osinga, S.A., Paudel, D., Mouzakitis, S.A., Athanasiadis, I.N.: Big data in agriculture: Between opportunity and solution. Agric. Syst. **195**, 103298 (2022). https://doi.org/10.1016/j.agsy.2021.103298, https://www.sciencedirect.com/science/article/pii/S0308521X21002511

16. Osman, A.M.S.: A novel big data analytics framework for smart cities. Futur. Gener. Comput. Syst. **91**, 620–633 (2019). https://doi.org/10.1016/j.future.2018.06.046, https://www.sciencedirect.com/science/article/pii/S0167739X17307446

17. Ouni, S., Fkih, F., Omri, M.N.: Toward a new approach to author profiling based on the extraction of statistical features. Soc. Netw. Anal. Min. **11**(1), 59 (2021). https://doi.org/10.1007/s13278-021-00768-6

18. Sandhu, A.K.: Big data with cloud computing: Discussions and challenges. Big Data Min. Anal. **5**(1), 32–40 (2022). https://doi.org/10.26599/BDMA.2021.9020016

19. Sokolova, M.: Big text advantages and challenges: classification perspective. Int. J. Data Sci. Anal. **5**(1), 1–10 (2018). https://doi.org/10.1007/s13278-021-00768-6

20. Song, H., xiu-ying Han, Montenegro-Marin, C.E., krishnamoorthy, S.: Secure prediction and assessment of sports injuries using deep learning based convolutional neural network. J. Ambient Intell. Hum. Comput. **12**(3), 3399–3410 (2021). https://doi.org/10.1007/s12652-020-02560-4

21. Souiden, I., Omri, M.N., Brahmi, Z.: A survey of outlier detection in high dimensional data streams. Comput. Sci. Rev. **44**, 100463 (2022). https://doi.org/10.1016/j.cosrev.2022.100463, https://www.sciencedirect.com/science/article/pii/S1574013722000107

22. T.K., B., Annavarapu, C.S.R., Bablani, A.: Machine learning algorithms for social media analysis: a survey. Comput. Sci. Rev. **40**, 100395 (2021). https://doi.org/10.1016/j.cosrev.2021.100395, https://www.sciencedirect.com/science/article/pii/S1574013721000356

23. Wang, Q., Su, T., Lau, R.Y.K., Xie, H.: Deepemotionnet: emotion mining for corporate performance analysis and prediction. Inf. Process. Manage. **60**(3), 103151 (2023). https://doi.org/10.1016/j.ipm.2022.103151, https://www.sciencedirect.com/science/article/pii/S0306457322002527

# Multistep Time Series Forecasting of Energy Consumption Based on Stacked Deep LSTM Network Architecture

Minyar Sassi Hidri[(✉)] [ID]

Computer Department, Deanship of Preparatory Year and Supporting Studies,
Imam Abdulrahman Bin Faisal University, Dammam, Saudi Arabia
mmsassi@iau.edu.sa

**Abstract.** The analysis of energy time series involves examining historical energy data and potentially considering external factors to make predictions. Various tasks fall within the broader field of energy time-series analysis and forecasting, with popular ones including forecasting electric load demand, predicting personalized energy consumption, and forecasting renewable energy generation. Given the impressive capabilities of Deep Learning (DL) models in numerous vision-related tasks, they have been applied to time-series forecasting. This paper explores whether and in what ways recently developed DL-based architectures, like Long Short-Term Memory (LSTM), can enhance performance in energy time-series forecasting tasks. To mitigate overfitting, the proposed model incorporates dropout layers to enhance its generalization capability.

**Keywords:** Deep learning · Stacked LSTM · Time Series · Forecasting · Energy Consumption

## 1 Introduction

Throughout the day, month, and year, there are continuous fluctuations in energy supply, demand, and pricing. Predicting these fluctuations is a precise method for mitigating uncertainties in the energy sector [1]. The dependability of forecasts in energy market sectors consistently serves as a key indicator for companies, investors, and governments when planning and making decisions [2,3].

Oil price [3,4], oil production and consumption [5], wind energy [6], electricity price [2,7], and electricity load [8] are just a few of the studies that have been done to forecast the energy sectors.

Furthermore, accurately predicting the direction of the energy market time series has always been difficult because of the complex, chaotic, volatile, nonlinear, and convoluted aspects of energy generation and pricing, especially in light of recent weather and energy price variations [6]. Consequently, earlier research has shown how important forecasting methods are to the energy industry.

N.-T. Nguyen et al. (Eds.): ICCCI 2024, CCIS 2165, pp. 132–143, 2024.
https://doi.org/10.1007/978-3-031-70248-8_11

Traditional time series forecasting methods rely on linear models, such as autoregressive integrated moving average (ARIMA) models [9]. These models use a set of equations to predict future values based on past values. They are often used in financial forecasting and are well suited to analyzing stationary data.

Deep Learning (DL) models, on the other hand, are more powerful than traditional time series forecasting methods. They are capable of learning complex patterns in data and can be used to make predictions on non-stationary data [10–12]. Additionally, DL models can be used to predict future values based on a longer history of past values, which can lead to more accurate forecasts. Using these models to predict future results based on past data is becoming increasingly popular across a variety of industries. One of the most promising applications of DL is in the area of time series forecasting, which involves predicting the future values of a given time series based on its past values. DL has emerged as the most cutting-edge technology that could provide an accurate time series forecast in various aspects.

Recent research has explored a number of techniques to optimize the performance of Recurrent Neural Networks (RNNs) for time series forecasting. These include the use of long short-term memory (LSTM) networks, which are better able to capture long-term dependencies in time series data. Other techniques include using regularization, which can help reduce overfitting, and using hyperparameter optimization to find the best model configuration.

Through this paper, we aim to explore the use of LSTM-based transfer learning for time series forecasting. We will use pre-trained models to initialize the weights of a new model, which allowed us to benefit from the knowledge acquired by the pre-trained model. This will help us reducing the amount of data and training time required to achieve good performance. 6-year hourly electricity consumption data has been used. To train the data, the stacked LSTM model was updated. In order to test the model and make it directly comparable to energy values in the data, we used the Mean Squared Error (MSE), the Mean Absolute Error (MAE), and the Root Mean Squared Error (RMSE) regression error metrics that help assess the performance of a regression model by quantifying the difference between predicted and actual values.

The reminder of the paper is organized as follows: Sect. 2 presents related work on deep neural networks for time-series forecasting. Section 3 highlights the stacking LSTMs for time series forecasting. Section 4 presents the model optimization. Section 5 presents the experiments and the results. Section 6 concludes the paper and highlights future directions.

## 2   Deep Neural Networks for Time-Series Forecasting: Related Work

DL has recently been used to a number of energy-related fields [13–15]. The Artificial Neural Network (ANN) model was used to anticipate oil output or extraction since it offers high precision and quick processing.

An effective DL framework was proposed in [16] to forecast power prices. The outcomes show that the DL models perform exceptionally well.

Based on physics-oriented and data-oriented time-series approaches for wind data augmentation, Chen et al. [17] applied data augmentation to predict wind power. A predicting algorithm which uses advanced augmentation methods is applied to five turbines. The data-oriented method performed better than the physics-oriented method, according to the results.

Yazici et al. [18] incorporated a variation mode decomposition to analyze and forecast crude oil prices (COP) on a daily and weekly basis. For forecasting on high-frequency modes, an AdaBoost (Adaptive Boosting) Random Forest (RF) was used to decompose COP data into multiple modes. A powerful framework for building forecasting models has been demonstrated with DL forecasting, offering an attractive alternative for developing sustainable oil price strategies. Each of the DL methods focuses only on extracting certain types of characteristics and has its own constraints even though they have introduced powerful approaches to improve prediction operations.

Energy consumption time series data is multifarious, and DL algorithms [19], which are well known for their capacity to extract complex patterns from complicated datasets, have shown to be very effective in managing this type of data. In order to provide more reliable predictions, these models are made to capture the dynamics of several time series simultaneously and take use of interdependencies between these series [20].

Consequently, DL models have found application in various time-series forecasting scenarios across diverse domains, such as retail [21], healthcare [22], biology [23, 24], aviation [25], energy [26], climate [27], automotive industry [28], and finance [29] to name a few.

## 3    Stacking LSTMs for Time Series Forecasting

For sequential data, RNNs are the most advanced method available. This is because it is the first algorithm with internal memory that can recall its input, which makes it ideal for machine learning issues involving sequential data. It is one of the algorithms that powers DL's incredible accomplishments over the last many years.

Information is looped through in an RNN. It considers the information it is receiving at the moment as well as the lessons it has learnt from earlier inputs before making a decision.

The memory of an RNN is usually short-term. Exploding gradients and disappearing gradients are the two main challenges that RNNs face. RNNs have an extension called LSTM networks, which essentially increases their memory. As a result, it is ideal for learning from significant events that occur over very long delays.

RNNs may retain their inputs for a considerable amount of time because to LSTMs. This is due to the fact that LSTMs store their data in a memory that functions similarly to a computer's memory in that it allows for the reading, writing, and deletion of data. This memory may be thought of as a gated cell. A gated cell is one that makes decisions about what information to keep or remove based on its perceived value, such as whether to open the gates or not.

Weights are used to assign importance, and the algorithm also learns these. This essentially indicates that it gradually discovers what information is significant and what is not.

Three layers make up an LSTM are: the input, forget, and output gates. These gates control whether to allow data to enter the system (input gate), discard data that isn't needed (forget gate), or allow data to affect the output at the current time step (output gate).

To solve the issue of vanishing gradients, LSTM was created to eliminate the problem with RNN analysis of long-term dependencies. There are multiple neural network modules in a Standard RNN, each with its own structure. Figure 1 module has a single *tanh* layer. It is discovered that these module architectures are basic.

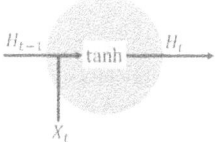

**Fig. 1.** RNN for time-series forecasting.

Although the chain of recurring modules in an LSTM and an RNN is the same, the LSTM module structure is somewhat more complex. Every module has four layers, as opposed to an RNN module's single layer.

Figure 2 shows the fundamental layout of an LSTM network. These modules, or memory blocks, consist of an input gate, a forget gate, an output gate, and the cell state. There's a distinct interaction at every level. Three gates regulate the information that gets added or withdrawn from the cell state. LSTM network maps an input sequence $X = (X_1, ..., X_t)$ to an output sequence $Y = (Y_1, ..., Y_t)$. First, it calculates the input gate activation vector and candidate values of the memory cell state $\tilde{C}_t$.

$$i_t = \sigma(W_i X_t + U_i H_{t-1} + b_i) \tag{1}$$

and

$$\tilde{C}_t = tanh(W_C X_t + U_C H_{t-1} + b_C) \tag{2}$$

where the logistic *sigmoid* function is denoted by $\sigma$. New information is stored in the cell state with the aid of the input gate activation vector. Each *tanh* layer generates a vector of new candidate values called $\tilde{C}_t$, which is then added to the state. Next, we compute the forget gate activation vector $f_t$ using Eq. (3).

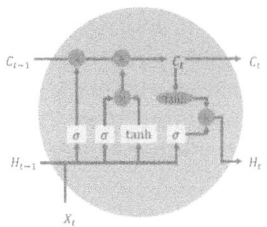

**Fig. 2.** LSTM for time-series forcasting.

$$f_t = \sigma(W_f X_t + U_f H_{t-1} + b_f) \tag{3}$$

The forget gate aids in both wiping out data from the cell state and in restarting the memory cells. The new state of memory cell $C_t$ is then computed using the computed values of $i_t$, $\tilde{C}_t$, and $f_t$. It is given in Eq. (4).

$$C_t = i_t \times \tilde{C}_t + f_t \times C_{t-1} \tag{4}$$

The cell state functions across the chain like a conveyor belt. The output gate activation vector $o_t$ is then calculated using [30] using the cell state vector $C_t$. It is given in Eq. (5).

$$o_t = \sigma(W_o X_t + U_o H_{t-1} + V_o C_t + b_o) \tag{5}$$

and may then be utilized to ascertain the LSTM's output vector in the manner shown in Eq. (6).

$$H_t = o_t \, tanh(C_t) \tag{6}$$

In addition to filtering the cell activations, the output gate aids in calculating the output using the cell state. Equations make use of the weight matrices $W_i$, $W_C$, $W_f$, $W_o$, $U_i$, $U_C$, $U_f$, $U_o$, and $V_o$. In the training phase, in addition to $b_i$, $b_C$, $b_f$, and $b_o$, which stand for bias vectors (Eq. (1) and Eq. (6)). According to [31], the LSTM structure that has been presented is effective in addressing the challenges related to vanishing gradients and may be used to long-term dependencies.

The process comprised a number of crucial components. The dataset was first imported and preprocessed, which involved normalizing the consumption and sorting the data by date. The data was then divided into testing and training sets. After building the RNN LSTM model with the proper input and output dimensions, training data was used.

Figure 3 shows the RNN LSTM stacking model.

To adapt the stacked LSTM network architecture for multi-step time series forecasting, we assume we are at time $t$, and we want to:

– Predict the values at times $(t + 1, ..., t + HORIZON)$

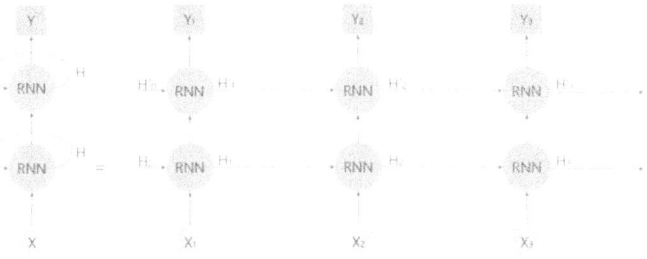

**Fig. 3.** RNN LSTM stacking.

– Conditional on the previous $T$ values of the time series.

The vector output approach is adopted because it is the simplest model, fastest to train and does not model dependencies between predicted outputs (see Fig. 4).

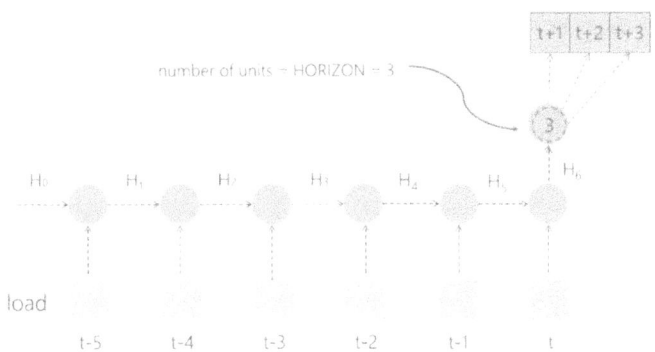

**Fig. 4.** Vector output model.

## 4    Model Optimization

During the training stage of neural networks, optimization algorithms are used to minimize the error (difference between the predicted output and the actual output). This optimization is done by readjusting or updating the weights in order to obtain more precise predictions. The backpropagation algorithm for updating the weights is described by the Algorithm 1.

Numerous optimization techniques have been proposed in the literature, and selecting the best algorithm for a DL model is essential [32,33].

Neural network optimization is most commonly done using gradient descent, one of the most widely used optimization methods. Depending on how much

---

**Algorithm 1.** Backpropagation algorithm for weight optimization in LSTM.

---

**Require:** Network, Training records, Learning rate.
**Ensure:** Network.
 1: Network ← Randomly initialize the weights;
 2: **while** Network does not converge **do**
 3:    **for** Example training data **do**
 4:       Network_Output ← Neural_Network_Output (Network, example);
 5:       Example_Err ← Target_Output - Network_Output;
 6:       $W_{j,i} \leftarrow W_{j,i} + \alpha \times \alpha_j \times Err_i \times g'(\text{input\_}Sum_i)$;
 7:       **for** Next layer of the network **do**
 8:          $\delta_j \leftarrow g'(\text{input\_}Sum_i) \sum_i W_{j,i} \Delta_i$;
 9:          $W_{k,j} \leftarrow W_{k,j} + \alpha \times \alpha_k \times \Delta_j$;
10:       **end for**
11:    **end for**
12: **end while**

---

data is utilized for gradient descent, there are three fundamental variations: mini-gradient descent, Stochastic Gradient Descent (SGD), and Batch Gradient Descent (BGD).

ADaptive Moment Estimation (ADAM) and Root Mean Square Propagation (RMSProp) [34] are two variants of SGD that have been introduced despite its established effectiveness and efficiency as an optimization technique for DL algorithms. Adam function is by far one of the most popular optimization functions for DL algorithms.

## 5    Experiments and Results

### 5.1    Model Architecture

Specifically designed to anticipate energy consumption, the stacked LSTM model predicts univariate time series. The sequentially structured model is implemented with the help of the Keras library. Because the LSTM model is built in a sequential manner, layers can be stacked one after the other. Three LSTM layers, dropout regularization, and a dense output layer are the main parts of the model. Each of the several layers that make up the LSTM model accurately captures patterns and temporal relationships present in the univariate time series data. The model architecture may be found in Table 1.

The architecture and parameter count are displayed in the model summary (see Table 1), which has 71,051 total parameters and 0 non-trainable parameters. The model's weights and biases are represented by the total parameters, whereas the parameters that may be changed during training are the trainable parameters. There are no fixed or frozen layers in the model if there are no non-trainable parameters.

**Table 1.** Model architecture.

| Layer (Type) | Output Shape | Param # |
|---|---|---|
| lstm (LSTM) | (None, 100, 50) | 10400 |
| dropout (Dropout) | (None, 100, 50) | 0 |
| lstm_1 (LSTM) | (None, 100, 50) | 20200 |
| lstm_2 (LSTM) | (None, 100,50) | 20200 |
| lstm_3 (LSTM) | (None, 50) | 20200 |
| dense (Dense) | (None, 1) | 51 |

Total params:71051 9277.54 KB)

Trainable params: 71051 (277.54 KB)

Non-trainable params: 0 (0.00 Byte)

## 5.2   Training and Validation

A "*ModelCheck point*" call back is used to guarantee that the top-performing model weights are maintained during training. This callback is set up with "*save_best_only=True*", which means that it will only save the model eights if the validation loss starts to drop. In order to avoid overfitting and capture the model's ideal performance on unknown data, this is an essential tactic.

We prepared the model using ADAM optimizer before starting the training procedure. Regression tasks are a good fit for this arrangement since it helps forecast continuous energy usage numbers.

Each of the 45 epochs that make up the training process represents a full run through the training dataset. The weights of the model are iteratively adjusted in order to minimize the mean squared error loss. Overfitting is monitored by the validation data, which is kept separate for evaluating generalization performance. The history variable, which contains the training history, offers information on how the model has changed throughout epochs. The model's steady convergence is shown by the presented loss values for the training and validation sets.

Keeping an eye on these patterns helps determine how effectively the model generalizes to new, untested data and adjusts to the training set.

Figure 5 shows actual vs train predictions.

Figure 6 shows actual vs validation predictions.

The model is becoming more adept at identifying patterns and correlations in the training data as the training and validation loss decrease with time (Fig. 7). By adding dropout layers and selecting the appropriate hyperparameters and architecture, the model is able to avoid fitting too closely to the training set and becoming too specialized.

An outstanding outcome is found when evaluating the trained LSTM model's performance on a test set. The evaluate function yielded an unexpectedly low test loss of 0.0279 MWh.

This low number indicates that the mean squared error between the model's predictions and the actual consumption data is successfully minimized by the

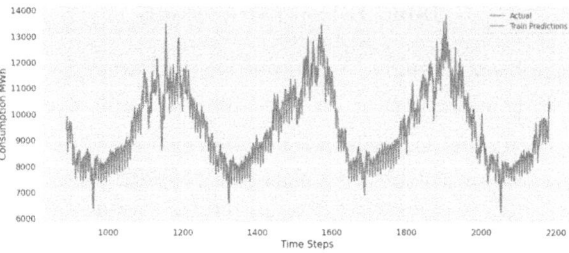

**Fig. 5.** Actual vs Train Predictions.

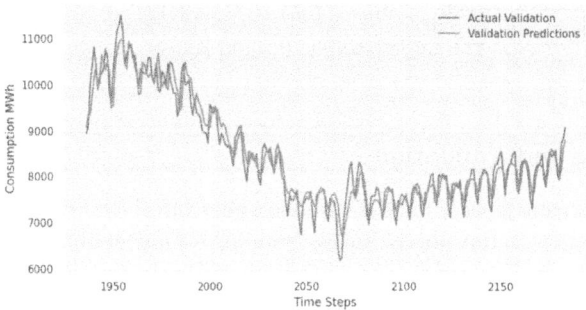

**Fig. 6.** Actual vs Validation Predictions.

model. The predict function is used to construct forecasts on the test dataset, which yield an MSE of 0.0279 MWh, an RMSE of 0.1670 MWh, and a MAE of 0.1072 MWh. All of these measures indicate how accurate and precise the model is at predicting energy use. The RMSE calculates the average error magnitude; a result of 0.1670 MWh indicates very minimal prediction mistakes. The model's performance is further supported by the MAE of 0.1072 MWh, which shows that forecasts often differ from real values by 0.1072 MWh.

To sum up, the model has strong predictive skills on the test set, and its potential for precise energy consumption forecasting is highlighted by the low values of assessment criteria.

## 5.3    Future Forecasting

The ability of the stacked LSTM model to produce precise energy consumption forecasts is shown in Fig. 8. When forecasting the initial test samples, the model performs best, nearly matching the actual values. This suggests that the model does a good job of identifying transient patterns in the data. We notice a little discrepancy between the model's predictions and the actual results when we go on to further test samples. The model's average MAE of 0.1072 MWh shows that it has good prediction performance. The model's predictions and the actual values differ by less than 11% on average.

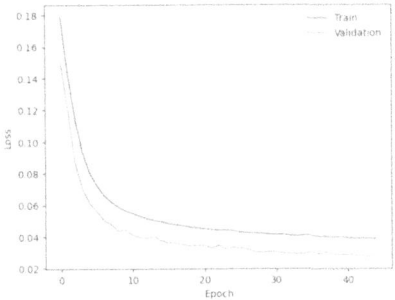

**Fig. 7.** Model Loss.

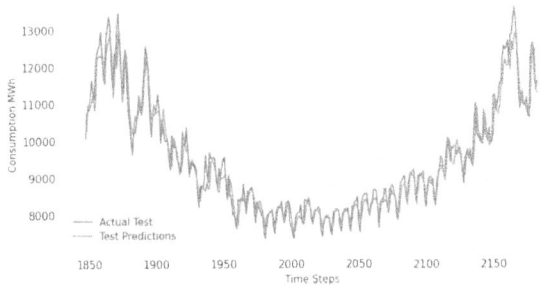

**Fig. 8.** Actual Test vs Test Predictions.

## 6    Conclusion

This work aimed to predict energy consumption using a stacked LSTM model, which was trained on a dataset containing 6 years of energy consumption.

The model was built successfully using four LSTM layers, one Dropout, and one output layer. This work used the *Adam* optimization algorithm and RMSE to test the model performance.

When using the stacked LSTM for short-term predictions, the results have demonstrated good performance. First, training, validation, and test datasets were used to test the model. It is important to note that the time steps-in this example, 50 days-have a significant impact on this judgment. The outcomes would vary if a different amount of timesteps were taken into consideration.

Expanding the research to alternative domains or datasets may confirm the applicability of the suggested methodology and enhance the overall comprehension of multivariate time series forecasting in various scenarios.

# References

1. Zhang, Y., He, M., Wen, D., Wang, Y.: Forecasting crude oil price returns: can nonlinearity help? Energy **262**, 125589 (2022)
2. He, J., Xu, B., Su, X., Tan, C.: A CNN-Relstm model based on a hybrid architecture of unidirectional and bidirectional LSTM for predicting stock prices. In: Proceedings of the 15th International Conference on Machine Learning and Computing, pp. 589–596 (2023)
3. Laib, O., Khadir, M.T., Mihaylova, L.: Toward efficient energy systems based on natural gas consumption prediction with LSTM recurrent neural networks. Energy **177**, 530–542 (2019)
4. Del Ser, J., et al.: Randomization-based machine learning in renewable energy prediction problems: critical literature review, new results and perspectives (2021)
5. Fu, W., Wang, K., Zhou, J., Xu, Y., Tan, J., Chen, T.: A hybrid approach for multi-step wind speed forecasting based on multi-scale dominant ingredient chaotic analysis, KELM and synchronous optimization strategy. Sustainability **11**(6), 1804 (2019)
6. Kottath, R., Singh, P.: Influencer buddy optimization: algorithm and its application to electricity load and price forecasting problem. Energy **263**(PC), 125641 (2023)
7. Haider, S.A., Sajid, M., Sajid, H., Uddin, E., Ayaz, Y.: Deep learning and statistical methods for short- and long-term solar irradiance forecasting for Islamabad. Renew. Energy **198**(C), 51–60 (2022)
8. Zini, M., Carcasci, C.: Machine learning-based monitoring method for the electricity consumption of a healthcare facility in Italy. Energy **262**(PB), 125576 (2023)
9. Harvey, A.C.: ARIMA Models, pp. 22–24. Palgrave Macmillan UK, London (1990)
10. Sassi Hidri, M.: Learning-based models for building user profiles for personalized information access. Interdiscip. J. Inf. Knowl. Manag. **19**, 010 (2024)
11. Ferjani, I., Sassi Hidri, M., Frihida, A.: SiNoptiC: swarm intelligence optimisation of convolutional neural network architectures for text classification. Int. J. Comput. Appl. Technol. **68**(1), 82–100 (2022)
12. Ferjani, I., Sassi Hidri, M., Frihida, A.: Multi-GPU-based convolutional neural networks training for text classification. In: Proceedings of the 2021 Intelligent Systems Conference, pp. 72–84 (2021)
13. Ying, C., Wang, W., Yu, J., Li, Q., Yu, D., Liu, J.: Deep learning for renewable energy forecasting: a taxonomy, and systematic literature review. J. Clean. Prod. **384**, 135414 (2023)
14. Benti, N.E., Chaka, M.D., Semie, A.G.: Forecasting renewable energy generation with machine learning and deep learning: current advances and future prospects. Sustainability **15**(9), 7087 (2023)
15. Sehovac, L., Nesen, C., Grolinger, K.: Forecasting building energy consumption with deep learning: a sequence to sequence approach. In: Proceedings of the IEEE International Congress on Internet of Things (ICIOT), pp. 108–116 (2019)
16. Ugurlu, U., Oksuz, I., Tas, O.: Electricity price forecasting using recurrent neural networks. Energies **11**(5), 1255 (2018)
17. Chen, H., Birkelund, Y., Zhang, Q.: Data-augmented sequential deep learning for wind power forecasting. Energy Convers. Manage. **248**, 114790 (2021)
18. Dar, L.S., Aamir, M., Khan, Z., Bilal, M., Boonsatit, N., Jirawattanapanit, A.: Forecasting crude oil prices volatility by reconstructing EEMD components using ARIMA and FFNN models. Front. Energy Res. **10**, 991602 (2022)

19. LeCun, Y., Bengio, Y., Hinton, G.: Deep learning. Nature **521**, 436–444 (2015)
20. Lim, B., Zohren, S.: Time-series forecasting with deep learning: a survey. Philos. Trans. Ser. A, Math. Phys. Eng. Sci. **379**, 20200209 (2021)
21. Böse, J.H., et al.: Probabilistic demand forecasting at scale. Proc. VLDB Endow. **10**(12), 1694–1705 (2017)
22. Kaushik, S., et al.: AI in healthcare: time-series forecasting using statistical, neural, and ensemble architectures. Front. Big Data (2020, in press)
23. Leise, T.L.: Analysis of nonstationary time series for biological rhythms research. J. Biol. Rhythms **32**(3), 187–194 (2017)
24. Lynn, L.: Artificial intelligence systems for complex decision-making in acute care medicine: a review. Patient Saf. Surg. **13**, 6 (2019)
25. Vonitsanos, G., Panagiotakopoulos, T., Kanavos, A., Tsakalidis, A.: Forecasting Air Flight Delays and Enabling Smart Airport Services in Apache Spark, pp. 407–417 (2021)
26. Martínez-Álvarez, F., Troncoso, A., Asencio-Cortés, G., Riquelme, J.C.: A survey on data mining techniques applied to electricity-related time series forecasting. Energies **8**, 13162–13193 (2015)
27. Mudelsee, M.: Trend analysis of climate time series: a review of methods. Earth-Sci. Rev., 310–322 (2019)
28. Nousias, S., Pikoulis, E.V., Mavrokefalidis, C., Lalos, A.S.: Accelerating deep neural networks for efficient scene understanding in multi-modal automotive applications. IEEE Access **11**, 28208–28221 (2023)
29. Berat Sezer, O., Gudelek, M.U., Ozbayoglu, A.M.: Financial time series forecasting with deep learning: a systematic literature review: 2005–2019. Appl. Soft Comput. **90**, 106181 (2020)
30. Hochreiter, S., Schmidhuber, J.: Long short-term memory. Neural Comput. **9**(8), 1735–1780 (1997)
31. Graves, A.: Supervised Sequence Labelling with Recurrent Neural Networks, vol. 385 of Studies in Computational Intelligence. Springer, Cham (2012). https://doi.org/10.1007/978-3-642-24797-2
32. Isaac, N., Saha, A.K.: A review of the optimization strategies and methods used to locate hydrogen fuel refueling stations. Energies **16**(5), 2171 (2023)
33. Hijma, P., Heldens, S., Sclocco, A., van Werkhoven, B., Bal, H.E.: Optimization techniques for GPU programming. ACM Comput. Surv. **55**(11), 1–81 (2023)
34. Rakshitha Kiran, P., Naveen, N.C.: Op-RMSprop (optimized-root mean square propagation) classification for prediction of polycystic ovary syndrome (PCOS) using hybrid machine learning technique. Int. J. Adv. Comput. Sci. Appl. **13**(6) (2022)

# On the Effect of Quantization on Deep Neural Networks Performance

Jihene Tmamna[1], Rahma Fourati[1,2](✉), and Hela Ltifi[1,3]

[1] REsearch Groups in Intelligent Machines, National Engineering School of Sfax, 3038 Sfax, Tunisia
jihen.tmamna@enis.tn, hela.ltifi@ieee.org
[2] Faculté des Sciences Juridiques, Economiques et de Gestion de Jendouba, Université de Jendouba, 8189 Jendouba, Tunisia
rahma.fourati@ieee.org
[3] Department of Computer Sciences, Faculty of Sciences and Techniques of Sidi Bouzid, University of Kairouan, Kairouan, Tunisia

**Abstract.** Neural network Quantization has emerged as an important technique for reducing the computational cost of deep neural networks (DNNs) and deploying them on resource-constrained devices. However, DNN models can exhibit vulnerabilities when exposed to various types of noise in real-world applications. While evaluating the impact of quantization on model performance is crucial, research on this topic is notably limited, often neglecting established principles of performance evaluation and consequently yielding inconclusive findings. To bridge this gap, we evaluated quantized models' performance under various noise types, including data perturbation, model parameter perturbation, and adversarial attacks, using the Tiny ImageNet dataset. The empirical results from this comprehensive evaluation present a valuable understanding of how quantized models perform across diverse scenarios, particularly when compared to the performance of the original models.

**Keywords:** Neural network quantization · Adversarial attacks · Uncertainty quantification · Sensitivity analysis

## 1  Introduction

Deep neural networks (DNNs) have showcased remarkable performance across various computer vision tasks, including speech recognition [5], natural language processing [2], and computer vision [9,34]. Nevertheless, these DNNs come with high memory and computational requirements, which hinder their deployment in resource-constrained environments like self-driving vehicles, Internet of Things (IoT) devices, robotics, and mobile phones [21]. To overcome these challenges, various DNN compression methods, including knowledge distillation [30,33], pruning [16,22], and quantization [23,24], have been proposed. Notably, among these methods, quantization has become a key approach for compressing DNNs

N.-T. Nguyen et al. (Eds.): ICCCI 2024, CCIS 2165, pp. 144–156, 2024.
https://doi.org/10.1007/978-3-031-70248-8_12

due to its ability to preserve network structure while achieving comparable performance. This is accomplished by minimizing memory usage and accelerating inference through the representation of model parameters with low-bit values instead of the standard 32-bit floating-point values.

Despite its efficacy in enabling the deployment of DNN models on limited-resource devices, quantization presents challenges in ensuring trustworthiness. This entails addressing aspects like robustness and reliability, especially when implemented in real-world scenarios [6, 15]. DNNs are notably vulnerable to perturbation examples, cunningly designed to remain imperceptible to human vision while having the ability to deceive DNNs. This vulnerability poses a substantial threat to the reliability of practical deep-learning applications. For example, perturbations in input data or model parameters can hurt model accuracy.

These vulnerabilities highlight the lack of reliability of quantized models, especially when applied in safety-critical domains like face recognition and autonomous driving, where they encounter various real-world perturbations. Consequently, a thorough evaluation of the reliability and robustness of quantized models becomes important before deployment. This assessment is designed to identify potential weaknesses, ensuring the models exhibit effective performance in complex and dynamic scenarios.

In recent years, the robustness of DNN models has been extensively evaluated by researchers [29, 32], using various adversarial attack methods for comprehensive assessments across tasks. However, limited attention has been given to the robustness evaluation of quantized models [3, 27]. Notably, this research is characterized by a narrow focus on adversarial attacks and a lack of diversity in noise sources. Consequently, the current literature fails to provide a thorough examination of the performance of quantized models, resulting in a significant gap in our comprehension of their strengths and vulnerabilities.

In this paper, a comprehensive performance evaluation benchmark is introduced. The robustness and reliability of quantized models are systematically assessed through the application of the BBPSO-Quantizer quantization method [24]. Three key techniques, namely adversarial attack, uncertainty quantification, and sensitivity analysis, are used in our analysis.

Our main contributions can be summarized as follows.

– This study represents a comprehensive evaluation of the performance of quantized models. The evaluation encompasses the BBPSO-Quantizer method and three classical architectures, assessing their performance across various noise types, including input data perturbation, model parameter perturbation, and adversarial attacks.
– Through extensive experiments, this study delves into the performance of quantized models, revealing valuable insights into their strengths and weaknesses. It provides a comprehensive analysis, comparing their performance with the original models across a variety of scenarios.

The subsequent sections of the paper are structured as follows: Sect. 2 provides an overview of the related work. Section 3 introduces the performance evaluation benchmark. Section 4 outlines the experimental results. Finally, Sect. 5 presents the conclusion.

## 2    Related Work

This section provides related works on network quantization and the performance evaluation of quantized models.

### 2.1    Neural Network Quantization

Quantization achieves compression of DNN models by reducing the number of bits needed to represent each weight and activation, thereby minimizing memory usage and accelerating model inference speed. Quantization involves two common operations: clipping and projection, which are applied to the model's parameters to obtain quantized values, with the quantization function defined as in [11]:

$$\overline{W} = \prod\nolimits_{Q(a,b)} clip(W, a) \tag{1}$$

where $\overline{W}$ represents the quantized value, and $W$ denotes the input value. $Q(a, b)$ refers to the quantization levels. The clipping function $clip(W, a)$ truncates whether it's 32-bit weight or activation to the ranges $[-a, a]$ and $[0, a]$, respectively. The projection function $\prod(.)$ plays a crucial role in mapping each clipped element $X$ into the space defined by $Q(a, b)$.

Quantization methods can be broadly categorized into two strategies: mixed precision quantization (MPQ) and fixed precision quantization (FPQ). In FPQ, the same bit width is assigned to all or most layers without considering their sensitivity to quantization errors. It may lead to suboptimal performance as the quantization levels are not tailored to the specific requirements of each layer.

On the other hand, MPQ considers the diverse sensitivities of layers to quantization errors, enabling the assignment of different bit widths to various layers. Therefore, critical layers may necessitate higher bit precision to maintain accuracy, whereas less critical layers could benefit from lower bit precision. Tailoring the quantization levels according to the specific requirements of each layer allows for the optimization of the average bit width, thereby minimizing accuracy degradation. Consequently, our emphasis is on MPQ methods, and we offer a brief review of commonly used approaches within this category.

A strand of research relies on predefined guidelines to allocate different precisions to various layers, considering the quantization sensitivity of each layer [1,28]. For instance, Zhewei *et al.* [28] used the Hessian matrix to guide the quantization process, mitigating the potential accuracy degradation associated with quantization. Other noteworthy endeavors have focused on automating the determination of the optimal bit width for network layers [17,18,24]. Specifically, [24] utilized a bare-bones particle swarm optimization approach to search for the optimal bit width for each layer.

## 2.2 Performance of Quantized Models

Several studies have exclusively focused on evaluating the robustness of floating-point models against adversarial attacks, exploring various attack strategies [4,13,25,26]. While numerous studies have made efforts to evaluate the robustness of floating-point models [14,31], the exploration of robustness in quantized networks is notably limited. Lin et al. [12] introduced a defensive quantization method to address this gap. Their focus is on mitigating the amplification of adversarial noise during propagation by controlling the Lipschitz constant of the network during the quantization process.

In contrast, our goal is to conduct a comprehensive evaluation of the performance of quantized models, taking into account various sources of noise. This encompasses the perturbation of model parameters, input perturbation, and adversarial attacks. This broader assessment aims to provide a more thorough understanding of the performance of quantized models across diverse scenarios.

# 3 Performance Evaluation

The methodology of performance evaluation is illustrated in Fig. 1.

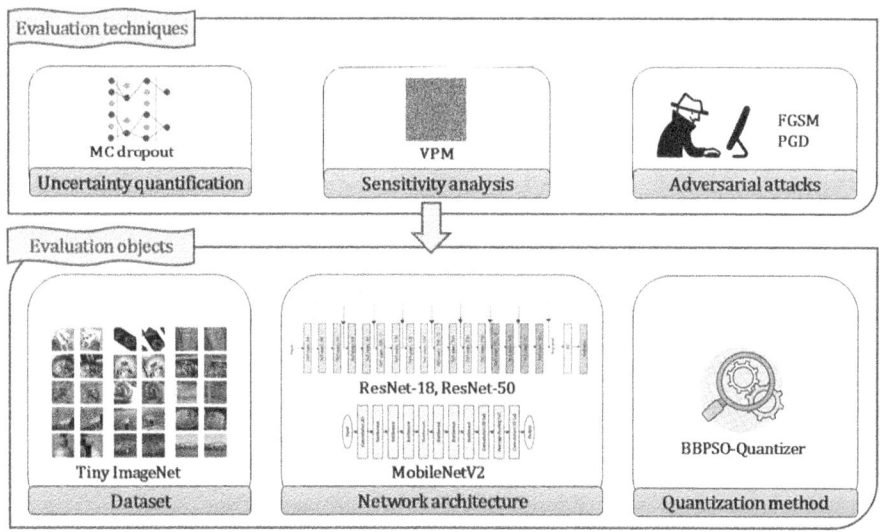

**Fig. 1.** Performance evaluation methodology

## 3.1 Evaluation Performance Techniques

Quantized models are commonly susceptible to various perturbations in real-world applications. To subject these models to perturbations, we used uncertainty quantification, sensitivity analysis, and adversarial attack techniques. It ensures a comprehensive assessment of the performance of quantized models.

**Uncertainty Quantification.** Uncertainty quantification involves measuring and analyzing uncertainties within model predictions, playing a pivotal role in evaluating reliability and confidence. Uncertainty quantification techniques can be classified into (i) aleatoric and (ii) epistemic uncertainties. Epistemic uncertainty arises from the inherent ambiguity surrounding model parameters, while aleatoric uncertainty explores the intrinsic variability and randomness within the data. In our study, our focus is directed towards investigating epistemic uncertainty.

To assess the impact of parameter perturbation on quantized models, we used Monte Carlo dropout (MC dropout). This technique extends dropout, a standard regularization method used during neural network training, into the prediction phase. MC dropout introduces stochasticity by preserving active dropout layers during prediction, allowing for the generation of an ensemble of predictions through multiple forward passes. The variability in these predictions serves as a measure of uncertainty. Hence, the uncertainty introduced by MC Dropout is quantified using predictive entropy as a key metric. This metric evaluates the level of uncertainty in a set of predictions, providing a precise measure of variability among the multiple predictions generated for each input.

**Sensitivity Analysis.** Deployed models in real-world applications often encounter diverse input perturbations. Sensitivity analysis aims to evaluate how perturbations in input impact the output of a model. It assesses the influence of individual input factors on model predictions, providing insights into the model's robustness and stability. To evaluate the impact of input perturbations on quantized models, Variable Perturbation Method (VPM) is used. It systematically perturbs input variables by adding a factor to each pixel in the input image.

**Adversarial Attacks.** Adversarial attacks involve introducing visually imperceptible perturbations to natural (clean) images to compromise a model's predictive capability. These attacks may access the model parameters and architecture design, called white-box attacks, or solely target the model output, called blackbox attacks. Our study focuses predominantly on white-box attacks to assess the adversarial robustness of models, leveraging their enhanced attack capabilities. Below, the relevant attack methods employed in our benchmark are introduced. The considered attacks considered in this work are as follows.

- **Fast gradient sign method (FGSM)** [4] represents a one-step attack method utilizing the $l_\infty - norm$. It involves a singular computation of the gradient of the loss function concerning the input, followed by the introduction of gradient noise to generate an adversarial example. Despite its comparatively less potent attack capability, FGSM stands out for its computational efficiency in creating adversarial examples.
- **Projected gradient descent (PGD)** [19] is recognized as one of the most effective attack methods, boasting a high success rate. It extends the FGSM approach by incorporating an iterative process with gradient projection at each step.

Adversarial attacks encompass diverse goals, ranging from untargeted to targeted attacks. In a targeted attack, the adversary manipulates an input to prompt the victim model to predict a predetermined incorrect class. Conversely, an untargeted adversarial attack seeks to induce any form of misclassification without specifying a particular false class. In this case, we focus on the untargeted attack scenario. Adversarial images are generated by executing a subset of popular white-box adversarial attacks, including FGSM [4], PGD [19]. The implementation of these attacks utilizes the Adversarial Robustness Toolbox (ART) library.

### Evaluation Objects

*Dataset:* To achieve the objective of yielding broadly relevant conclusions concerning quantized models in the field of computer vision, our research focuses on fundamental image classification tasks and adheres to established quantization literature. The selection of the Tiny ImageNet dataset [10], known for its extensive collection of images and classes, further enhances its suitability as a benchmark for evaluating model quality compared to commonly used small-scale datasets such as CIFAR-10 [8] and CIFAR-100 [8]. With a substantial repository comprising 100,000 training images and 10,000 validation images across 200 different classes, Tiny ImageNet provides a robust foundation for our performance evaluation.

*DNN Architectures:* In our study, we incorporate three architectures: ResNet-18 [7], ResNet-50 [7], and MobileNetV2 [20]. ResNet-18 and ResNet-50 are established backbone architectures known for their effectiveness across various computer vision tasks. These architectures rely on residual blocks, incorporating skip connections to address the vanishing gradient problem. MobileNetV2, on the other hand, is a lightweight network that utilizes depthwise separable convolutions, separating spatial and channel-wise convolutions to reduce computational load while preserving performance.

*Quantization Methods:* Within our study, we concentrate on the proposed quantization method: BBPSO-Quantizer [24]. The BBPSO-Quantizer performs its quantization on model weights and activation values. For each architecture, the performance against perturbations generated on Tiny ImageNet is evaluated.

## 4    Experiments

This section conducts an in-depth performance evaluation of the quantized models by encompassing three crucial dimensions: uncertainty quantification, sensitivity analysis, and adversarial attacks.

## 4.1   Evaluation Metrics

To assess the influence of adversarial attacks and sensitivity analysis, we relied on the model's accuracy. Simultaneously, the impact of uncertainty quantification was measured using predictive entropy $H$ as shown in Eq. (2).

$$H[y|x, D_{train}] = -\sum_c p(y = c|x, D_{train})logp(y = c|x, D_{train}) \qquad (2)$$

where y is the label, x is the input image, $D_{train}$ is the training data, c is the class and p is the probability.

## 4.2   Clean Accuracy

In Table 1, the clean accuracies of the quantized models are displayed. The majority of these models maintain accuracy levels comparable to those of 32-bit pretrained models. The quantization process involves the use of a bit-width vector s = {2,3,4}. Notably, the first layer and the last layer, considered critical, are quantized to 8 bits.

**Table 1.** Clean accuracy of original and quantized models. W and A are the bitwidths assigned to weights and activations respectively. M indicates mixed precision quantization.

| Model | W | A | Accuracy | Compression ratio | Average operation bit-width |
|-------|---|---|----------|-------------------|------------------------------|
| MobileNetV2 original | 32 | 32 | 60.41% | 1 | 32 |
| BBPSO-Quantizer-3 | M | M | 60.01% | 10.43 | 3.04 |
| ResNet-18 original | 32 | 32 | 63.39% | 1 | 32 |
| BBPSO-Quantizer-3 | M | M | 63.01% | 10.85 | 3.08 |
| ResNet-50 original | 32 | 32 | 65.50% | 1 | 32 |
| BBPSO-Quantizer-3 | M | M | 65.72% | 10.85 | 3.04 |

## 4.3   Uncertainty Quantification

In this section, we delve into the exploration of uncertainty quantification techniques to gain deeper insights into the reliability and robustness of predictions made by quantized models. Our objective is to offer insights into the reliability of their predictions and address the pivotal question: Does quantization improve reliability compared to the original models?

Table 2 presents the results of the uncertainty quantification for different models. Based on these findings, we can conclusively address the question posed at the beginning of this section. Quantization, by reducing the model's complexity, not only enhances prediction accuracy but also improves the model's reliability. This trend is evident across all models examined. For instance, in the case of MobileNetV2, the predictive entropy decreased from an average of 0.014 for the original model to 0.008, indicating a lower level of prediction uncertainty. Figure 2 visually depicts the distribution of entropy for quantized and original MobileNetV2. The study emphasizes quantization's benefits in reducing model complexity, improving prediction accuracy, and reducing uncertainty levels, thereby resulting in more robust and efficient models.

**Table 2.** Uncertainty quantification analysis

| Model | Accuracy | compression ratio | Average predictive entropy |
|---|---|---|---|
| MobileNetV2 original | 60.41% | 1 | 0.014 |
| MobileNetV2 quantized | 60.01% | 10.43 | **0.008** |
| ResNet-18 original | 63.39% | 1 | 0.14 |
| ResNet-18 quantized | 63.01% | 10.85 | **0.05** |
| ResNet-50 original | 65.50% | 1 | 0.17 |
| ResNet-50 quantized | 65.72% | 10.85 | **0.06** |

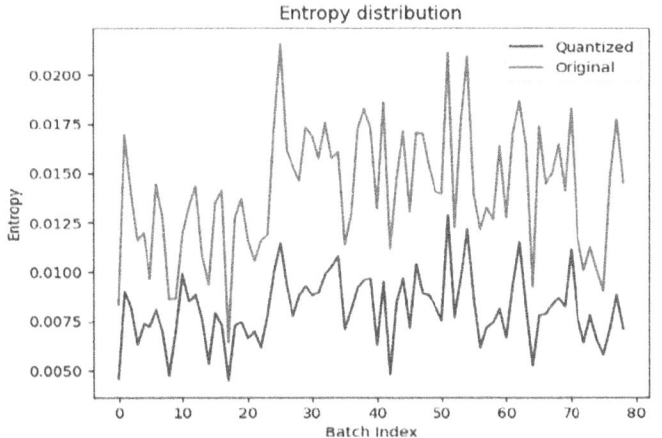

**Fig. 2.** Entropy distribution of MobileNetV2

## 4.4   Sensitivity Analysis

In this section, we will explore how quantization responds to changes in input features, aiming to evaluate its impact on the sensitivity analysis of the model. To assess the model's sensitivity to variations in input data, the Variable Perturbation Method (VPM) is used, which systematically perturbs input variables by introducing a factor to each pixel in the input image. The test dataset is used for analysis with a perturbation factor of 0.1, indicating a 10% change in the original data. This factor was applied to assess its impact on the predictions of both the original and quantized models. The objective is to gain insights into how model quantization influences sensitivity and robustness.

The study evaluates the impact of VPM on a model using accuracy as a metric. Table 3 compares the original and quantized models' accuracy based on input variable changes. The quantized models show greater robustness and less sensitivity to changes in input variables.

**Table 3.** Sensitivity analysis results

| Model | Original Accuracy | Perturbed accuracy | Accuracy change |
|---|---|---|---|
| MobileNetV2 original | 60.41 % | 58.01% | −2.40 |
| MobileNetV2 quantized | 60.01 % | 59.54 % | **−0.46** |
| ResNet-18 original | 63.39 % | 62.03% | −1.36 |
| ResNet-18 quantized | 63.01 % | 62.20% | **−0.81** |
| ResNet-50 original | 65.50% | 63.43% | −2.04 |
| ResNet-50 quantized | 65.72% | 64.71% | **−1.01** |

To further illustrate the significance of quantization, the T-SNE visualization of the quantized and original MobileNetV2 over perturbed data is presented in Fig. 3.

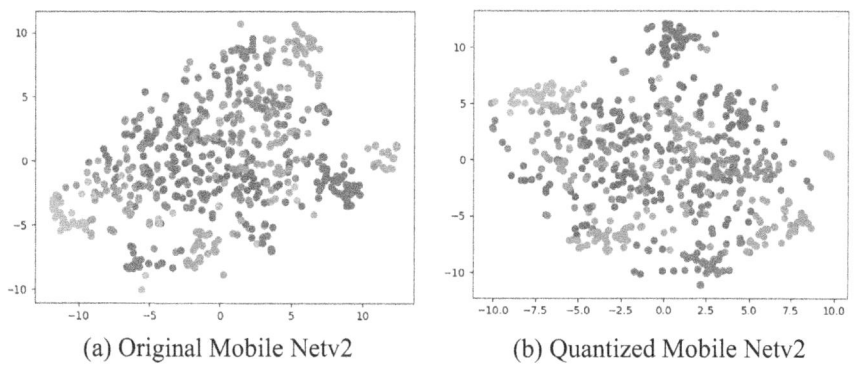

(a) Original Mobile Netv2                (b) Quantized Mobile Netv2

**Fig. 3.** t-SNE Visualization of perturbed dataset

Each point on the graph represents a perturbed data sample after processing through the model. The t-SNE algorithm has grouped similar data samples into clusters positioned in various regions of 2D space. These clusters correspond to different classes in the dataset. As depicted in Fig. 3, the distances between clusters for the quantized MobileNetV2 are more well-separated than in the original model, indicating that the quantized model can effectively distinguish between different classes, making it more robust than the original model.

## 4.5   Adversarial Attacks

This section delves into the analysis of how quantization affects the model's resilience to adversarial attacks. The central question we aim to address is: Do quantized models exhibit enhanced security compared to the original models when confronted with adversarial attacks?

For the generation of adversarial images, a subset of popular white-box adversarial attacks, including FGSM [4] and PGD [19] are used. The execution of these attacks is facilitated through the Adversarial Robustness Toolbox (ART) library, utilizing the default parameters for each attack as specified by ART. Comprehensive details of these parameters can be found in Table 4.

**Table 4.** White-box attacks details.

| Attack | Epsilon | Epsilon step | Max iteration |
|--------|---------|--------------|---------------|
| FGSM | 0.2 | – | 100 |
| PGD | 0.3 | 0.1 | 100 |

Table 5 presents the results of two white-box attacks on the Tiny ImageNet dataset. The robust accuracy, defined as the percentage of samples correctly classified by the model, is reported in Table 5 for each attack. These results demonstrate that quantized models exhibit greater robustness against FGSM and PGD attacks compared to the original models. For example, the quantized MobileNetV2 achieves 20.92% robustness against FGSM attacks, whereas the original MobileNetV2 achieves only 12.97%. Based on these findings, we can answer the original question posed at the beginning of this section. Overall, we

**Table 5.** Adversarial test accuracies for the origin model and the quantized model on Tiny ImageNet.

| Model | FGSM | PGD | Original Accuracy |
|-------|------|-----|-------------------|
| MobileNetV2 original | 12.97% | 0.31% | 60.41% |
| MobileNetV2 Quantized | 20.92% | 3.01% | 60.01% |
| ResNet-18 original | 12.78% | 0.91% | 63.39% |
| ResNet-18 Quantized | 17.62% | 2.05% | 63.01% |
| ResNet-50 original | 16.61% | 1.13% | 65.50% |
| ResNet-50 Quantized | 25.88% | 3.01% | 65.72% |

have shown that quantization can enhance robustness and advance security in the field of DNNs when facing adversarial attacks.

## 5   Conclusion

In this paper, we introduce a benchmark designed to assess the robustness and reliability of quantized models across various perturbations using various techniques, encompassing uncertainty quantification, sensitivity analysis, and adversarial attacks. The benchmark evaluates three classical architectures and the recent quantization method, BBPSO-Quantizer, considering low bit widths. Our findings indicate that lower-bit quantized models exhibit heightened performance compared to floating-point models.

**Acknowledgment.** The research leading to these results has received funding from the Ministry of Higher Education and Scientific Research of Tunisia under grant agreement number LR11ES48.

## References

1. Bablani, D., Mckinstry, J.L., Esser, S.K., Appuswamy, R., Modha, D.S.: Efficient and effective methods for mixed precision neural network quantization for faster, energy-efficient inference. arXiv preprint arXiv:2301.13330 (2023)
2. Bahdanau, D., Cho, K., Bengio, Y.: Neural machine translation by jointly learning to align and translate. arXiv preprint arXiv:1409.0473 (2014)
3. Bernhard, R., Moellic, P.A., Dutertre, J.M.: Impact of low-bitwidth quantization on the adversarial robustness for embedded neural networks. In: 2019 International Conference on Cyberworlds (CW), pp. 308–315. IEEE (2019)
4. Goodfellow, I.J., Shlens, J., Szegedy, C.: Explaining and harnessing adversarial examples. arXiv preprint arXiv:1412.6572 (2014)
5. Graves, A., Mohamed, A.R., Hinton, G.: Speech recognition with deep recurrent neural networks. In: 2013 IEEE International Conference on Acoustics, Speech and Signal Processing, pp. 6645–6649. IEEE (2013)
6. Guo, J., et al.: A comprehensive evaluation framework for deep model robustness. Pattern Recogn. **137**, 109308 (2023)
7. He, K., Zhang, X., Ren, S., Sun, J.: Deep residual learning for image recognition. In: Proceedings of the IEEE Conference on Computer Vision and Pattern Recognition, pp. 770–778 (2016)
8. Krizhevsky, A., Hinton, G., et al.: Learning multiple layers of features from tiny images (2009)
9. Krizhevsky, A., Sutskever, I., Hinton, G.E.: ImageNet classification with deep convolutional neural networks. Commun. ACM **60**(6), 84–90 (2017)
10. Le, Y., Yang, X.: Tiny ImageNet visual recognition challenge. CS 231N **7**(7), 3 (2015)
11. Li, Y., Dong, X., Wang, W.: Additive powers-of-two quantization: an efficient nonuniform discretization for neural networks. arXiv preprint arXiv:1909.13144 (2019)
12. Lin, J., Gan, C., Han, S.: Defensive quantization: when efficiency meets robustness. arXiv preprint arXiv:1904.08444 (2019)

13. Liu, A., et al.: Perceptual-sensitive GAN for generating adversarial patches. In: Proceedings of the AAAI Conference on Artificial Intelligence, vol. 33, pp. 1028–1035 (2019)
14. Liu, A., Liu, X., Yu, H., Zhang, C., Liu, Q., Tao, D.: Training robust deep neural networks via adversarial noise propagation. IEEE Trans. Image Process. **30**, 5769–5781 (2021)
15. Liu, A., et al.: Towards defending multiple lp-norm bounded adversarial perturbations via gated batch normalization. Int. J. Comput. Vision, 1–18 (2023)
16. Liu, Y., Wu, D., Zhou, W., Fan, K., Zhou, Z.: EACP: an effective automatic channel pruning for neural networks. Neurocomputing **526**, 131–142 (2023)
17. Liu, Z., Zhang, X., Wang, S., Ma, S., Gao, W.: Evolutionary quantization of neural networks with mixed-precision. In: ICASSP 2021-2021 IEEE International Conference on Acoustics, Speech and Signal Processing (ICASSP), pp. 2785–2789. IEEE (2021)
18. Lou, Q., Liu, L., Kim, M., Jiang, L.: AutoQB: AutoML for network quantization and binarization on mobile devices. arXiv preprint arXiv:1902.05690, vol. 2(8) (2019)
19. Madry, A., Makelov, A., Schmidt, L., Tsipras, D., Vladu, A.: Towards deep learning models resistant to adversarial attacks. arXiv preprint arXiv:1706.06083 (2017)
20. Sandler, M., Howard, A., Zhu, M., Zhmoginov, A., Chen, L.C.: MobileNetV2: inverted residuals and linear bottlenecks. In: Proceedings of the IEEE Conference on Computer Vision and Pattern Recognition, pp. 4510–4520 (2018)
21. Tmamna, J., Ayed, E.B., Ayed, M.B.: Deep learning for internet of things in fog computing: survey and open issues. In: 2020 5th International Conference on Advanced Technologies for Signal and Image Processing (ATSIP), pp. 1–6. IEEE (2020)
22. Tmamna, J., Ayed, E.B., Ayed, M.B.: Neural network pruning based on improved constrained particle swarm optimization. In: Mantoro, T., Lee, M., Ayu, M.A., Wong, K.W., Hidayanto, A.N. (eds.) Neural Information Processing: 28th International Conference, ICONIP 2021, Sanur, Bali, Indonesia, 8–12 December 2021, Proceedings, Part VI 28, pp. 315–322. Springer, Cham (2021). https://doi.org/10.1007/978-3-030-92310-5_37
23. Tmamna, J., Ayed, E.B., Ayed, M.B.: Automatic quantization of convolutional neural networks based on enhanced bare-bones particle swarm optimization for chest X-ray image classification. In: Nguyen, N.T., et al. (eds.) International Conference on Computational Collective Intelligence, pp. 125–137. Springer, Cham (2023). https://doi.org/10.1007/978-3-031-41456-5_10
24. Tmamna, J., Ayed, E.B., Fourati, R., Hussain, A., Ayed, M.B.: Bare-bones particle swarm optimization-based quantization for fast and energy efficient convolutional neural networks. Expert Syst. **41**(4), e13522 (2024)
25. Wang, J., Liu, A., Yin, Z., Liu, S., Tang, S., Liu, X.: Dual attention suppression attack: generate adversarial camouflage in physical world. In: Proceedings of the IEEE/CVF Conference on Computer Vision and Pattern Recognition, pp. 8565–8574 (2021)
26. Wei, Z., Chen, J., Wu, Z., Jiang, Y.G.: Cross-modal transferable adversarial attacks from images to videos. In: Proceedings of the IEEE/CVF Conference on Computer Vision and Pattern Recognition, pp. 15064–15073 (2022)
27. Xiao, Y., Zhang, T., Liu, S., Qin, H.: Benchmarking the robustness of quantized models. arXiv preprint arXiv:2304.03968 (2023)
28. Yao, Z., et al.: HAWQ-V3: dyadic neural network quantization. In: International Conference on Machine Learning, pp. 11875–11886. PMLR (2021)

29. Yi, C., Yang, S., Li, H., Tan, Y.P., Kot, A.: Benchmarking the robustness of spatial-temporal models against corruptions. arXiv preprint arXiv:2110.06513 (2021)
30. Yim, J., Joo, D., Bae, J., Kim, J.: A gift from knowledge distillation: fast optimization, network minimization and transfer learning. In: Proceedings of the IEEE Conference on Computer Vision and Pattern Recognition, pp. 4133–4141. IEEE (2017)
31. Zhang, C., et al.: Interpreting and improving adversarial robustness of deep neural networks with neuron sensitivity. IEEE Trans. Image Process. **30**, 1291–1304 (2020)
32. Zhang, T., Xiao, Y., Zhang, X., Li, H., Wang, L.: Benchmarking the physical-world adversarial robustness of vehicle detection. arXiv preprint arXiv:2304.05098 (2023)
33. Zhao, H., Sun, X., Dong, J., Chen, C., Dong, Z.: Highlight every step: knowledge distillation via collaborative teaching. IEEE Trans. Cybern. **52**(4), 2070–2081 (2020)
34. Zhao, Z., Zhang, J., Xu, S., Lin, Z., Pfister, H.: Discrete cosine transform network for guided depth map super-resolution. In: Proceedings of the IEEE/CVF Conference on Computer Vision and Pattern Recognition, pp. 5697–5707 (2022)

# Natural Language Processing

# Three-Stage Extraction of Spatial Relationships Using Markers

Michał Olek and Maciej Piasecki[✉]

Department of Artificial Intelligence, Wrocław University of Science and Technology, Wrocław, Poland
maciej.piasecki@pwr.edu.pl

**Abstract.** We present SpaRel – a system for Spatial Relation Extraction (SRE), which is focused on intra-sentence Spatial Relation Extraction. It was evaluated on three datasets: English (CLEF2017 mSpRL, SpaceEval) and Polish (PST2.0). We propose a novel procedure for joint relation elements discovery and classification. It significantly reduces the number of possible triplets checked for representing relation instances. This results in significant improvement in the quality and speed of the relation extraction process. To represent instances of relation instances, we expanded approaches to Relation Extraction, but we tried to extract limited amount of the necessary intermediate information. We propose a special procedure for resolving issues of one spatial indicator involved in many relations. The best results were achieved with the DeBERTa v3 (large) language model. We also studied SRE from the point of evaluation reliability and reproducibility of the previous approaches. Several issues related to the SpaceEval dataset were identified and their potential impact on results was discussed. Problems with one of the SOTA systems were identified and discussed. In a broader context, we have identified and discussed all the factors that hamper comparison of results in Relation Extraction.

**Keywords:** spatial relations · contextual embeddings · relation extraction

## 1 Introduction

Spatial Relation Extraction (henceforth SRE) is a special case [25] of Relation Extraction, and it maps utterances onto spacial locations and structures, e.g. in geographical information harvesting from documents or creating maps from descriptions in texts etc. SRE searches for spatial relations (SR) between entities mentioned in text. Usually, a SR is represented as a triple $\langle SI, TR, LD \rangle$ whose elements corresponds to *Spatial Roles* played by the mentioned entities in a given SR, namely: **SI** – *Spatial Indicator*, a *spatial trigger*, i.e. an expression signalling a SR, **TR** – *Trajector* represents an entity whose location is being described, **LD** – *Landmark* refers to an entity towards which the location is described. Not always all roles are explicitly expressed in a SR, quite often one of them is missing, e.g. in the relation: $\langle$ on the right, room, - $\rangle$ LD is empty. There are also relations with more than 3 roles present, e.g. *Path* and *Motion Indicator* elaborating dynamic aspects of a TR. The same expressions may play different roles in

© The Author(s), under exclusive license to Springer Nature Switzerland AG 2024
N.-T. Nguyen et al. (Eds.): ICCCI 2024, CCIS 2165, pp. 159–172, 2024.
https://doi.org/10.1007/978-3-031-70248-8_13

different relations, e.g. a TR in one relation and an LD in another one, like "table" in the sentence:

S1: *"Flowers stood on the table in the second room on the right"*

where in ⟨on, flowers, table⟩ it is LD and in ⟨in, table, room⟩ it is TR.

Concerning the scope, SRs may be represented by expression tuples located in: a sentence, a paragraph, or a document. Here, we focus on the most frequent SRE representation in corpora where SRs are labelled within sentences. Some approaches to SRE assume that a training corpus has already been tagged with Spatial Roles and the goal is to arrange and classify them as relation instances, while other works take a more realistic task with raw text on input and the goal of identifying and assigning Spatial Roles to expressions in text. Sometimes the latter task is characterised as Joint Learning for SRE and in our work we focus on this type of task. It is worth to mention that expecting raw text on input does not prevent different forms of language processing inside. [1] proposed a conceptual general scheme for the Relation Extraction task – relevant for SRE as a kind of RE. Starting from raw text, first potential relation elements, often entity mentions, together with their types (optional) are identified. They can be Named Entities – Named Entity Recognition (NER) or, more broadly, Mention Detection (MD) – or just language expressions (single or multiword), sometimes limited to specific categories, like nominals, pronouns. In SRE this is called *Spatial Role Labelling* (SRL). Next *Relation Identification (RI)* and *Relation Classification (RC)* are done (Fig. 1).

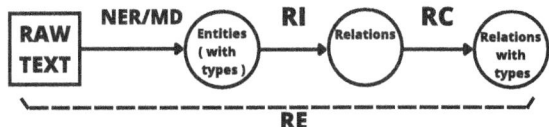

**Fig. 1.** Relation Extraction pipeline. NER: Named Entity Recognition; MD: Mention Detection; RI: Relation Identification; RC: Relation Classification.

In SRE, the former corresponds to identification all tuples of relation elements (language expressions) that are potentially in some SR. The second is a multi-class classification assigning a relation type to a tuple. Sometimes the two steps are combined and a tag "no_relation" is assigned to tuples not representing any relation. In some approaches, processing stops on Relation Identification and the results from this stage are final. It is worth noticing that we are looking for relations somehow expressed in a sentence, not inferred from common knowledge as done, e.g., by [3].

In this work, we present the *SpaRel* system for SRE from raw text, based on contextual embeddings as semantic representation. Extraction is done as a three-stage process. First possible SIs are found, then possible TRs and LDs for each candidate SI separately and finally *SpaRel* tries to find a relation associated with each possible SI, TR and LD triplet. It is common that the last part is done by using additional markers as [36]. Usually such markers are used only in the last step but we apply them in all stages.

We also solve a situations with more than one relation per a single SI. Our approach can be easily adapted to different natural languages, and was evaluated on the English and Polish datasets. Moreover, we addressed the issues of reproducibility of the previous approaches as a non-fully-expected side effect of the performed evaluation and comparison.

Our key contributions is the extension of marker-based technique to the two earlier stages – the first, where only SIs are located, and the second, where TRs and LDs are found. Moreover, our approach applied in the second stage in the case of many TRs and LDs associated to one single SI, significantly improves the quality of learning.

In addition, we propose an approach to the representation of SR triples which utilises contextual embeddings vectors in a way expanding methods proposed for the general task of Relation Extraction (in our case adapted to element triples, instead of pairs). We used the DeBERTa language model [9] for English and HerBERT [21] for Polish. Finally, during comparison and experiments we analysed and discussed several issues related to the reproducibility of the previous approaches.

## 2    Related Works

Initial work on SRE was done by [28], but interest in SRE started with SemEval-2012 Task 3 [13], since then continues. The following SemEval 2013 Task 3 [11] and SemEval 2015 Task 8 [25] expanded the problem of SRs with fine-grained semantics and dynamic SRs. However, the latter is rarely uptaken.

X-Space [29] used additional knowledge resources: WordNet [7] to facilitate SRL and PropBank [24] to improve SRE. [5] proposed Sieve-Based approach combining ideas of *tree kernels* and *multi-pass sieves*. SpRL-CWW of [22] uses CRF classifier for SRL and SVM one for SRE. [6] proposed an ensemble-based UTD that addressed the complexity of extracting parameters of dynamic relations by decomposing them into smaller parts. The BERT [4] model was used for SRL by [31]. It uses R-BERT [34] for SRE. A hybrid HMCGR of [33] uses a generative approach and reflexivity evaluation mechanism to improve the accuracy based on the reflexivity relations. Several works were done on CLEF2017-mSpRL [15]. [35] proposed rule-base system ASSA based on grammar and cognitive linguistic theories to identify spatial roles and relations. LIP6 of [20] attempted multimodal task but showed that text-based solution outperformed the bi-modal one. A few works on SRE were done for Polish [19] and [17].

## 3    Datasets

Two textual datasets annotated with SRs have been proposed for English, namely "CLEF2017: Multimodal Spatial Role Labeling (mSpRL) Task" [15] and SpaceEval of SemEval'2015 [25]. CLEF2017-mSpRL contains touristic pictures along with their textual descriptions, but we focus on the texts only. The annotated data is a subset of the IAPR TC-12 image Benchmark [8]. The texts are annotated with spatial roles according to spatial role labelling scheme [14]. Descriptions are not always limited to spatial information and phrases, not individual words, are annotated with roles.

SpaceEval is a dataset used in the SRE task of the SemEval family, especially SemEval-2015 Task 8, and is a subset of ISO-SpaceBank [27]. It contains travel narratives annotated with three types of SRs following ISO-Space (2012). It includes spatial static and dynamic relations. Dynamic relations are enriched with additional optional parameters but we only consider the static ones. A similar dataset for Polish – "Polish Spatial Texts 2.0" (PST) [18] offers an interesting perspective, as Polish is significantly different than English. It contains travel blog texts annotated with spatial expressions according to [23] based on: (SpatialML [16], SpatialRole Labelling from SemEval-2013 Task 3 [12] and ISO-Space1.4 [26]) (Fig. 1).

**Table 1.** Statistics for text datasets annotated with spatial relations

|  |  | Sentences | Relation instances | Spatial indicators |
|---|---|---|---|---|
| CLEF | train | 600 | 761 | 666 |
|  | test | 613 | 939 | 795 |
| Space Eval | train | 1082 | 1675 | 2504 |
|  | val. | 281 | 517 | 757 |
|  | test | 237 | 394 | 555 |
| PST | total | 4326 | 2035 | 1625 |

## 4    Approach Used

Each dataset was converted to a slightly extended version of the CONLLU format (the last column encodes a SR and the token role). *SpaRel* searches for SRs within a single sentence in three stages. Firstly, it only locates SIs, secondly, for each SI the sets of potential TRs and LDs are created. In third stage each feasible combination of SI, TR and LD is linked into a SR instance. The same approach is used across all stages, only slightly altered to match different targeted inputs and outputs.

### 4.1    Selection of SIs

We check each single word by creating a separate sample for it. The sentence is tokenised by BPE (from BERT) [30] and then additional special token is introduced before and after the (subword) token sequences corresponding to the analysed single word. DeBERTa v3 for English and HerBERT for Polish are applied to represent modified sentences. Then token embedding of the first subtoken of a given word is the input to a classifier (Fig. 2). This stage results in mapping of a sentence into a set of words, potential SIs. For S1 example sentence (Sect. 1) we should obtain: {on, in, on the right}.

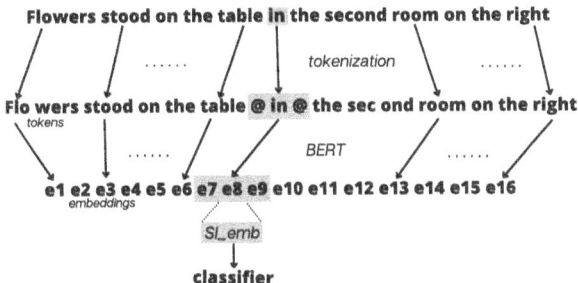

**Fig. 2.** Finding SIs - checking word "in". (Tokenisation is exemplary). Whole set of samples for sentence is obtained by "walking" markings '@' all over the words in the sentence

## 4.2 SI-Driven Selection of TRs and LDs

Assuming SIs found so far, we try to identify their corresponding TRs and LDs. Samples representing SIs and their candidate TRs/LDs – all words from a sentence, except the candidate SIs (from the first stage) – are generated for processing from a sentence. For each candidate SI, a sample based on the tokenised sentence including a pair of special tokens is inserted around the (subword) token sequences corresponding to the SI (as in the previous step), and another special tokens around the (subword) token sequences corresponding to the analysed single word (Fig. 3).

Next, modified sentences are processed with the language model and the input vectors for a classifier are composed by concatenating the embeddings of the first subtoken of the SI and the first subtoken of the word being analysed. The classifier is separate from the one used in the previous step. It is also worth noticing that a single word can be a TR for one sample, and LD or just a common word for another sample. The model trained for this stage is a separate from the one of the first step, and is trained on gold annotations but the final output is produced using the list of possible SIs from the previous step. The training is described in more details in Sect. 5. This step outputs for a pair (sentence,SI) two sets of potential TRs and LDs for the given SI, e.g.

```
{on->{TRs->{Flowers}, LDs->{table}}, in->{TRs->{Flowers,table}, LDs->{room}},
 on the right->{TRs->{room}, LDs->{ }} }
```

Here, the system made one error: the word "Flowers" was assigned as a possible TR to SI "in" but should be only assigned to the first "on". This is not a problem yet, but it is only the third step that finally decides on a relation represented by a triple.

## 4.3 Building Sample Representation

We create candidate triplets and classify them according to relations. From the pair sets from the previous step all possible triplet combinations are generated, taking SIs as seeds and adding TRs and LDs found to be associated with them. By repeating this procedure for all SIs in a sentence we obtain all potential triplets for the sentence, e.g.

```
{{on,Flowers,table}, {in,Flowers,room}, {in,table,room}, {on the right,room,-}}
```

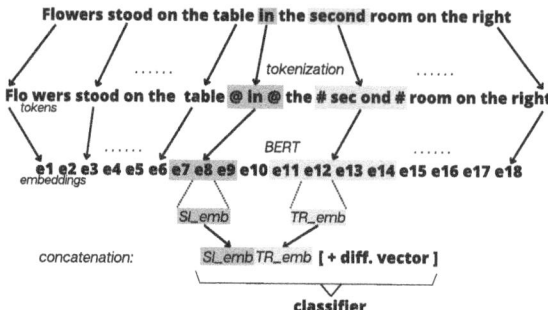

**Fig. 3.** Finding TRs and LDs - checking word "second" for SI fixed as "in". Whole set of samples for SI "in" is obtained by "walking" markings '#' all over the words in the sentence (except SI).

After all potential triples for a given sentence have been found, they are verified by a classifier. A sentence is tokenised and special tokens are inserted around the (subword) token sequences corresponding to SI, TR and LD: different special tokens for different roles. A similar solution is in [36], but we expanded it beyond the binary relations only, and, as mentioned earlier, adapted the method not only to classify relations, but also to find SIs and next to find the rest of components of SRs. Modified sentences are processed with the selected language model. Token embeddings corresponding to first subtoken of each role are concatenated into sample vectors preserving the order: $\langle SI, TR, LD \rangle$. Next, in a similar way to techniques used in lexico-semantic relation recognition, e.g. [10], the vector is expanded with three element-wise difference vectors, namely: $SI - TR$, $TR - LD$, $LD - SI$. The idea behind such representation is to emphasise the relation of the two Spatial Roles to the contextual vector of the relation lexical marker $SI$, and also to better characterise the relation between the components $TR - LD$. The resulting concatenated vector is the input to the classifier (Fig. 4).

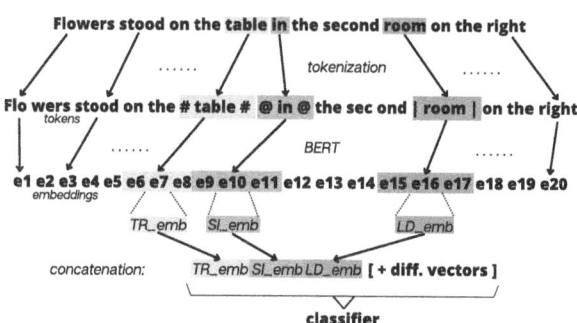

**Fig. 4.** Third stage: building sample representation for a triple <in, table, room>. (example tokens).

### 4.4 Classification

We used an MLP neural network (two fully connected layers, and the dropout: 0.1). For each of the three stages we used a separate classifier model:

1. the classifier decides whether a word is a first word of SI, a subsequent word of SI or just common word.
2. it decides if a word is just a common word, a start word of TR or LD, or a subsequent word of TR or LD – five different labels in total.
3. it decides what type of relation is to be assigned to a given triple. The types depend on the dataset being used. All datasets contain type "no_relation".

## 5 Discussion and Extension

Usually, when a marker-based approach is used, it is implemented only in one stage (corresponding to our 3-rd one) where triples already found are to be classified for a relation. We generalised this approach to earlier stages also.

The first step of training is quite straightforward: for each sentence we generate as many samples as there are words in a sentence and label them according to golden relations: whether a word between inserted special tokens is part of SI in any relation.

There are several further issues related to training in the second stage. We decided to base training on the SIs from the annotated training data rather than results of the first step. This saves us from re-training the second stage each time the outcome of stage one changes, which is a significant time-saver. Originally the training samples were based on annotated relations: a full set of samples was generated per each relation in sentence. However, problems occur when model has to deal with finding TRs and LDs for SIs that occur in more than one relation in a sentence. It is quite frequent, as illustrated in Fig. 5.

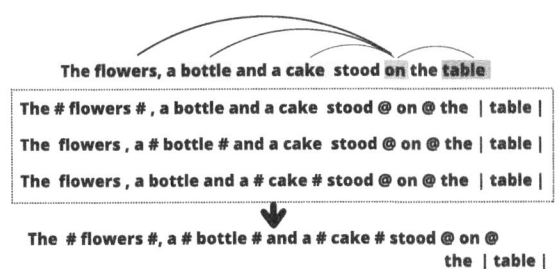

**Fig. 5.** SI with many relations. Merging info for many relations with a particular SI to just one info

In Fig. 5, unfortunately, there are three different relations in the dataset and for some words the model receives conflicting information whether a given word is or is not a TR for that SI. In this case, there are, e.g., three exactly the same samples generated for the word 'flower', but only in one of them it is labelled as TR. We tested the possibility that the model would nevertheless learn and take into account all possible outcomes, but experiments showed that the opposite happens: conflicting samples cancel each other out and the model learns nothing in such cases.

So we extended the approach so that for any SI there is only one info about all its relations. We obtain this by merging all info from all relations for the same SI into just one record that contains multiple TRs and LDs. And only from this one info the raw samples are generated and since all possible meaningful TRs and LDs are mentioned in it, there will be no conflicting input data.

Another problem results from this extension, but very rare (5 instances in whole SpaceEval train): when there are two relations for the same SI and including exactly the same word in different roles (in the first relation as TR and LD in the second). There can be only one simple label per sample, so it is impossible to encoded such cases (Fig. 6).

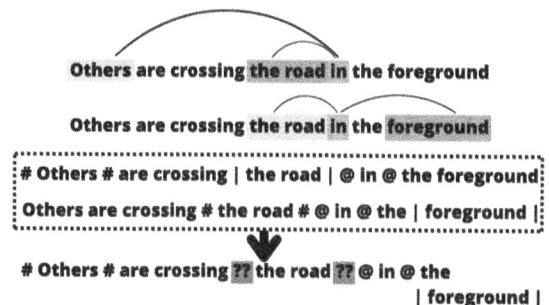

**Fig. 6.** SI with many relations. Merging conflict

## 6   Experiments

SRE is a specific kind of Relation Extraction and shares with it all the problems related to measuring performance, e.g. [32]. Results depend on the whole pipeline: Spatial Role Labelling (in the full task), Relation Identification and Relation Classification – so the scores can be reported to any component or their combinations. In addition, all papers report on F1, but closer inspections reveals that some refer to F1-micro and some to F1-macro. In some papers, recognition of "no_relation" is considered as True Positive and in others not. Thus, we may calculate the final result in 16 slightly different ways.

Further on, different requirements for the detection of relation arguments may be taken into account [32]: not only relation occurrence, but its arguments. Because the recognition of a relation is not only the recognition of its existence or type, but also the recognition of its arguments. There are three possible evaluation settings [2]: *strict* – all boundaries and argument types must be correct, *boundaries* – boundaries are compared, but the argument type is not considered, and *relaxed*: NER is reduced to Entity Classification, i.e. predicting a type for each token, a multi-token entity is considered correct, if at least one token is correctly typed.

As a result, there are 48 different ways for score calculation. In SRE domain type can be equalled to the triple position, but the question is whether we require the argument boundaries to be given precisely or one token selected per element is enough.

As [1] noticed the lack of information whether RI and RC steps are performed, or if gold Spatial Roles are used leads to "blurriness that hampers fair evaluation". And that "these details are utterly important". Often such important details are not reported in papers and they can be only learned from the source code inspection (if it is available). Sometimes authors make mistakes comparing results with different settings and, as [32] describe, it is worrisome that mistakes are invisible in the article and can only be detected in code. Often the code is not made available, causing such mistakes to go undetected.

We prefer F1-micro to F1-macro as the latter can be boosted by targeting less populated classes. We decided to not treat "no_relation" as a proper (positive) class, because this could result in increasing the score by generating more negative samples.

### 6.1  Evaluation Settings

For all experiments we use metric F1-micro that does not count correct guesses for "no_relation" as true positives. The requirement for arguments detection is of the type *Relaxed*. The code is based on [36], but very significantly changed and extended.

### 6.2  CLEF2017-mSpRL

For this dataset we made experiments including Spatial Role Labelling as a part of the task and measured results for both RI and RC.

### 6.3  SpaceEval

For this dataset we made two series of experiments: first is with our "standard" setup which includes Spatial Role Labelling as a part of the task. Second – in which SRL is assumed to be already done (existing annotation is utilised) – is performed additionally just to obtain result that can be compared with other systems that uses this way.

Additionally for SpaceEval in the first phase, when recognising SIs, the system does not only recognises, if a word is a SI, but also decides, if it is an SI of the type "motion" or "spatial_signal", since this info is provided in dataset and used by other approaches.

### 6.4  PST

To compare with [17] we tried to reproduce the setup from it using the test set selected by the authors, i.e. only SR instances with all three elements: 1474 (1200 train, and 274 test) out from 2035. The evaluation was made only on the training data (it is a rule based solution) and 375 out of 1200 were correctly classified. Assuming no false positives, it gives F1 = 47.61. We prepared dataset with only full relations included, and divided it to: 1166 samples in train and 310 in test subsets. After training model on the train dataset it was evaluated on test dataset. Only Relation Identification was performed, as [17] did. They assumed Spatial Role Labeling already done but we stuck to our default setting of the more difficult task (where SRL is part of the task) to fully test.

# 7   Results

SpaRel achieved 74.55 ($\sigma = 1.17$) for RI task and 71.17 ($\sigma = 0.52$) for RC task (averaged over 5 evaluation iterations). Table 2 cites the results of [35] where system ASSA was introduced, and shows comparison with other systems discussed in Sect. 2.

**Table 2.** Comparison of results (F1-micro values) for CLEF2017-mSpRL.

| Model | RI | RC |
|---|---|---|
| SpaRel | **74.55** | **71.17** |
| ASSA | 62.28 | – |
| LIP6 | 56.41 | 46.29 |
| Baseline | 56.67 | – |

SpaRel achieved 65.45 when SRL is a part of a task ($\sigma = 0.75$) and its comparison to other results is shown in Table 3. When SRL is not a part of the task the result is 70.4 ($\sigma = 2.12$) and the column RCgSR denotes other works which we know that used gold Spatial Roles. For each series we evaluated model 5 times and the results are the averages. The case of HMCGR system is described in details below.

**Table 3.** Comparison of results (F1-micro values) for SpaceEval2015.

| Model | RCgSR | RC |
|---|---|---|
| SpaRel | **70.42** | **65.5** |
| HMCGR (declared 70.9 RC) | 64.40 | – |
| BERT-based SpIE | – | 61.2 |
| UTD: Ensemble-Based SpRE | 62.00 | – |
| SpRL-CWW | – | 56.1 |
| X-Space | – | 53.0 |
| Sieve Based | – | 41.8 |

*Issues with Replicating Results for HMCGR.* We tried to replicate the results reported for system HMCGR in [33], but when we ran the original source code it achieved much different, lower result from 70.9 reported in the paper. On closer inspection, we also found that in spite of the full RE pipeline declared in the paper, there is an assumption that Spatial Role Labelling has been already performed in the source code and this annotation is available. This changes the whole task implemented to a different (and much easier) task than the one described in the paper. The entire module described in [33] (named CTE) and responsible for this part of the task is missing from the code.

The module (called RFX) announced to perform the interesting and novel Reflexivity Evaluation part of the task is also missing in the code. We contacted the authors and asked them to share the full code, but received a reply that due to lack of time and an overabundance of tasks they are unable to make the entire code available. Thus we based the comparison on the results reproduced (to some extent) by us for HMCGR.

Admittedly, F1-micro results are reported in [33], but the included source code calculates F1-macro score: 65.76, in fact. We recalculated the result using the F1-micro measure and obtained: 64.4.

It is difficult to estimate what the score would have been, if both modules had been in place. Although the RFX module would have increased the score, the introduction of the CTE module would have lowered it considerably – since it would have meant performing Spatial Role Labelling on its own, rather than using the ready-made gold annotations of Spatial Roles. The authors reported an F1-micro score of 89.1 for the Spatial Role Labelling task alone by the CTE module application. From the data reported further on in [33], it appears that the RFX module would not make up for this loss: in the ablation study when the system used only the GEN module the score was 56.5 and when RFX module was added the score increased by 2.3 p.p.

So we left out speculations and decided to include the score, which is reported by the original script only in the F1-micro version: 64.40 (the one that uses gold Spatial Roles).

The results are presented in Table 4 (KBRSE is from [17]). However, the results are not directly comparable. Our score was achieved for a much more difficult task as it required SRL. And also they tested on the train set (since their rule based system did not need training phase) and SpaRel was trained on the train set and tested on the test set.

**Table 4.** Comparison of results (F1-micro values) for the PST2.0 dataset.

| Model | RIgSR | RI |
|-------|-------|-------|
| SpaRel | – | 68.31 |
| KBRSE | 47.61 | – |

## 8    Ablation Study

We chose the RI task of CLEF2017, because the differences between the tested variants should be most visible. We first tested what effect on the result has the vector generation method providing input to the classifier. Originally, only the first subtoken embedding is taken from each element (SI, RT or LD) found. We tested what is influence of taking the average of embeddings of all subtokens instead. It turned out that the results are lower by 2-3 p.p. When extension from Sect. 5 is not used then the final score is lower by 10 p.p: all the cases with more than one relation to the same SI, cannot be correctly identified.

## 9   Limitations

The conflicting roles for a word in the relations for the same SI (Sect. 5) are not possible to be solved by the method. However, such cases are very infrequent. Our method is able to cope with relation instances of less than 3 elements, but in Step 3 candidates should be also pairs, not only triples. This would increase the computational complexity. In a similar way, cases of more than three components could be handled. The method generates first all possible candidates, next to filter them by classification. This is an obstacle, but it could be reduced by some heuristically driven filtering or pre-ordering of candidates. All three classifiers from the three steps are trained separately and do not share parameters. This is not clearly a limitation, an area for further improvements. Some kind of knowledge passing could help selecting relations instances that fit together well. Our method is limited to sentence boundaries, like most works in SRE.

## 10   Conclusion

We presented the SpaRel system for SRE based on a mechanism of introducing additional tokens-markers into processed sentences and a heuristically orchestrated stack of classifiers. It performed very well for both Relation Identification and Relation Classification tasks for English and Polish. The results seem to be very high in comparison to SOTA (state-of-the-art), when we take into account all doubts with the reproducibility of the SOTA results that we experienced. Further works include dealing with the mentioned known limitations and balancing between performance and efficiency of processing.

**Acknowledgements.** This work has been supported by the Ministry of Science and Higher Education as a part of CLARIN-PL project.

## References

1. Bassignana, E., Plank, B.: What do you mean by relation extraction? A survey on datasets and study on scientific relation classification. In: Proceedings of the 60th ACL: Student Research Workshop, pp. 67–83. ACL (2022)
2. Bekoulis, G., Deleu, J., Demeester, T., Develder, C.: Adversarial training for multi-context joint entity and relation extraction. In: Proceedings of the 2018 EMNLP, pp. 2830–2836. ACL (2018)
3. Bouraoui, Z., Camacho-Collados, J., Schockaert, S.: Inducing relational knowledge from BERT. In: The 34th Conference on Artificial Intelligence, AAAI, pp. 7456–7463. AAAI Press (2020)
4. Devlin, J., Chang, M.W., Lee, K., Toutanova, K.: BERT: pre-training of deep bidirectional transformers for language understanding. arXiv: abs/1810.04805 (2019)
5. D'Souza, J., Ng, V.: Sieve-based spatial relation extraction with expanding parse trees. In: Proceedings of the 2015 EMNLP, pp. 758–768. ACL, September 2015
6. D'Souza, J., Ng, V.: UTD: ensemble-based spatial relation extraction. In: Proceedings of the 9th SemEval 2015, pp. 862–869. ACL, June 2015
7. Fellbaum, C.: WordNet: An Electronic Lexical Database. Bradford Books (1998)

8. Grubinger, M., Clough, P., Müller, H., Deselaers, T.: The IAPR TC12 benchmark: a new evaluation resource for visual information systems. In: Workshop Ontoimage, October 2006
9. He, P., Liu, X., Gao, J., Chen, W.: DeBERTa: decoding-enhanced BERT with disentangled attention (2020). https://doi.org/10.48550/ARXIV.2006.03654
10. Kanclerz, K., Piasecki, M.: Deep neural representations for multiword expressions detection. In: Proceedings of the 60th ACL: Student Research Workshop, pp. 444–453. ACL (2022)
11. Kolomiyets, O., Kordjamshidi, P., Bethard, S., Moens, M.: Semeval-2013 task 3: spatial role labeling. In: *SEM 2013 - 2nd Joint Conference on Lexical and Computational Semantics, pp. 255–262. ACL (2013)
12. Kolomiyets, O., Kordjamshidi, P., Moens, M.F., Bethard, S.: SemEval-2013 task 3: spatial role labeling. In: 2nd Joint Conference on Lexical and Computational Semantics (*SEM), vol. 2: Proceedings of the 7th (SemEval 2013), pp. 255–262. ACL (2013)
13. Kordjamshidi, P., Bethard, S., Moens, M.: Semeval-2012 task 3: spatial role labeling. In: Proceedings of the 6th SemEval, pp. 365–373. *SEM 2012 - 1st Joint Conference on Lexical and Computational Semantics, ACL (2012)
14. Kordjamshidi, P., van Otterlo, M., Moens, M.F.: Spatial role labeling: task definition and annotation scheme. In: LREC (2010)
15. Kordjamshidi, P., Rahgooy, T., Moens, M.F., Pustejovsky, J., Manzoor, U., Roberts, K.: CLEF 2017: multimodal spatial role labeling (MSPRL) task overview. In: Jones, Gareth J.F., et al. (eds.) Experimental IR Meets Multilinguality, Multimodality, and Interaction, pp. 367–376. Springer, Cham (2017)
16. Mani, I., et al.: SpatialML: annotation scheme, resources, and evaluation. Lang. Resour. Eval. **44**, 263–280 (2010)
17. Marcińczuk, M., Oleksy, M., Wieczorek, J.: Evaluation of knowledge-based recognition of spatial expressions for Polish. In: Nguyen, N.T., Hoang, B.H., Huynh, C.P., Hwang, D., Trawiński, B., Vossen, G. (eds.) ICCCI 2020. LNCS (LNAI), vol. 12496, pp. 682–693. Springer, Cham (2020). https://doi.org/10.1007/978-3-030-63007-2_53
18. Marcińczuk, M., Oleksy, M., Wieczorek, J.: PST 2.0 – corpus of Polish spatial texts. In: Proceedings of the 12th LREC, pp. 2167–2174. ELRA, May 2020
19. Marcińczuk, M., Oleksy, M., Wieczorek, J.: Towards recognition of spatial relations between entities for polish. Cogn. Stud.|Études cognitives **16**, 119–132 (2016)
20. Moussa, A., Fournier, S., Mahmoudi, K., Espinasse, B., Faiz, S.: Mixing static word embeddings and RoBERTa for spatial role labeling. Procedia Comput. Sci. **207**, 2950–2957 (2022). Proceedings of the 26th International Conference on KES2022
21. Mroczkowski, R., Rybak, P., Wróblewska, A., Gawlik, I.: HerBERT: efficiently pretrained transformer-based language model for Polish. In: Proceedings of the 8th Workshop on Balto-Slavic NLP, pp. 1–10. ACL (2021)
22. Nichols, E., Botros, F.: SpRL-CWW: spatial relation classification with independent multiclass models. In: Proceedings of the 9th SemEval 2015, pp. 895–901. ACL, June 2015
23. Oleksy, M., Marcińczuk, M., Bernaś, T., Wieczorek, J., Kocoń, J.: KPWr annotation guidelines - spatial expressions (2.0) (2019). http://hdl.handle.net/11321/719, CLARIN-PL
24. Palmer, M., Kingsbury, P., Gildea, D.: The proposition bank: an annotated corpus of semantic roles. Comput. Linguist. **31**, 71–106 (2005)
25. Pustejovsky, J., Kordjamshidi, P., Moens, M.F., Levine, A., Dworman, S., Yocum, Z.: SemEval-2015 task 8: SpaceEval. In: Proceedings of the 9th SemEval 2015, pp. 884–894. ACL (2015)
26. Pustejovsky, J., Moszkowicz, J.L., Verhagen, M.: A linguistically grounded annotation language for spatial information. Trait. Autom. des Langues **53**, 87–113 (2012)
27. Pustejovsky, J., Yocum, Z.: Capturing motion in ISO-SpaceBank. In: Proceedings of the 9th Joint ISO - ACL SIGSEM Workshop on Interoperable Semantic Annotation, pp. 25–34. ACL (2013)

28. Roberts, K., Harabagiu, S.: UTD-SpRL: a joint approach to spatial role labeling. In: *SEM 2012: The 1st Joint Conference on Lexical and Computational Semantics, pp. 419–424. ACL (2012)

29. Salaberri, H., Arregi, O., Zapirain, B.: IXAGroupEHUSpaceEval: (X-space) a WordNet-based approach towards the automatic recognition of spatial information following the ISO-space annotation scheme. In: Proceedings of the 9th SemEval 2015, pp. 856–861. ACL (2015)

30. Sennrich, R., Haddow, B., Birch, A.: Neural machine translation of rare words with subword units. In: Proceedings of the 54th ACL (Vol. 1: Long Papers), pp. 1715–1725. ACL (2016)

31. Shin, H.J., Park, J.Y., Yuk, D.B., Lee, J.S.: BERT-based spatial information extraction. In: Proceedings of the Third International Workshop on Spatial Language Understanding, pp. 10–17. ACL (2020)

32. Taillé, B., Guigue, V., Scoutheeten, G., Gallinari, P.: Let's stop incorrect comparisons in end-to-end relation extraction! In: Proceedings of the 2020 EMNLP, pp. 3689–3701. ACL (2020)

33. Wang, F., Li, P., Zhu, Q.: A hybrid model of classification and generation for spatial relation extraction. In: Proceedings of the 29th COLING, pp. 1915–1924 (2022)

34. Wu, S., He, Y.: Enriching pre-trained language model with entity information for relation classification. CoRR abs/1905.08284 (2019). http://arxiv.org/abs/1905.08284

35. Xu, C., Dietz Saldanha, E.A., Gromann, D., Zhou, B.: A cognitively motivated approach to spatial information extraction. In: Proceedings of the Third International Workshop on Spatial Language Understanding, pp. 18–28. ACL, November 2020

36. Zhou, W., Chen, M.: An improved baseline for sentence-level relation extraction. In: Proceedings of the 2nd Conference of the Asia-Pacific Chapter of the ACL and the 12th IJCNLP (Vol. 2: Short Papers), pp. 161–168. ACL (2022)

# A Quadruplication Multilingual and Multilevel Topic Seeding Approach Towards a Bottom-Up Graph Generation and Enhancement

Amani Mechergui[1,2]($\boxtimes$), Wahiba Ben Abdessalem Karaa[1,2], and Sami Zghal[3]

[1] High Institute of Management of Tunis, Tunis University, Tunis, Tunisia
amenimechergui4T@gmail.com
[2] RIADI Laboratory, National School of Computer Science, Manouba University, Manouba, Tunisia
[3] LIPAH-LR11ES14, University of Jendouba, FSJEGJ, Jendouba, Tunisia

**Abstract.** Multilevel topic-seeding models serve as a connection that links unstructured data to a well-structured knowledge representation in a **K**nowledge **G**raph (**KG**). However, these approaches face challenges in representing KGs for multilingual term clustering, which involves refining the combination of predefined hierarchy, unstructured multilingual data, and structural knowledge. We propose "Quadruplicated **Multi**lingual and **Multi**level **L**atent **D**irichlet **A**llocation" (**4Multi²LDA**) that integrates topic seed terms into multilevel modeling. Our objective is to generate distinct clusters by adjusting term clusters according to a predefined hierarchy. 4Multi²LDA involves monolingual/multilingual distributional term clustering over **C**ore **C**oncepts (**CCs**) across different domains, forming upper KGs. This process contributes to the semi-automatic learning and supplementation of a **B**ottom-**U**p **U**niversal **U**pper **K**nowledge **G**raph (**BU³KG**) from unstructured data. In the initial training phase, we focus on establishing "the bottom level" of BU³KG. During the second training phase, we enhance the KG by constructing its "upper level" . In the third and fourth training phases, we aim to further enrich the constructed bottom and upper levels of the KG with new seed terms. The findings demonstrate that 4Multi²LDA exhibits outstanding performance in term clustering over CCs. It surpasses the performance of monolingual and multilingual baselines trained on four datasets.

**Keywords:** Multilevel topic-seeding · multilingual term clustering · quadruplicated Latent Dirichlet Allocation · topic seed terms · core concept · bottom-up universal upper knowledge graph

## 1 Introduction

Probabilistic **T**opic **M**odels (**TM**), exemplified by **L**atent **D**irichlet **A**llocation (**LDA**) [1], have achieved notable efficacy in text mining and analysis. The objective of TM involves the derivation of per-document topic proportions and the identification of latent topics within the target corpus through the examination of word co-occurrences within

© The Author(s), under exclusive license to Springer Nature Switzerland AG 2024
N.-T. Nguyen et al. (Eds.): ICCCI 2024, CCIS 2165, pp. 173–192, 2024.
https://doi.org/10.1007/978-3-031-70248-8_14

individual documents. LDA serves a dual role by not only aiding in the formation of concepts but also establishing a connection between unstructured data and structured knowledge representation in **K**nowledge **G**raphs (**KG**s). However, the central aim of the LDA objective function is to maximize the probability of observed data, a tendency that prioritizes high-frequency words while potentially neglecting infrequent ones. Additionally, the lack of prior knowledge may lead to the allocation of infrequent words to irrelevant topics. Consequently, this can result in the formation of topics that are not intuitive or meaningful to end users, presenting a practical challenge. This issue is particularly evident in multilevel (hierarchical structure) contexts, where the modeling of a large number of topics and their relevance becomes more complicated [2, 3]. To overcome these limitations, multilevel topic-seeding models are employed. These models introduce topic-seed words and take into account latent topic hierarchies during the creation of the corpus. This motivation stems from the understanding that topics have natural correlations. For instance, in Fig. 1, 'ontology' represents a higher-level topic that encompasses the concept of 'technique'. By incorporating this multilevel seeding approach, the models can strategically give preference to particular topics or concepts right from the start, thereby enhancing the accuracy of topic inferences. Nonetheless, the use of multilevel topic-seeding models in conjunction with KGs for multilingual clustering poses a significant challenge. This difficulty stems from integrating diverse data from different languages and incorporating the structural knowledge embedded within the KG. The complexity lies in addressing the intricacies of various languages and the intricate connections within the KG. Failing to bridge this gap could restrict the models' ability to extract meaningful and non-overlapping topics from multilingual datasets. To overcome the mentioned shortcomings, we propose an innovative model referred to as *"Quadruplicated **Multi**level and **Multi**lingual LDA" (**4Multi²LDA**)*. It constitutes an innovative five-step Bayesian generative cross-domain approach that incorporates prior domain knowledge into a multilevel topic-seeding model. As exemplified in Fig. 1, users with domain expertise often prefer a predefined topic hierarchy (depicted on the right), visually represented as the knowledge graph (on the left). This graph organizes topics (pink and yellow nodes) and words (orange nodes) into a tree-like taxonomy (on the right). Centered around the target corpus and seed words (positioned in the middle of Fig. 1) and complemented by the provided topic hierarchy (on the right), the proposed approach will: (1) Establish a semantic connection between words and their corresponding topic nodes (on the right) to guide the exploration of non-overlapping topics, and (2) reveal novel topic hierarchies not explicitly captured in the provided topic tree (on the right). The overarching goal is to semi-automatically generate and further enhance a **B**ottom-**U**p **U**niversal **U**pper **K**nowledge **G**raph (**BU³KG**). We aim to generate high-quality and non-overlapping multilingual term clusters that align with the predefined **C**ore **C**oncepts (**CC**s) composing the **U**pper **K**nowledge **G**raphs (**UKG**s) of a BU³KG. More precisely, the UKGs are models that contain CCs and the connections between them in a BU³KG. Additionally, through this pipeline, our focus lies in automating the task of term clustering for CC classes, emphasizing concept formation. These classes consist of sets of terms linked to CCs, encompassing synonyms, hyponyms, and terms connected through semantic relationships. Unlike existing topic-seeding approaches, our approach involves selecting seed terms associated with CCs and allocating a CC label to each topic. To achieve this, we use cross-domain multilingual data and a predefined hierarchy. We conduct four iterations of algorithm training: In the initial training, we integrate topics with their subsumed terms at the lowest level of the $BU^3KG$. Subsequently, during the second training, we refine and label the cluster terms generated from the first training based on the CCs specified in the predefined UKGs. This facilitates their interconnection and organization into 'upper' level clusters within the $BU^3KG$ structure. In the third and fourth training iterations, our goal is to employ new seed terms to further enhance both the constructed bottom and upper clusters of the $BU^3KG$. Our contributions can be outlined as follows: (1) Our proposed four-trained pipeline not only constructs but also refines a multilingual $BU^3KG$. As far as we know, in the field of graph mining, we introduce the first approach that constructs a universal upper KG by training the model twice using a bottom-up strategy and subsequently enhances it with the same two initial training stages along with new seeds. (2) We present an innovative approach for term clustering, encompassing both monolingual and multilingual contexts. To the best of our knowledge, existing topic-seeding models are typically used for document clustering and have not been extended to address term clustering in both monolingual and multilingual settings within graph mining. (3) The 4Multi²LDA methodology also facilitates the exploration of synonym and hypernym relationships. As a result, this offers the advantage of improving computational efficiency by cultivating a more nuanced representation of information within KGs. (4) The proposed model exhibits noteworthy flexibility and efficiency when trained on ontology, fisheries, and medicine datasets. It establishes a robust baseline for knowledge-based TM. The paper's organization is detailed as follows: Sect. 2 provides an overview of relevant literature concerning TM and KGs. Section 3 presents our methodology for aligning a multilingual topic-seeding model with a predefined hierarchy. Section 4 illustrates the sequence of experiments conducted across four corpora. Finally, Sect. 5 provides a conclusion and outlines potential future works.

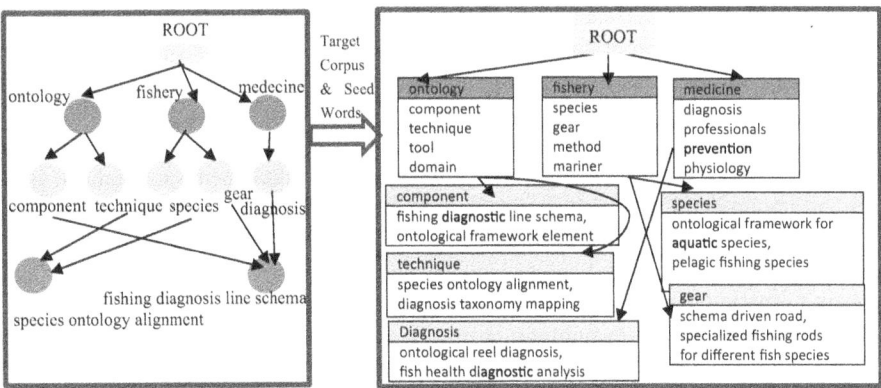

**Fig. 1.** The motivation behind the proposed approach.

## 2  Related Work

TM approaches aim to autonomously reveal and extract hidden thematic structures within a collection of documents. They operate under the assumption that documents are mixtures of topics, and each topic is characterized by a distribution of words. In the realm of *multilingual TM*, the primary objective is to simultaneously identify both language correspondences and latent topics across diverse linguistic datasets. The authors of [4] introduce an LDA model for unaligned text, utilizing a stochastic expectation maximization inference procedure. To further improve it, the proposed model of [5] departs from the assumption of substantial comparability between documents in diverse languages. This model employs a learning mechanism for weighted LDA topic connections, establishing links between cross-lingual topics based on the similarity of their defining dominant words. Diverging from the approaches outlined in [4] and [5], Xie et al. [6] examine topics within both monolingual and multilingual datasets. They employ BERT (Bidirectional Encoder Representations from Transformers) multilingual sentence embeddings to formulate LDA distributions. Their findings validate the investigation of topic evolution without necessitating machine translation or relying on subjective author translations. However, they fail to consider the integration of topics with other fields of study. Concurrently, the research discussed in [7] generates topics using embedded data for Indonesian customer complaints. In this process, LDA assigns probabilistic topic assignment vectors, while BERT manages sentence embedding, attaining the highest coherence score. A recent approach in [8] enhanced the adaptability of multilingual LDA models by integrating alternative priors such as generalized Dirichlet and Beta-Liouville distributions through variational inference, along with the utilization of the Jaccard index. Some other studies, such as [9], broaden the scope of multilingual context by integrating graph partitioning. Bansal et al. [9] introduce a system for personalized hashtag recommendation in low-resource languages, utilizing graph-based deep neural networks with attention mechanisms. The paper highlights the integration of linguistic knowledge and graphs to enhance hashtag recommendations in a multilingual context. On the other hand, TM may encounter challenges, such as the difficulty in achieving interpretable and relevant topics without some form of guidance. To overcome such an issue, *the seed-guided TM approach* is employed, involving the supply of relevant seed terms to guide the modeling process. The work outlined in [10] enhances the stability and semantic coherence of neural TM through the incorporation of seed words at both the word and document levels. Mustapha et al. [11] utilized predefined topics and relationships to establish connections among entities but fell short of comprehensive identification of all associated terms within CCs. Xu et al. [12] introduced a doubly-trained LDA technique that not only classifies terms into subdomains but also incorporates prior knowledge to facilitate concept formation. Analogous to [11], Xu et al. [12] have not investigated the implications of omitting domain-meaningful terms. To overcome this limitation, Huang et al. [13] propose a dual approach involving the utilization of CCs from a domain ontology as prior knowledge for learning domain-specific clusters and adjusting term clustering with seed knowledge-based LDA models to incorporate these CCs. This adaptation aims to identify the pertinent concept labels for each cluster, enhancing the accuracy and relevance of the clustering process. Nevertheless, the potential issue with seed-guided TM is the risk of bias introduced by the initial seeds, potentially restricting the diversity of identified topics and reducing model flexibility. Consequently, an initial specification of seeds may prompt the emergence of a need for a multilevel organization of topics, directing the model toward general or specific topics. The goal of *multilevel TM* is to deduce latent hierarchies within the dataset, capturing correlations among topics and constructing multilevel topic structures. The authors in [2] integrate tree and text embeddings within a spherical space to concurrently represent the structure of the category tree and the generative processes of the corpus. They prove advantageous for weakly-supervised multilevel text classification tasks. However, further investigation may be needed to assess the model's applicability to other types of structured data. Duan et al. [14] present a semantic graph-guided TM that views words and topics as Gaussian distributions and employs the Kullback-Leibler (KL) divergence to regulate the structure of the pre-defined topic tree. Nonetheless, this KL divergence is time-consuming, limiting its application in large-scale knowledge scenarios. Similar to [14], Meng et al. [15] adopt a graph adaptive technique to investigate missing links in the topic tree derived from the target corpora, resulting in an improved document representation. The final revised topic tree is derived by merging the original structure of the provided topic tree with the additional structures learned from the current corpus. This integration enhances the overall structure and completeness of the topic tree based on the information obtained from the specific dataset. However, their concentration is limited to a particular type of knowledge, intended for classic KGs structured around entity relationships. Likewise, Pei et al. [16] exclusively address semi-supervised entity alignment by leveraging both labeled and unlabeled data to refine the structural arrangement of KGs. Recently, Mechergui et al. [17] utilized a twice-trained LDA agglomerative approach on a generic Core KG. They incorporate cross-domain CCs' prior knowledge using two techniques for topic-seed terms: seed key knowledge injection and entity masking. However, the task of labeling proves time-consuming, necessitating a significant level of expertise in the field to ensure accurate term categorization. In general, current multilevel topic-seeding approaches face notable challenges, resulting in overlapping topics. Firstly, the issue of ensuring semantic coherence within a multilingual cluster remains unresolved, given that the terms within a cluster may not consistently relate to the same concept, thereby reducing the meaningfulness of the cluster. Secondly, these approaches do not effectively address the relevance of clusters within the targeted knowledge domain. Consequently, a cluster may fall outside the intended scope, representing terms that refer to an irrelevant concept for a predefined domain.

## 3    Methodology

In Fig. 2, we outline the fundamental steps for monolingual and multilingual term clustering across a $BU^3KG$. The architecture's pipeline comprises five steps: 1) Data processing, 2) Initial training, 3) Second training, 4) Third training, and 5) Fourth training. In the subsequent sections, we will offer a comprehensive explanation of each component and the entire training process.

## 3.1  Step 1: Data Preparation

The data preparation (Step 1 in Fig. 2) comprises three stages. It involves "term retrieval" and the generation of two input datasets essential for training our model which are the document-term matrix and the CC-seed sets. To be more specific, our approach involves the utilization of original documents from three distinct domains, namely English, French, and Arabic languages. This enables an exploration of the multilingual cross-domain effects. Firstly, In Step 1.1, the unstructured data passes through a set of pre-processing phases, including dependency parsing, text segmentation, part-of-speech tagging, and lemmatization of terms. The vocabulary is then formed based on the lemmas derived from these processed terms. Notably, terms with lemmas appearing fewer than three times in the corpus or those listed as 'stop words' are excluded. The outcome terminology constitutes the content of *"a term dataset"*. In Step 1.2, our objective is to construct *"the document-term matrix"*, serving as a quantitative depiction of the relationship between documents and the previously prepared "term dataset." Each entry within this matrix represents the normalized term frequency within a document. This frequency is determined by dividing the number of occurrences of a term in a document by the total number of terms in that document. In Step 1.3, our emphasis is on preparing the CC seed collection, designated as *"CC seeds 1"*, which directly contributes to the quality of topics. Unlike seeded LDA [18] which relies on information gain and necessitates labeling each document with a CC, we propose a unique configuration for CC seed collection within a topic. Specifically, we generate "the CC seeds 1" by exploring the DBpedia knowledge, encompassing hyponyms and synonyms associated with CCs. We assume that all terms linked to a CC will be influenced by its synonyms and hyponyms, thereby ensuring a more nuanced representation of the concept. Subsequently, the retrieved terms experience processing, with each term being augmented by appending an underscore. Finally, domain specialists validate these terms, producing *"the authenticated CC seed set"* to guarantee the authenticity and relevance of the employed terminology.

## 3.2  Step 2: Initial Training

The initial model training procedure is divided into three sub-stages: 1) Model training, 2) document-centered topic retrieval, and 3) topic-centered term retrieval. Firstly, the goal of Step 2.1 is to initiate model training. To seamlessly integrate existing knowledge and prevent overfitting, a practical approach involves implementing a sampling procedure from 'the authenticated CC-seed set'. This method aims to achieve a balance and avoid excessive adaptation to specific training data. We implement a priori topic labeling by establishing a direct correspondence between the number of topics and the authenticated CC-seed set. This labeling approach ensures that each topic is explicitly associated with a CC, guided by its corresponding CC seed collection throughout the training process. The goal is to achieve more meaningful, non-overlapping, and high-quality clusters. The algorithm utilizes a two-phase strategy, commencing with Gibbs sampling on the document-term matrix. It progresses from a phase without prior knowledge to a phase where the authenticated CC seeds are incorporated. It continuously adjusts the topic distribution within the matrix following each iteration, producing *a sequence of topic distributions* specific to a given topic. These distributions illustrate how topics are distributed across documents and the distribution of terms within each topic. Next, our focus turns to analyzing the output of this initial model training. More specifically, in Step 2.2, referred to as 'The Document-centered Topic Retrieval,' we examine the topic-document distributions. This step involves extracting all dominant topics from document-topic distributions, sorting them by their significance within the document, and composing a set of *"documents' primary topics"* with their highest scores. Subsequently, the procedure of Step 2.3 labeled 'topic-centered term retrieval', initiates by systematically examining each term within the distributions of topic terms. For each term, the algorithm evaluates its significance and verifies its alignment with one of the top topics identified through "the documents' primary topics" and the information outlined in the topic-term distributions. If a term's relevance exceeds the threshold (T_aggregated = 0.1) and it is associated with one of the "documents' primary topics", it will be integrated into an existing cluster. In cases where no appropriate cluster is identified, the algorithm dynamically establishes a new cluster to incorporate the term. Following this, to guarantee the non-overlapping nature of term topics, we systematically iterate through the set of term clusters, comparing each cluster with every other cluster. In instances where a common term appears in multiple clusters, we evaluate its relevance to each respective cluster. If a term is more relevant to one cluster than another, we eliminate it from the less relevant cluster. This iterative process persists until all common terms are assigned to the topics where they hold the highest relevance. However, during this process, an imbalance phenomenon in term cluster sizes becomes evident, with some clusters being several hundred times larger than others.

Therefore, a restriction is imposed on the size of each cluster, limiting it to a maximum of 100 terms. In cases where a cluster surpasses this threshold, we systematically eliminate low-weight terms from the cluster. This approach ensures that each topic maintains a balanced size, preventing disparities in size among clusters. The terms obtained as a result are stored in a dataset referred to as *"topical terms"*. Adhering to the bag-of-words assumption, the utilization of this technique in TM brings numerous benefits. Firstly, a marked reduction in the size of the vocabulary is achieved, resulting in a noteworthy decrease in computational complexity. Additionally, the technique contributes to the enhancement of the statistical importance of hypernym terms, as their frequency experiences a substantial increase during the training phase. The generated topics, along with their associated subsumed terms, constitute *the lowest level of the $BU^3KG$*. They are visually represented as descendants or terminal nodes within the universal upper KG, providing comprehensive coverage of specific concepts embedded within the graph.

### 3.3  Steps 3: Second Training

In Step 3.1 referred to as "model thinning", we enhance the seed terms and fine-tune hyperparameters. We integrate a selection of high-probability CC seeds collection, denoted as *"CC seeds 2"*, to augment the existing set of *"topical terms"*. Additionally, we adjusted the Beta parameter, influencing the distribution of words within topics as indicated in Eq. (1). The adjusted Beta value ($\beta\_adjusted$) reflects the modifications, with $\beta\_base$ serving as the baseline Beta value for word distribution in topics. The weighting factor $\alpha$, set at 0.01, indicates the impact of adjustments to both seed

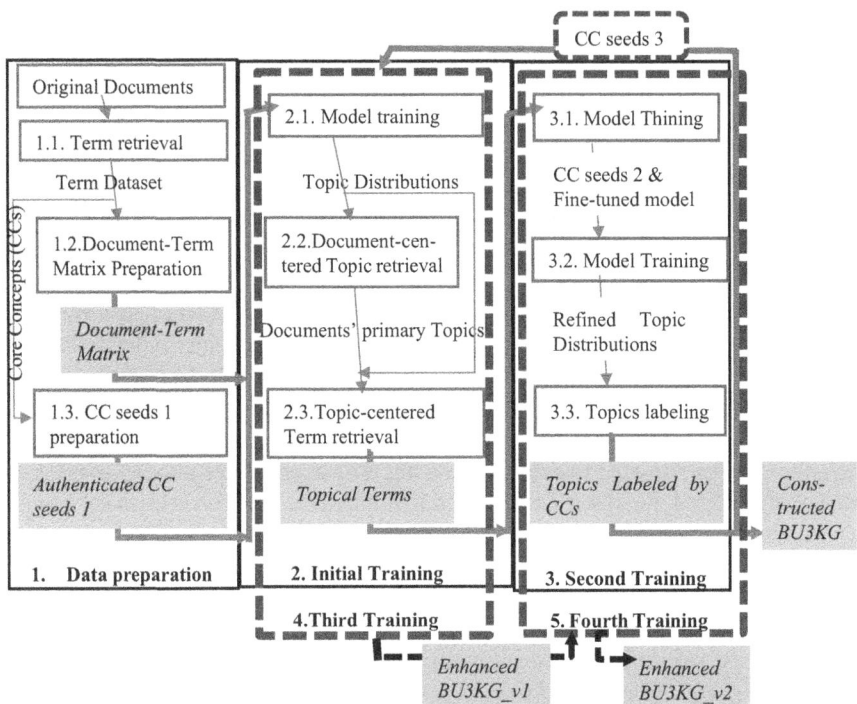

**Fig. 2.** Quadruplicated Multilingual and Multilevel LDA Steps over a Bottom-Up Universal Upper Graph.

terms and hyperparameters on Beta. Furthermore, β_seed serves as a probability distribution vector that incorporates seed terms into the model, predominantly comprising high-probability seed terms. Consequently, this *fine-tuned model setting* encompasses the collective influence of both seed terms and hyperparameter adjustments.

$$\beta\_adjusted = \beta\_base + \alpha * \beta\_seed \qquad (1)$$

Stage 3.2 consists of training the model for the second round to generate high-quality and enhanced topics. More precisely, the algorithm systematically determines the optimal topic associated with each seed term from 'CC seeds 2' by maximizing its probability within the existing topic-term distribution. Indeed, as the construction advances, the graph captures the connections and interdependencies among the various terms. Through successive iterations of assigning topics to terms, the algorithm enhances the topic-term distribution, culminating in the generation of *"refined topic distributions"*. It is a cohesive and structured representation of semantic information, organized in a unified and multilevel manner. In the subsequent Step 3.3, designated as "topics labeling", the goal is to assign the most fitting label to *"refined topic distributions"*. For each topic within these distributions, we initiate variables to monitor the highest probability and the corresponding CC label, facilitating a systematic approach to characterizing the refined topics. We methodically traverse each CC label within the "CC labels collection," calculating the probability of its association with the current topic. Upon finding a calculated probability that surpasses the existing maximum, we update the maximum probability and assign the corresponding CC label. This mechanism ensures the identification of the label with the highest probability for each topic. Following this, we integrate the labeled topics into a newly organized dataset, designated as *"topics labeled by CCs"*. These higher-level topics collectively constitute the upper level of the $BU^3KG$ by representing the higher-level concepts (CCs) of the UKGs within this $BU^3KG$. Furthermore, the refined topical terms within *"topics labeled by CCs"* contribute to the enriched lowest level of the $BU^3KG$, forming specific concepts. Hence, the outcome of the second training is the creation of a *"constructed BU3KG"*, comprising 36 CCs, as expressed in Eq. (2). In the context of Eq. (2), "V" represents a collection of vertices that encompass both abstract concepts and concrete entities. The interconnected "Concepts" rely on essential abstract notions to comprehend complicated relationships. The introduction of "Interconnections" in $BU^3KG$ captures semantic relations and dependencies, thereby improving knowledge acquisition. The "Structured Hierarchy" organizes concepts at multiple levels to facilitate efficient data retrieval. Within 'Highlighted Domains,' emphasis will be placed on three specific areas: ontology, fishery, and medicine. To exemplify, within the UKG related to the fishery field, there are 13 CCs: *Regulation and monitoring (FR), Populations (FP), Zones (FZ), Sustainability (FS), Ecosystem (FE), Ship (FH), Mariner (FA), Technical Characteristic (FN), Biological Characteristic (FB), Gear (FG), Timeframe (FT), Method (FM)*, and *Ship Owner (FO)*. All these CCs are connected to the *Ecosystem (FE)*. In the UKG related to the ontology field, there are 11 CCs: *seMantic (OM), Evaluation (OE), Component (OC), Domain (OD), User (OU), tooL (OL), Step (OS), Resource (OR), Process (OP), Technique (OT)*, and *Ontology (OO)*. For the medicine UKG, there are 12 CCs: *medical sCience (MC), preVention (MV), Epidemiology (ME), Medical Research (MR), Treatment (MT), healthcare proFessionals (MF), Patient (MP), Diagnosis (MD), patHology (MH), pharMacology (MM), phySiology (MS)*, and *Anatomy (MA)*.

$$C_{BU3KG} = \{V, \text{Concepts}, \textit{Interconnections, Structured Hierarchy, Highlighted Domains}\}$$
$$(2)$$

### 3.4   Step 4 and 5: Third and Fourth Training

The third and fourth training phases aim to reinforce the presence of terms associated with CC seeds within the corpus, allowing for an in-depth analysis of its influence on our approach. Thus, in the third training, we enhance *"the topics labeled by CCs"* by introducing a third collection of high-probability CC seeds, labeled as *"CC seeds 3"*. Consequently, these data will function as input for the iterative repetition of the initial training pipeline. This iterative process will augment the lower level of the $BU^3KG$ with additional knowledge, resulting in *"enhanced BU3KG_v1"*. The objective of the fourth training is to continue refining this representation by further augmenting the upper levels of the $BU^3KG$. We will employ the *"enhanced BU3KG_v1"* as input data in the iterative process of the second training phase. The outcome of this Step is an *"enhanced BU3KG_v2"* expressed in Eq. (3), demonstrating how the "constructed BU3KG" enhances to encapsulate newly discovered relations and introduced terms. In Eq. (3), "enhanced_BU3KG_v2" denotes the final enriched BU3KG, "enhanced_BU3KG_v1" signifies the supplemented BU$^3$KG developed during the third training, and the function "enhance_BU3KG(Topics_T4, Terms_T4, New_Terms, New_Cons)" incorporates information from the fourth training. Specifically, Topics_T4 denotes the set of topics produced from the third training, Terms_T4 represents the set of associated terms corresponding to each topic in Topics_T4, New_Terms indicates the set of recently introduced terms, and New_Cons designates the set of newly discovered connections. The additional operator (+) serves as a symbol for merging insights from the fourth training into the third training.

$$enhanced\_BU3KG\_v2 = enhanced\_BU3KG\_v1 + enhance \\ \_BU3KG(Topics\_T4,\ Terms\_T4,\ New\_Terms,\ New\_Cons) \tag{3}$$

## 4 Experimentation

In this section, we perform both qualitative and quantitative evaluations to illustrate the effectiveness of our model. The qualitative analysis encompasses two series of experiments, with the initial set comprising seven experiments and the subsequent set consisting of three. Concurrently, the quantitative analysis involves a third set, which encompasses twenty-eight distinct experiments.

### 4.1 Experimentation Settings

**Datasets.** We have four datasets: **FISH, ONTO, MED**, and HOM, extracted respectively from the **FISH**eries, **ONTO**logy, and **MED**icine domains. The merging of these three datasets, referred to as 'FISH-ONTO-MED,' results in a new dataset called **HOM**. These datasets include academic scientific publications available online, along with articles sourced from Wikipedia. They comprise numerous seed terms aligned with our objectives and were annotated with CCs by domain experts. Each of these datasets exhibits an even distribution of documents across three languages (English, French, and Arabic). The FISH corpus comprises 900 papers, the ONTO corpus includes 600 papers, and the MED corpus consists of 300 papers. The HOM corpus comprises 1,800 documents. The HOM corpus distinguishes itself as having the most substantial volume of textual information among the various corpora, featuring 52,200 sentences and a total of 783,000 terms. We highlight that 80% of our datasets will be dedicated to the training phase. Conversely, the remaining 20% of our datasets will be specifically assigned for testing purposes.

**Gold Standard.** The gold standard is a vocabulary baseline that serves as the optimal term cluster judge since it reflects the outcome that an ideal algorithm should attain. Domain experts create a gold standard, consisting of a carefully reviewed list of terms, for each corpus through manual efforts. Their focus is on utilizing structural methodologies rooted in the internal structure of terms for classification. Professionals guarantee that each processed term is semantically annotated with a CC concept. We consider the CC label *"Other" (O)* to denote terms that do not have a direct connection or relevance to our specified domains. The experts individually annotated the topics and subsequently merged their findings. In cases where differences of opinion arose among the experts, discussions were undertaken to harmonize their perspectives and arrive at a consensus. Table 1 presents a quantitative evaluation of the final gold standards, organized according to the CC classes characterizing the $BU^3KG$. The CC annotation process involves categorizing terms into three distinct groups: *meaningful* (i.e., terms found in the gold standard), *meaningless* (i.e., terms absent from the gold standard), and the combination of both, denoted as *"full terms"*. In the FISH corpus, *13 meaningful CCs* contribute to a total of 349,669 meaningful terms. The ONTO corpus has *11 meaningful CCs*, resulting in 230,024 meaningful terms. The MED dataset comprises *12 meaningful CCs*, contributing to 113,292 meaningful terms. The HOM corpus, with *36 meaningful CCs* and 90,015 meaningless terms, totals 783,000 full terms.

**Seed Terms.** The formulation of seed terms involved a collaborative effort by three groups of university professors, each specializing in our designated domains. These academic experts identified and selected the most frequently utilized terms for each CC, transforming them into a standardized format for inclusion in the final corpus. As depicted in Fig. 1, each set of *"CC seeds 1," "CC seeds 2," and "CC seeds 3"* correspond to 100, 200, and 300 seed terms for each of the CC, respectively. Specifically, for the HOM corpus with 36 CCs, these sets comprise 3,600, 7,200, and 10,800. In the FISH corpus with 13 CCs, the counts are 1,300, 2,600, and 3,900. The ONTO corpus, consisting of 11 CCs, has counts of 1,100, 2,200, and 3,300, while the MED corpus, comprising 12 CCs, has counts of 1,200, 2,400, and 3,600 seed terms in each respective set.

**Baselines.** To showcase the effectiveness of our approach and ensure a fair comparison, we evaluate it against four categories of multilevel baselines: *1) monolingual TM, 2) multilingual TM, 3) monolingual KGs*, and *4) multilingual KGs*. In the first set: (1) *JoSH* [2] is a multilevel topic mining method that leverages knowledge by utilizing category hierarchy as side information. It employs the embedding TM to learn the spherical tree and text embedding. (2) *CC-Seeded LDA* [13] proposes a dual approach involving the

**Table 1.** Labeling CCs and Quantifying Terms within the Final Gold Standards.

| Datasets | CC labeling | | Terms Counting | |
|---|---|---|---|---|
| | Meaningful | meaningless | meaningful | full |
| FISH | FB (52,266); FP (46,531); FR (43,376); FM (33,084); FZ (30,854); FG (28,260); FC (25,412); FS (22,099); FE (18,945); FT (16,601); FH (15,479); FA (12,794); FO (3,968) | O (41,831) | 349,669 | 391,500 |
| ONTO | OC (42,061); OT (31,597); OR (27,364); OS (25,988); OP (22,127); OL (20,240); OM (17,306); OU (15,192); OO (12,608); OD (10,708); OE (4,833) | O (30,976) | 230,024 | 261,000 |

(*continued*)

**Table 1.** (*continued*)

| Datasets | CC labeling | | Terms Counting | |
|---|---|---|---|---|
| | Meaningful | meaningless | meaningful | full |
| MED | MA (19,606); MM (17,848); MS (15,978); MD (11,708); MH (10,509); MR (8,422); MF (8,039); MP (5,611); MT (6,998); ME (3,273); MV (2,904); MC (2,396) | O (17,208) | 113,292 | 130,500 |
| HOM | FB (52,266); FP (46,531); FR (43,376); FM (33,084); FZ (30,854); FG (28,260); FC (25,412); FS (22,099); FE (18,945); FT (16,601); FH (15,479); FA (12,794); FO (3,968); OC (42,061); OT (31,597); OR (27,364); OS (25,988); OP (22,127); OL (20,240); OM (17,306); OU (15,192); OO (12,608); OD (10,708); OE (4,833); MA (19,606); MM (17,848); MS (15,978); MD (11,708); MH (10,509); MR (8,422); MF (8,039); MP (5,611); MT (6,998); ME (3,273); MV (2,904); MC (2,396) | O (90,015) | 692,985 | 783,000 |

utilization of CCs from a domain ontology as prior knowledge for learning domain-specific clusters and adjusting term clustering with seed knowledge-based LDA models to incorporate these CCs. As for the second set, *LDA-BERT* [7] is applied for TM of Indonesian customer complaints. This involves generating TM from embedded data, where LDA assigns probabilistic topic assignment vectors, and BERT handles sentence embedding. In the third set: (1) *TopicKGA* [15] is a graph-generative approach designed to uncover interpretable taxonomies by simultaneously capturing documents and prior knowledge. It adopts a graph adaptive technique to investigate missing links in the topic tree derived from the target corpora. (2) $T^2AggLDA\_GSCKG$ [17] is a cross-domain approach. It employs a Twice-Trained Agglomerative framework utilizing LDA over a Generic Semantic Core KG. The method incorporates cross-domain CCs' prior knowledge using two techniques for topic-seed terms: seed key knowledge injection and entity masking. In the subsequent set, *TAGALOG* [9] introduces a personalized hashtag recommendation system for multilingual content and low-resource Indic languages, utilizing graph-based deep neural networks and knowledge graphs. In all cases of baselines, we utilize their official codes and default settings retrieved from their respective release repositories.

**Evaluation Protocol.** We employed a metric combination approach that integrates weighted metrics to evaluate the outcomes of our model. In the qualitative assessment, we systematically examine the graphical representation of generated topics to discover connections and identify overlapping themes. Additionally, we analyze the 25 most probable terms associated with each topic, arranged in descending order based on their topic-specific collapsed probabilities. The objective of this evaluation is to assess the significance of the acquired relationships between topics and terms. For the quantitative assessment, we evaluate the similarity between a single topic or multiple topics and CC classes, relying on expert annotations available in the gold standard. We use precision as a metric to measure the accuracy of term associations. Precision serves to express the ratio of accurately identified terms to the total number of terms identified by a system. Essentially, precision quantifies the system's capability to accurately identify items, disregarding instances where it may have missed correct elements. Formally, precision is defined as:

$$Precision = \frac{Quantity\ of\ terms\ correctly\ recognized\ /\ classified}{Quantity\ of\ terms\ recognized\ /\ classified} \tag{4}$$

**Model Settings.** We integrate the collapsed Gibbs sampling algorithm into our model to compute the topic-term distribution, thereby enhancing the generation of topics through a bag-of-words representation. To ensure fair comparisons with baseline methods, we establish consistent hyperparameters for all approaches as outlined below: (1) Fixing the iteration number parameter at 50, which denotes the frequency of algorithmic iterations through the entire dataset during the training process. (2) Assigning $\beta$ a value of 0.01 to investigate a term's association with a topic with a 1% weight. (3) Setting $\alpha$ to 1/k, where $\alpha$ influences the number of topics present in a document. It is important to highlight that the optimal selections for both $\alpha$ and K exhibit a robust correlation, as evidenced in reference [19]. (4) Determining the quantity of LDA topics (K) to be recognized within a dataset. We expect a term cluster to encapsulate a specific and detailed concept rather than a mixed one. Therefore, the determination of the value of K was guided by the CC count in each UKG. More precisely, in qualitative sets of experiments, the topic allocation is specified as K = 37 because our focus is on the HOM dataset. However, in the quantitative set, the designated values of k are established as follows: k = 13 for experiments 1, 5, 9, 13, 17, 21, 25; k = 11 for experiments 2, 6, 10, 14, 18, 22, 26; k = 12 for experiments 3, 7, 11, 15, 19, 23, 27; and k = 37 for experiments 4, 8, 12, 16, 20, 24, 28. (5) Given that LDA models involve a degree of randomization, running the algorithm multiple times enhances result stability and reliability. Therefore, for all methods, we conduct 5 runs of the algorithms.

### 4.2  Experimental Results and Discussion

**Qualitative Analysis of 4Multi²LDA.** Table 2 presents the first series of experiments, showcasing the varied compositions of cluster (topic) number 22 generated by both baseline models and our proposed model. Our analysis specifically emphasizes the 25 most frequently occurring terms within each cluster. Our optimal scenario involves identifying a single clear CC within each cluster. Conversely, the least favorable situation arises when a topic comprises overlapping CCs.

Our approach produces 37 clusters that faithfully capture the 36 CCs linked to the HOM corpus along with the "Others CC," resulting in an exceptional purity rate of 99.99%. As an example, in Experiment 7 of Table 2, the resulting cluster specifically corresponds to the "FM" CC. Thanks to our approach, we not only eliminated the occurrence of overlapping CC topics but also achieved the highest level of expert annotation. Moreover, the extracted term clusters are not only purer but also maintain a balance between compactness and separateness. More precisely, they are marked by strong semantic coherence and conceptual consistency, closely mirroring the Gold Standard. This aligns with the ideal clustering scenario. This approach presents other significant advantages: 1) Our model achieves computational efficiency. In contrast to classic multilingual models where inference times might be slow due to the need to process data in multiple languages, our model shares the performance characteristics of monolingual models, resulting in faster processing times. 2) Considering that LDA algorithms are based on the principle of randomness, multiple launches are typically required to stabilize the results. Notably, our model has consistently produced optimal results from the initial model training. The described approach's success across different conditions and datasets highlights its versatility and robustness, making it a valuable tool in various domains and research contexts. We agree that the success can be attributed to the quadruplication approach which enhances the quality of clustering. Specifically, its application enables a more thorough exploration of the dataset and complex relationships within it by generating multiple instances, thereby increasing the chance of capturing nuanced and intricate topics that might be disregarded in a single clustering attempt. This

**Table 2.** Qualitative Evaluation of Topic 22 Extraction Across Different Models Using the HOM Dataset.

| Models | Experiments | Composition and labeling of generated topics across CC classes | CC Encapsulation | |
|---|---|---|---|---|
| | | | Before | After |
| JOSH | 1 | FB (5.4%); FP (2.7%); FR (2.7%); FM (2.7%); FZ (2.7%); FG (2.7%); FC (2.7%); FS (2.7%); FE (2.7%); FT (2.7%); FH (2.7%); FA (2.7%); FO (2.7%); OC (2.7%); OT (2.7%); OR (2.7%); OS (2.7%); OP (2.7%); OT (2.7%); OM (2.7%); OU (2.7%); OO (2.7%); OD (2.7%); OE (2.7%); MA (2.7%); MM (2.7%); MS (2.7%); MD (2.7%); MH (2.7%); MR (2.7%); MF (2.7%); MP (2.7%); MT (2.7%); MV (2.7%); MC (2.7%); O (2.7%) | 36 | 30 |
| CC-Seeded LDA | 2 | FB (8.31%); FP (2.7%); FR (2.7%); FM (2.7%); FZ (5.4%); FG (2.7%); FC (2.7%); FS (2.7%); FE (2.7%); FT (2.7%); FH (2.7%); FA (2.7%); FO (2.7%); OC (2.7%); OT (2.7%); OR (2.7%); OS (2.7%); OP (2.7%); OT (2.7%); OM (2.7%); OU (2.7%); OO (2.7%); OD (2.7%); OE (2.7%); MA (2.7%); MM (2.7%); MS (2.7%); MD (2.7%); MR (2.7%); MP (2.7%); MT (2.7%); MC (%); O (5.4%) | 33 | 28 |

*(continued)*

**Table 2.** (*continued*)

| Models | Experiments | Composition and labeling of generated topics across CC classes | CC Encapsulation | |
| --- | --- | --- | --- | --- |
| | | | Before | After |
| LDA-BERT | 3 | FB (5.4%); FP (5.44%); FR (2.7%); FM (2.7%); FZ (2.7%); FG (2.7%); FC (2.7%); FS (2.7%); FE (2.7%); FT (2.7%); FH (2.7%); FA (2.7%); FO (2.7%); OC (2.7%); OT (2.7%); OR (2.7%); OS (2.7%); OP (2.7%); OT (2.7%); OM (2.7%); OU (2.7%); OO (2.7%); OD (2.7%); OE (2.7%); MA (2.7%); MM (2.7%); MS (2.7%); MD (2.7%); MH (2.7%); MR (2.7%); MF (2.7%); MT (2.7%); ME (2.7%); MV (2.7%); O (2.7%) | 35 | 30 |
| Topic KGA | 4 | FB (2.7%); FP (5.4%); FR (2.7%); FM (5.4%); FZ (2.7%); FG (2.7%); FC (2.7%); FS (2.7%); FE (2.7%); FT (2.7%); FH (2.7%); FA (2.7%); FO (2.7%); OC (2.7%); OT (2.7%); OR (2.7%); OS (2.7%); OP (2.7%); OT (2.7%); OM (2.7%); OU (2.7%); OO (2.7%); OD (2.7%); OE (2.7%); MA (2.7%); MM (2.7%); MS (2.7%); MD (2.7%); MH (2.7%); MR (2.7%); MF (2.7%); MP (2.7%); MT (2.7%); ME (2.7%); O (2.7%) | 35 | 29 |
| $T^2$AGGLDA_GSCKG | 5 | FB (22.16%); FP (8.31%); FR (13.85%); FM (5.54%); FZ (13.85%); FS (5.54%); FH (11.08%); FA (2.7%); OC (2.7%); O (13.85%) | 10 | 7 |
| TAGALOG | 6 | FB (2.7%); FP (2.7%); FR (2.7%); FM (2.7%); FZ (2.7%); FG (2.7%); FC (2.7%); FS (2.7%); FE (2.7%); FT (2.7%); FH (2.7%); FA (2.7%); FO (2.7%); OC (2.7%); OT (2.7%); OR (2.7%); OS (2.7%); OP (2.7%); OT (2.7%); OM (2.7%); OU (2.7%); OO (2.7%); OD (2.7%); OE (2.7%); MA (2.7%); MM (2.7%); MS (2.7%); MD (2.7%); MH (2.7%); MR (2.7%); MF (2.7%); MP (2.7%); MT (2.7%); ME (2.7%); MV (2.7%); MC (2.7%); O (2.7%) | 37 | 32 |
| 4Multi$^2$LDA | 7 | FM (100%) | – | – |

comprehensive exploration ensures a richer representation of the underlying semantic structure.

**Qualitative Analysis of 1st and 2nd versus 3rd and 4th Training.** We present the visual representation of the learned topic hierarchies of 1st and 2nd model training sessions in comparison to 3rd and 4th training sessions on the HOM dataset in Fig. 3(a–b), respectively. Indeed, Fig. 3 illustrates the outcomes of the second series of experiments. In each topic box, the top bar displays the pre-specified concept name, while the bottom content provides a list of associated terms. Topics within dashed boxes (free topics) were incorporated by 3rd and 4th training sessions based on the target corpus. The width of the arrows signifies the strength of connections, while distinct colors indicate various layers. Our observations produce significant insights: (1) The extracted terms demonstrate a remarkable degree of relevance to their respective topics, providing a clear and comprehensive representation of the underlying concepts. (2) The incorporation of CC knowledge guides the formation of highly interpretable connections between topics across adjacent layers, resulting in intuitively understandable topic taxonomies. (3) Notably, in an evaluation of the adaptability of our 3rd and 4th training approaches, we introduced several free topics (depicted as dashed boxes in Fig. 3(b)) to each layer of the existing topic tree. These additional topics, not initially considered in the predefined knowledge, serve as a test of the algorithm's capacity to extract concepts from the target corpus. Our algorithm convincingly showcases its ability to identify and incorporate these missing concepts, leading to a refined prior graph that aligns with the specific characteristics of the current corpus.

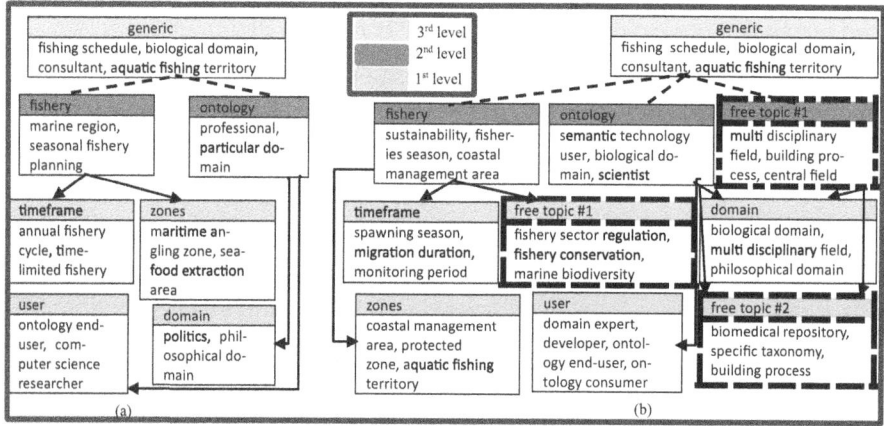

**Fig. 3.** The multilevel arrangement of topics from 1st, 2nd, 3rd, and 4th training sessions of 4Multi²LDA on the HOM dataset.

**Qualitative Comparison.** We aim to provide a clear and transparent qualitative comparison between the baseline models and our proposed model based on the outcomes of the initial series of experiments. Experiments 1 to 6 detailed in Table 2 provide insights into the outcomes of baseline methodologies, showcasing clusters marked by a lack of robust semantic coherence. We highlight the observation that the baseline topics do not exhibit apparent connections to any CC a priori. The range of overlapping CC classes within clusters extends across from 10 to 37 CCs. This results in a deficiency of evident conceptual cohesion. For instance, consider topics characterizing TAGALOG, JOSH, LDA-BERT, Topic KGA, and CC-Seeded LDA, each containing 37, 36, 35, 35, and 33 CCs, respectively. This lack of direct association with specific CCs is unfavorable for our intended purpose. It is crucial to assign a low precision value close to zero to these clusters, given the limited semantic relevance associated with these diverse topics. Nevertheless, a notable improvement is observed in the consecutive topics. For instance, the approach of [17] yielded a reduced set of 10 CCs. This particular cluster exhibits precision results that are evaluated more satisfactorily for our research, attaining a value of

0.7. It is crucial to emphasize that, while multilingual methods may surpass monolingual methods in specific scenarios, this does not necessarily indicate superior quality in their cluster partitions. Examine the case of LDA-BERT in Experiment 3, highlighting its superior performance when compared to JOSH in Experiment 1. In contrast, the quantitative analysis of the HOM dataset in Table 3 shows that JOSH in Experiment 4 (0.1394) demonstrates superior performance compared to LDA-BERT (0.1205) in Experiment 12. Moreover, a substantial influence from terms associated with the fishery and ontology fields is evident in the majority of clusters produced by all baseline methods. The most common dominant and overlapping set of CCs across all baseline models include *"FB," "FP," "FR," "OC," "FM," "OT"*, and *"FZ"*, leading to the omission of certain concepts and thus resulting in partitions of poor quality. In light of the aforementioned observations, we ensure that the clustering generated by baseline models deviates significantly from our objective, given the absence of an outstanding precision score for the most frequently occurring terms characterizing at least some topics. Furthermore, certain topics share common semantics or CC classes. To address this, we opted to employ encapsulation, grouping two or more classes into a single encapsulated class. This decision was informed by the qualitative study presented in Table 1, which illustrates the co-occurrence of these CCs in the corpus. Consequently, by consolidating topics with shared semantics, we have witnessed a notable enhancement in the quality of clustering generated by these baseline models. For instance, the topic from reference [9] initially comprised 37 overlapping CCs, but through the encapsulation, it was reduced to only 32 CCs. Similarly, the topics related to [2, 7, 13, 15], and [17] have experienced reductions in overlapping CCs to 30, 30, 29, 28, and 7, respectively. Nonetheless, our method doesn't necessitate encapsulation. It outperforms all baseline models, surpassing even the robust approach of [17]. When comparing the implementation results of experiments 1 to 6 with experiment 7, it becomes evident that significant differences exist, with our methodology yielding the most favorable results.

**Quantitative Analysis.** Table 3 presents the third series of experiments, with the two highest values for each column highlighted in bold. Each generated topic from Experiments 25 to 28 encapsulates the distinct vocabulary of a single CC class. This leads to a clear interpretation of these topics that aligns with our expectations. As a result, relevant precision values were attained. To illustrate, 4Multi$^2$LDA achieved 0.9896, 0.9853, 0.9611, and 0.9962 for experiments 25, 26, 27, and 28, respectively. We emphasize a progressive improvement in our model's outcomes, starting from the smallest dataset (MED) with a score of 0.9611 to the largest HOM dataset, where it attains a notable score of 0.9962. This pattern indicates that our semi-supervised approach prioritizes an expanded multilingual lexicon, thereby strengthening the occurrences of terms characterizing CCs. We can infer that there is a correlation between the corpus size and the performance of term clustering. Building upon Table 1, we emphasize a disparity in the range of occurrence differences among the CCs characterizing *meaningful terms* in all datasets. Illustratively, the variance observed in the HOM dataset falls within the predetermined range [580; 49,870]. In a deeper exploration, the analysis of CCs reveals that the extent of "FB" surpasses "OE" by a factor of 11, "ME" by 16, and "MC" by 22. This observation can be attributed to the fact that the meaningful vocabulary in FISH

(349,669) surpasses that in ONTO (230,024) by a factor of 1.52 and exceeds the vocabulary in MED (113,292) by a factor of 3.08. This limitation contributes to an *'inequitable data distribution'*, thereby influencing the quality of the derived topics. However, in 4Multi$^2$LDA, the bridging of this gap has been achieved through the implementation of *CC seeds 1, 2,* and *3.* The performance demonstrates improvement in tandem with the increasing number of CC seeds, guiding the algorithm toward alignment with our goals. Regarding the baseline models, there is also a noticeable absence of significant improvement from Experiment 1 to 24. Without CC encapsulation, the range of scores falls between 0.0277 and 0.4197. Upon the incorporation of CC encapsulation, their scores increase, ranging between 0.0318 and 0.4426. In both scenarios, we agree that these values are considerably below our targeted objectives, especially when addressing the MED dataset. As an illustration, within Experiment 3, the unsupervised JOSH attains a score of 0.0277. Following the CC encapsulation, this value increases to 0.0318, designating it as the least favorable scenario. In parallel, the unsupervised algorithms, including [13, 15], and [9], achieve the following respective scores 0.1732, 0.0714, and 0.0352. In the semi-supervised category, LDA-BERT achieves a significantly very low score of 0.0551, whereas T$^2$AGGLDA_GSCKG outperforms with a value of 0.2531. On the other hand, the baseline results also demonstrate the quantitative superior performance of monolingual models over the multilingual ones. For instance, when trained on HOM, the monolingual [17], algorithm achieved a score of 0.4197, contrasting with the 0.1178 and 0.1205 achieved by multilingual [9], and [7], respectively. Notably, apart from our approach, multilingual models may face challenges in handling languages with shared structures or vocabulary, resulting in potential interference and confusion. In contrast, monolingual models effectively may avoid such issues, leading to more accurate TM. In general, we observe some enhancements when examining the baseline findings from MED, ONTO, FISH, to HOM datasets. However, the overall outcomes emphasize the limitations of baseline algorithms in handling multilingual datasets. They still do not meet our satisfaction criteria, except for the approach of [17], which consistently produces results closely aligned with our requirements. Consequently, it is considered the primary high-performing competitor to our approach. Even so, across datasets of varying sizes, the results of our methodology demonstrate its superior performance compared to all other approaches.

The findings emphasize the potential of our methodology to fulfill various objectives. This includes autonomously identifying and extracting meaningful terms to structure and enhance a multilingual BU$^3$KG. Added to that, it ensures annotation expense prevention, KG integration, interoperability, scalability, reasoning, reusability, and sharing. Furthermore, it demonstrates domain-agnosticism, and consistency, thereby collectively enhancing the efficacy of KG representation and engineering in diverse domains. However, a thorough qualitative examination of the outcomes reveals certain areas that could benefit from enhancement. Moreover, the time-intensive nature of annotating work is influenced by the reliance on experts' prior knowledge. To illustrate, with an average annotation time of 36 min for 200 terms, extending to 1000 terms would necessitate 180 min (3 h), and for 5000 terms (the vocabulary size in a single corpus), a single annotator would require 900 min (15 h).

**Table 3.** Quantitative Evaluation of Various Models on Four Datasets: Precision Score Analysis.

| Models | Experiments | Before/After encapsulation of CCs in multilingual datasets | | | | | | | |
|---|---|---|---|---|---|---|---|---|---|
| | | FISH | | ONTO | | MED | | HOM | |
| | | Before | After | Before | After | Before | After | Before | After |
| JOSH | 1-2-3-4 | 0.1015 | 0.1081 | 0.0836 | 0.0930 | 0.0277 | 0.0318 | 0.1394 | 0.1909 |
| CC-Seeded LDA | 5-6-7-8 | 0.2159 | 0.2412 | 0.1907 | 0.2242 | 0.1732 | 0.1841 | 0.2423 | 0.2861 |
| LDA-BERT | 9-10-11-12 | 0.1097 | 0.1174 | 0.0966 | 0.1127 | 0.0551 | 0.1065 | 0.1205 | 0.1730 |
| Topic KGA | 13-14-15-16 | 0.1746 | 0.2218 | 0.1332 | 0.1882 | 0.0714 | 0.1386 | 0.2270 | 0.2938 |
| $T^2$AGGLDA_GSCKG | 17-18-19-20 | **0.3508** | **0.3977** | **0.2831** | **0.3046** | **0.2531** | **0.2889** | **0.4197** | **0.4426** |
| TAGALOG | 21-22-23-24 | 0.1062 | 0.1145 | 0.0898 | 0.1097 | 0.0352 | 0.0647 | 0.1178 | 0.1403 |
| 4Multi$^2$LDA | 25-26-27-28 | **0.9896** | – | **0.9853** | – | **0.9611** | – | **0.9962** | – |

# 5 Conclusion and Future Work

In this manuscript, we introduce the Quadruplicated Multilingual and Multilevel LDA, a novel semi-supervised methodology designed for the structuration and augmentation of BU$^3$KG. This generic methodology signifies an advanced model founded on the principles of topic seeding and multilevel TM. Its overarching focus is directed toward the generation of concepts (topics), each encapsulating terms linked to a distinct CC following a predefined hierarchy. We address challenges in current clustering-based methods, specifically overcoming difficulties in representing unstructured multilevel and monolingual/multilingual data across clusters, enhancing semantic coherence within these clusters, and labeling them. The experimental findings affirm that our proposition outperforms the baseline models on four datasets characterized by imbalanced CC class distribution and an appropriate allocation of clusters for each CC. The enhanced performance of our model is attributed to the quadruplicated approach, coupled with the inclusion of CC seeds 1, 2, and 3. In future research, a noteworthy focus should be on the removal of meaningless terms due to their direct influence on the dynamics of LDA models and their contribution to the reduction of False Positives. Furthermore, employing our methodology on an imbalanced dataset without prior knowledge of the size of CC classes poses a challenging task. In such cases, optimization algorithms can be employed to select the appropriate number of topics for each CC.

**Disclosure of Interests.** The authors have no competing interests to declare that are relevant to the content of this article.

# References

1. Blei, D.M.: Probabilistic topic models. Commun ACM **4**, 77–84 (2012)
2. Meng, Y., Zhang, Y., Huang, J., Zhang, Y., Zhang, C., Han, J.: Hierarchical topic mining via joint spherical tree and text embedding. In: Proceedings of the 26th ACM SIGKDD Conference on Knowledge, p. 10, 23–27 August 2020

3. Duan, Z., et al.: Sawtooth factorial topic embeddings guided gamma belief network. In: Proceedings of the 38th International Conference on Machine Learning, PMLR, vol. 139, pp. 2903–2913 (2021)
4. Jordan Boyd-Graber, D.B.: Multilingual topic models for unaligned text. In: UAI (2009)
5. Yang, W., Boyd-Graber, J., Resnik, P.: A multilingual topic model for learning weighted topic links across corpora with low comparability. In: Conference on Empirical Methods in Natural Language Processing and the 9th International Joint Conference on Natural Language Processing, Hong Kong, China, pp. 1243–1248, 7 November 2019
6. Xie, Q., Zhang, X., Ding, Y., Song, M.: Monolingual and multilingual topic analysis using LDA and BERT embeddings. J. Informetrics **14**, 101055 (2020)
7. Mutiara Auliya Khadija, W.N.: Enhancing Indonesian customer complaint analysis: LDA topic modelling with BERT embedding. SINERGI **28**, 153–162 (2024)
8. Maanicshah, K., Manouchehri, N., Amayri, M., Bouguila, N.: Novel topic models for parallel topics extraction from multilingual text. In: Nguyen, N.T., et al. (eds.) Intelligent Information and Database Systems. ACIIDS 2023. LNCS, vol. 13996, pp. 297–309. Springer, Cham (2023). https://doi.org/10.1007/978-981-99-5837-5_25
9. Bansal, S., Gowda, K., Kumar, N.: Multilingual personalized hashtag recommendation for low resource Indic languages using graph-based deep neural network. Expert Syst. Appl. **236**, 121188 (2024)
10. Lin, Y., Gao, X., Chu, X., Wang, Y., Zhao, J., Chen, C.: Enhancing neural topic model with multi-level supervisions from seed words. In: Findings of the Association for Computational Linguistics, pp. 13361–13377, ACL 2023 (2023)
11. Ben Mustapha, N., Aufaure, M.A., Baazaoui Zghal, H., Ben Ghezala, H.: Modular ontological warehouse for adaptative information search. In: Abelló, A., Bellatreche, L., Benatallah, B. (eds.) Model and Data Engineering. MEDI 2012. LNCS, vol. 7602, pp. 79–90. Springer, Cham (2012). https://doi.org/10.1007/978-3-642-33609-6_9
12. Xu, Z., Harzallah, M., Guillet, F., Ichise, R.: Modular ontology learning with topic modelling over core ontology. Procedia Comput. Sci. **159**, 562–571 (2019)
13. Huang, H., Harzallah, M., Guillet, F., Xu, Z.: Core-concept-seeded LDA for ontology learning. Procedia Comput. Sci. **192**, 222–231 (2021). 25th International Conference on Knowledge-Based and Intelligent Information & Engineering
14. Duan, Z., et al.: TopicNet: semantic graph-guided topic discovery. In: 35th Conference on Neural Information Processing Systems (NeurIPS 2021), Sydney, Australia, 27 October 2021
15. Wang, D., et al.: Knowledge-aware Bayesian deep topic model. In: 36th NeurIPS 2022, New Orleans, LA, USA, 20 September 2022
16. Pei, S., Yu, L., Hoehndorf, R., Zhang, X.: Semi-supervised entity alignment via knowledge graph embedding with awareness of degree difference. In: WWW 2019: The World Wide Web Conference, pp. 3130–3136, 13 May 2019
17. Mechergui, A., Ben Abdessalem Karaa, W., Zghal, S.: Twice-trained agglomerative clustering approach using topic modeling over generic semantic core knowledge graph (2023)
18. Jagarlamudi, J., Daumé III, H., Udupa, R.: Incorporating lexical priors into topic models. In: Proceedings of the 13th Conference of the European Chapter of the Association for Computational Linguistics, pp. 204–213 (2012)
19. Griffiths, T.L., Steyvers, M.: Finding scientific topics. Proc. Natl. Acad. Sci. **101**, 5228–5235 (2004)

# Question Answering System to Answer Questions About Technical Documentation

Szymon Olewniczak$^{(\boxtimes)}$ ⓘ, Michał Maciszka ⓘ, Kamil Paluszewski ⓘ,
Grzegorz Pozorski ⓘ, Wojciech Rosenthal ⓘ, and Łukasz Zaleski ⓘ

Department of Computer Architecture, Faculty of Electronics, Telecommunications
and Informatics, Gdańsk University of Technology, Gdańsk, Poland
szyolewn@pg.edu.pl,
{s180522,s180194,s180169,s180458,s180390}@student.pg.edu.pl

**Abstract.** This article ventures into the realm of specialized AI systems for question answering, with a specific focus on programming languages, using Rust as the case study. Our research harnesses the capabilities of BERT, a leading model in natural language processing, to explore its effectiveness in interpreting and responding to complex, domain-specific queries. We have developed a novel dataset, derived from Rust's detailed documentation, which surpasses the usual input size for language models. This dataset serves as a foundation for evaluating BERT's performance in a domain-specific context, providing a new resource for testing question-answering systems and shedding light on their strengths and limitations in processing specialized technical information. In this paper, we proposed a solution based on retrieval-reader architecture, the fine-tuned RoBERTa model with the usage of the mentioned dataset, and conducted typical tests for said problem. It is shown, that domain-specific question-answering remains a challenging problem.

**Keywords:** Question Answering · Information Retrieval · AI · Chatbot · Natural Language Processing · Documentation

## 1 Introduction

The integration of machine learning, and more recently deep learning, has revolutionized QA systems. Models like BERT [5] and GPT-3 [2], employing neural networks trained on vast datasets, have significantly enhanced the accuracy and contextual understanding of these systems.

QA systems can be broadly categorized into factoid, descriptive, and conversational types, each with unique capabilities and applications. Factoid systems answer straightforward factual questions (e.g. [3,15]). Descriptive systems provide detailed responses, often synthesizing information from multiple sources (e.g. [9]). Furthermore, conversational systems, like ChatGPT, maintain dialogue context over multiple interactions (e.g. [12,13]).

© The Author(s), under exclusive license to Springer Nature Switzerland AG 2024
N.-T. Nguyen et al. (Eds.): ICCCI 2024, CCIS 2165, pp. 193–205, 2024.
https://doi.org/10.1007/978-3-031-70248-8_15

Models like BERT and its derivatives have transformed NLP, offering deep bidirectional representations and fine-tuning capabilities that enable a nuanced understanding of complex texts. Our experiments focus on fine-tuning the RoBERTa-based model to work on a custom-created dataset. The objective of the task is to determine, given an input sequence and a query, whether the answer to the query exists in at least one sentence of the pool of retrieved text parts.

The challenge in this realm lies in transcending the conventional boundaries of QA systems, which traditionally rely on diverse information sources for answer retrieval. In contrast, this study ventures into leveraging a singular, extensive knowledge source - the detailed documentation of a programming language - to probe the efficacy of language models in domain-specific contexts. This approach aligns with recent trends in open-domain QA systems like BERTserini [18], which integrate robust information retrieval techniques with BERT-based models (e.g. RoBERTa - [10]) to extract answers from extensive text corpora.

This article explores this domain, emphasizing the adaptation of advanced language models to interpret and respond to domain-specific queries in programming languages, with Rust as a focal point.

Our exploration contributes to the broader discourse on QA systems' capabilities in handling specialized technical information, a niche yet increasingly relevant area in AI research. It aims to extend the understanding of how sophisticated language models like BERT can be fine-tuned and applied to specific domains, particularly in programming languages, to facilitate efficient information retrieval and accurate response generation.

This paper makes the following contributions: a dataset of over 1000 questions and answers based on Rust Book, fine-tuning of the RoBERTa model with the usage of this dataset, and a system for factoid QA based on retriever-reader architecture created with Haystack library.

## 2    Related Work

Currently, many systems are dealing with question-answering. They use different approaches. One way is Information Retrieval based systems. [15] is an example of this kind of system. It uses the dataset based on the Text Retrieval Conference QA dataset. The TREC dataset was divided into 5500 manually labeled questions in the training set and 500 in the test set with 6 coarse class labels and 50 fine class labels. The solution uses a bidirectional long short-time memory model and recurrent neural network. The results of this solution contain BM25, Single-Layer LSTM, Single-Layer BLSTM, and Three-Layer BLSTM and combine BM25 and Three-Layer BLSTM features. The most successful was the last one. Combined BM25 and Three-Layer BLSM achieved 0.7134 value of MAP metrics and 0.7913 value of MRR. Some question-answering systems do not focus on one specific model and use different models and compare them. [1] uses models such as LSTM, BLSTM, RNN, ELECTRA and BERT. The most effective was the ELECTRA and BERT models based on transformers. The BERT model achieved the following results: Precision = 93.27, Recall = 93.71, F1 Score = 93.49 and ELECTRA model reached following results: Precision = 89.87, Recall = 90.98, F1 Score = 90.42. Both systems show, that sequential models are not the most successful approach and it is worth considering transformer models. [17] is an example of the model using sequential knowledge graph approaches. The advantage of this idea over traditional Information Retrieval approaches is enabling the model to understand the context and query intention. It achieved 0.22 MRR and 0.19 MAP, but those metrics do not show that this solution can find answers for up to 87% of questions.

An area of research connected to QA that recently gained a lot of attention is retrieval-augmented generation (RAG). RAG systems can generally be described as systems that use a generative LLM on top of a QA pipeline. Metrics like faithfulness are used to evaluate such systems [7].

QA models can specialize in different areas. For example [8], although also uses the BERT model, it deals with answering questions about medicine, not technical documentation. On the MEDIQA-2019 dataset it achieved 79.41 accuracy, 90.00 MRR and 84.02 precision. [18] is the system, which architecture of that system is based on Anserini retriever and BERT reader. The dataset used here is large and based on indexed articles on Wikipedia. The usage of chatbot created in this system is not focused on one specific area and can answer any question. The recall metric used here showed good results and achieved more than 0.8 value of recall. [6] also seems worth mentioning here as it shows the practical usage of a QA system. It revolves around a specific domain, in that case, a specific REST API provided by the user. It provides an interesting idea for developing the research in the future so it could create a knowledge base of specific documentation pointed out by the user.

## 3   Dataset

We have created a new dataset for training and evaluating QA models based on the official Rust programming language book (https://doc.rust-lang.org/stable/book/) in SQuAD 2.0 format which is standard in the field. The creation of this dataset has been motivated by the fact that few publicly available datasets are specific to QA for code documentation. By releasing the new dataset, we hope to contribute to this area of research. Our dataset consists of 1068 questions with answers and is available under the following URL: https://github.com/MichalMaciszka/rust_docs_qa/tree/main/annotation.

We have used the Haystack annotation tool (https://docs.haystack.deepset.ai/docs/annotation) to create examples for our dataset. All 105 chapters of the language book have been evenly split between five annotators, who then devised questions based on each chapter's content.

Depending on the question, the answer may vary between one word and multiple sentences in length. Our approach to formulating examples in the dataset is that they should resemble queries made by a programmer who is researching a particular topic. Table 1 shows example training questions with answers.

**Table 1.** Example questions and answers.

| Question | Answer |
|---|---|
| What is the purpose of trait? | their specific purpose is to allow abstraction across common behaviour |
| Should I use 'static' lifetime in error messages? | In such cases, the solution is fixing those problems, not specifying the 'static' lifetime |
| What is the difference between shadowing and mutability? | Shadowing is different from marking a variable as 'mut' because we'll get a compile-time error if we accidentally try to reassign to this variable without using the 'let' keyword |

**Table 2.** Dataset split.

| Subset | Number of examples |
|---|---|
| Train | 854 |
| Validation | 107 |
| Test | 107 |

The dataset has been randomly split into three subsets described in Table 2. The dataset is stored in JSON format, one file per subset. The JSON files are compatible with Haystack processing. Notable fields are:

1. "context"—larger fragment of text. In our dataset it corresponds to a particular chapter from the language book.
2. "qas"—table of questions with answers for the specified context. Each entry of "qas" is an object with "question", "id", "answers" and "is_impossible" fields. Each question in the dataset has exactly one answer, which means that "is_impossible" is set to "false" in all examples in the dataset.
3. "question"—question in textual format.
4. "text"—answer in textual format.
5. "answer_start" - position of the first symbol of the answer in the context text.

## 4    Proposed Solution

The main idea of the proposed solution was inspired by popular lately retriever-reader architecture (e.g. [3]). In this approach, the retriever is used for filtering documents most valuable in the context of a given query. The reader is supposed to choose the best fragments from given documents in the context of the same query. As those elements need a knowledge base to operate on we also had to handle the loading and preprocessing of Rust Book.

The architecture of the proposed solution is presented in Fig. 1. The applied approach is based on two pipelines. Each pipeline consists of nodes that are executed sequentially. The first pipeline handles data preparation, such as loading files, reading them as text and preprocessing them by cleaning ambiguous white spaces and empty lines, but also by splitting large files into smaller documents. The second pipeline is the core of this solution. It is based on said retriever-reader approach. Retriever uses a simple BM25 algorithm [14] to classify documents based on the query given by the user. The reader is a transformer-based node that chooses the most suitable fragment of the provided context for the given query. During the development, we fine-tuned the RoBERTa-base-squad-2 model (https://huggingface.co/deepset/roberta-base-squad2). To create our solution we used the Haystack library delivered by Deepset AI (https://github.com/deepset-ai/haystack).

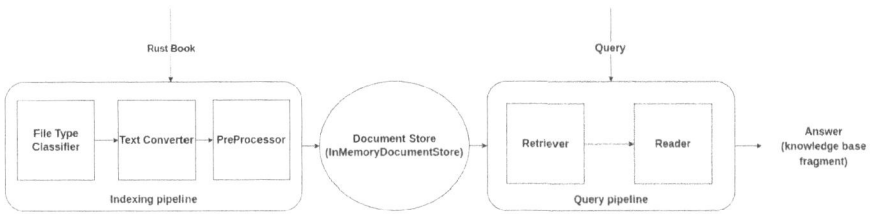

**Fig. 1.** Architecture of proposed solution

## 4.1 Indexing Pipeline

The purpose of the indexing pipeline is to provide data to the query pipeline. It consists of the following nodes:

1. File type classifier - filters files in a given folder based on their extension. In this case, we loaded markdown files.
2. Text converter - converts text files to documents that can be processed. In this case, we used *TextConverter*, even though there is a converter for markdown files because we wanted to keep formatting, as it can provide some information.
3. PreProcessor - cleans data and splits it into sensible units. Thanks to *PreProcessor* class we could easily clean files of empty lines, unnecessary white spaces and split documents that each had maximum $N$ words.
4. Document store - saves the processed data. In this case, we used the simplest *InMemoryDocumentStore* to store raw data on a drive. Additionally, it allows us to use an algorithm from the BM25 family while querying through it.

## 4.2 Query Pipeline

The query pipeline plays a crucial part in our system. We used a two-staged approach. It consists of two nodes: the retriever and the reader. In this pipeline, the given question is processed, and the most suitable documents and parts of the documentation are selected.

**Retriever.** The first stage of a typical query pipeline is information retrieval. Its purpose is to filter the knowledge base so the reader can give a proper answer and context to an output. Retriever, using chosen techniques, such as BM25 or dense embedding vectors of text produced by neural networks, provides a ranking of contexts connected with a given query. We have chosen *BM25Retriever*, which is based on BM25, a classic TF-IDF (term frequency-inverse document frequency) approach as it was giving the best results (as shown in Sect. 6, Table 4). This algorithm takes the query and based on that gives a score for every document in the knowledge base. Top $N$ documents are given as input for the next node.

**Reader.** Reader is a component that required the most attention in our solution. Based on lately popular approaches (e.g. BERTserini - [18]) we decided to fine-tune the BERT-based model with our dataset (described in Chap. 3, in SQuAD 2.0 format). RoBERTa-base model tuned with the SQuAD 2.0 dataset was used as a base of training (mentioned earlier RoBERTa-base-squad-2). We fine-tuned this model for 20 epochs, with a learning rate of 1e-5. As a loss function, we used Kullback-Leiber divergence given with the following formula:

$$L(y_{pred}, y_{true}) = y_{true} \cdot \log(\frac{y_{true}}{y_{pred}}) \tag{1}$$

Next, based on validation metrics, we chose a model from the 11th epoch. As we compared manually this model with those from later epochs by running this program and asking questions from the validation split we observed, that later models tended to shorten the response excessively, the final models even limited it to one character. The model of our choice was then used as a part of the whole pipeline. As an input, it gets documents classified by retriever. As an output, it returns an answer in the form of a fragment selected from a document and context - the whole document.

## 5   Results

As part of the system's quality assessment, we decided to test the system described above on a test split of a dataset we created (Sect. 3). The quality measurement was conducted in two stages. In the first stage, the retriever component is tasked with finding five text fragments within the Rust programming language book for which there is the highest likelihood of finding a matching passage. In the second stage of evaluation, the alignment of the answers extracted by the reader from the previously selected fragments by the retriever is assessed, comparing them with the answers present in the dataset.

### 5.1   Retriever Evaluation

The retrieved fragments are then evaluated using a series of metrics. One of them is Mean Average Precision (MAP) [4] which assesses the effectiveness of a system by calculating the average precision at each position where relevant items are found in the ranked list of results for multiple queries. Another metric used by us is Mean Reciprocal Rank (MRR) which is calculated by taking the average of the reciprocals of the ranks at which the first relevant items are found across multiple queries. MRR emphasizes the importance of retrieving relevant items early in the ranked list, providing a concise measure of the overall effectiveness of a retrieval system. The subsequent metric implemented is Normalized Discounted Cumulative Gain (nDCG) [11]. This metric quantifies the quality of ranked lists by considering both the relevance and position of items. It normalizes the Cumulative Gain by discounting it based on item position, providing a concise measure of retrieval system effectiveness.

In the realm of question-answering systems, precision gauges the accuracy of provided answers, representing the ratio of correct responses to the total claimed by the system. On the other hand, recall measures the system's effectiveness in retrieving and delivering answers to all pertinent questions, expressed as the ratio of correctly answered questions to the total number of relevant questions. In the context of retriever, a correct answer is understood as the existence of at least one sentence of the desired system answer in a pool of retrieved text parts.

### 5.2   Reader Evaluation

The evaluation metrics for sentence retrieval quality are as follows: exact match, which can only take one of two values: 1, when the system's provided answer is the same as in the dataset, or 0 - otherwise. The subsequent metric is the F1 score [19], which in Information Retrieval indicates how much the found answer overlaps with the one existing in the dataset. The more the texts overlap, the higher the value of the metric. Both metrics were measured for the best matching part of text from the retriever as well as for the best answers for each question from the reader based on all five fragments retrieved in the first step.

Table 3 contains the results of Reader tests, which we consider as the performance of the whole system.

**Table 3.** Values of metrics for reader.

| Metric | Value for Top1 | Value for Top5 |
|--------|----------------|----------------|
| Exact match | 0.262 | 0.274 |
| F1 | 0.406 | 0.474 |

In 27% of questions, our Question Answering system returns the same answer as the one in the dataset (assuming that the retriever extracts five contexts). The F1 score stands at 0.474. The results will be described in detail in the following sections.

## 6   Solution Variants

Modular architecture allows us to easily change nodes within our solution. As part of our studies, we decided to compare the performance of two types of Retrievers - the simple *BM25Retriever* described above and a more advanced *EmbeddingRetriever* based on our chosen transformer model. Table 4 provides results of the evaluation of both types of Retriever components of the system.

As can be seen in the results, as we tried to apply *EmbeddingRetriever*, we observed that results didn't improve, but answering took much longer as this method was much more complex. Potentially, however, this node could produce much better results than the used retriever if we fine-tuned it on data specific to our solution. As we didn't want to focus on this part of our system specifically we decided to stay with a simple BM25 algorithm, but the usage of different

**Table 4.** Values of metrics for retriever.

|  | BM25 | Embedding |
|---|---|---|
| MAP | 0.255 | 0.205 |
| MRR | 0.259 | 0.210 |
| NDCG | 0.275 | 0.232 |
| Precision | 0.086 | 0.074 |
| Recall (multi hit) | 0.331 | 0.306 |

retrievers along with choosing the most suitable model can become the beginning of further research. Currently, the retriever component can be recognized as the bottleneck of the system, being a subject of future improvements.

## 7    Qualitative Analysis

Table 5 illustrates an example of context retrieval in full-text documentation by a retriever. The first column presents a sample question for which the retriever successfully located the desired text snippet, while the second column provides an example question for which the retriever failed in this task. The "gold context" row represents the ground truth, indicating the fragment that was the correct answer.

For the first question, context 2 precisely aligns with the desired content. It is noteworthy that each context contained words like "Rc," "reference," and "count," which are characteristic of the question. However, only the second response included the word "get," crucial for understanding the specific aspect of reference count we want to retrieve. Other contexts relate to how this counter behaves when calling other functions, not how to obtain its value.

In the case of the second question, the retriever focused on the keyword 'panic.' Likely, the question did not contain a sufficient number of distinctive keywords for the retriever to adequately narrow down the set of matching fragments.

Table 6 presents several sample questions from the test set, along with the ground truth, the answer returned by the reader based on the fragment identified by the retriever as the best match (along with its F1 score result), and the answer with the highest F1 score among all selected retriever fragments (5).

In some cases, the first matched fragment turned out to be exactly the one containing the same answer as the ground truth (this situation can be observed in the first two rows of the table). It also happened that the first returned answer was not the same as the ground truth, although it correctly addressed the question. This can be observed in the third question in the table - the first answer is a correct response to the question, but it is not an exact match (F1 score is only 0.64). The same answer as in the dataset was found based on the third best match from the retriever.

There were also situations where only a part of the desired answer was returned (question 4 in the table). Sometimes it turned out that theoretically a

**Table 5.** Sample contexts extracted by retriever

| Question | How to get Rc's current reference count? | How does 'panic!' macro work? |
|---|---|---|
| Gold context | each point in the program where the reference count changes, we print the reference count, which we get by calling the 'Rc::strong_count' function. This function is named 'strong_count' rather than 'count' because | it has the type 'Result<T, E>' for recoverable errors and the 'panic!' macro that stops execution when the program encounters an unrecoverable error. This chapter covers calling 'panic!' first and then talks about |
| Context 1 | the count goes down by 1. We don't have to call a function to decrease the reference count like we have to call 'Rc::clone' to increase the reference count: the implementation of | has died, the test is marked as failed. In Chap. 9, we talked about how the simplest way to panic is to call the 'panic!' macro. Enter the new test as a |
| Context 2 | each point in the program where the reference count changes, we print the reference count, which we get by calling the 'Rc::strong_count' function. This function is named 'strong_count' rather than 'count' because | to do with 'r' wouldn't work correctly. So how does Rust determine that this code is invalid? It uses a borrow checker. ### The Borrow Checker The Rust compiler has a *borrow checker* that |
| Context 3 | count. When looking for performance problems in the code, we only need to consider the deep-copy clones and can disregard calls to 'Rc::clone'. ### Cloning an 'Rc <T>' Increases the Reference Count Let's change our | to cause a panic in practice: by taking an action that causes our code to panic (such as accessing an array past the end) or by explicitly calling the 'panic!' macro. In both |
| Context 4 | threads. When 'Rc <T>' manages the reference count, it adds to the count for each call to 'clone' and subtracts from the count when each clone is dropped. But it doesn't use any concurrency | will be the parameter that we pass to 'expect', rather than the default 'panic!' message |
| Context 5 | in 'leaf' has a strong count of 1 and a weak count of 0, because the variable 'leaf' is now the only reference to the 'Rc <Node>' again. All of the logic that manages | to 'panic!")[to-panic-or-not-to-panic]<!– ignore –> section later in this chapter. Next, we'll look at how to recover from an error using 'Result'. [to-panic-or-not-to-panic]: |

more matching answer with the dataset answer could not be considered correct by a human due to the absence of important words in the response. An example of such a situation is the question about the 'mod' declaration.

There were also cases where the answer extracted from the first match was unrelated to the context of the question, and one of the later answers satisfactorily addressed the question (in the question about the 'unwrap' method of the 'Result' class, the first answer refers to the purpose of using the 'Result' class, not what the 'unwrap' method does. Only the third answer contains the requested information).

For some questions, the situation occurred where the returned answer did not directly answer the question, but the context of the answer was correct, and inference could be made based on it. Examples of such situations are questions 7 and 8 in the table - the system, when asked about the default random number generator, does not explicitly respond that Rust does not have such a generator by default but mentions that the Rust team provides 'randcrate.' Similarly, when asked about the number of release channels, it does not directly answer "3" but provides a list of available channels.

However, there were questions with which the system struggled. Examples of such questions are the last two rows in the table. The system attempted to handle this problem in various ways - for instance, by trying to return a fragment containing a phrase from the question. Such answers were characterized by a small but nonzero F1 value. It also happened that in case of the inability to find an answer to the question, the system returned a period as the answer. It was also noticed that the system might have had difficulties with questions where the expected answers were special characters (operators "!", "]", "/", etc.).

**Table 6.** Answers for example questions from test set.

| Question | Ground truth | Top1 answer (F1 Score) | Answer with highest F1 Score from Top5 (F1 Score) | Rank of best match answer |
|---|---|---|---|---|
| What does status code 200 mean? | The status code 200 is the standard success response. | The status code 200 is the standard success response (F1 = 1.000) | The status code 200 is the standard success response (F1 = 1.000) | 1 |
| Do rust strings support indexing? | Rust strings don't support indexing. | Rust strings don't support indexing. (F1 = 1.000) | Rust strings don't support indexing. (F1 = 1.000) | 1 |
| How to create a string from a string literal? | We can also use the function 'String::from' to create a 'String' from a string literal. | You can create a 'String' from a string literal using the 'from' function (F1 = 0.640) | We can also use the function 'String::from' to create a 'String' from a string (F1 = 1.000) | 3 |
| What is an iterator? | The iterator pattern allows you to perform some task on a sequence of items in turn. An iterator is responsible for the logic of iterating over each item and determining when the sequence has finished. | The iterator pattern allows you to perform some task on a sequence of items in turn. (F1 = 0.636) | The iterator pattern allows you to perform some task on a sequence of items in turn. (F1 = 0.636) | 1 |
| How many times I should load a file using a 'mod' declaration in my module tree? | Note that you only need to load a file using a 'mod' declaration *once* in your module tree. | once* in your module tree. (F1 = 0.476) | Note that you only need to load a file using a (F1 = 0.720) | 5 |
| What is the purpose of unwrap method of Result? | If the 'Result' value is the 'Ok' variant, 'unwrap' will return the value inside the 'Ok'. If the 'Result' is the 'Err' variant, 'unwrap' will call the 'panic!' macro for us. | The purpose of these 'Result' types is to encode error-handling information. (F1 = 0.118) | If the 'Result value is the 'Ok' variant, 'unwrap' will return the value inside the 'Ok'. (F1 = 0.667) | 3 |

(*continued*)

**Table 6.** (*continued*)

| Question | Ground truth | Top1 answer (F1 Score) | Answer with highest F1 Score from Top5 (F1 Score) | Rank of best match answer |
|---|---|---|---|---|
| Does Rust have a random number generator by default? | Rust doesn't yet include random number functionality in its standard library | the Rust team does provide a ['rand' crate][randcrate] with (F1 = 0.200) | random number generator (F1 = 0.286) | 2 |
| How many release channels does rust have? | three | * Nightly * Beta * Stable Most Rust developers primarily use the stable channel, but those who want to try out experimental new features may use nightly or beta. | All answers have F1 = 0 | – |
| What are integration tests? | using only the public interface and potentially exercising multiple modules per test | unit tests and integration tests. (F1 = 0.125) | unit tests and integration tests. (F1 = 0.125) | 1 |
| What is the meaning of a '!' operator? | Macro expansion | . (F1 = 0.000) | Note that the '*' operator is replaced with a call to the 'deref' method and (F1 = 0.000) | 5 |

# 8   Discussion

In this paper, we described our Question Answering system that helps programmers query through technical documentation, in our case specifically through Rust Book - a manual for Rust programmers. During the research, we created a dataset of over 1000 pairs of questions and answers in SqUAD 2.0 format by manually annotating Rust Book with the usage of the Haystack Annotation Tool. Based on recent advancements in Natural Language Processing we decided to use retriever-reader architecture and the RoBERTa model as a base of our training. For that, we used the Haystack library, which helps us with creating pipelines consisting of nodes typical for QA problems. After the training, we have put together our system which consists of two pipelines: indexing and query. The indexing pipeline processes data and creates a knowledge base for the system. The query pipeline represents the retriever-reader architecture mentioned earlier. We evaluated our system based on typical QA metrics, such as F1-Score for the Reader and Mean Average Precision or Recall for the Retriever. Even though their values are not the best, in manual testing we discovered, that answers usually revolve around the real ones. This shows that both the creation and evaluation of domain-specific question-answering systems are difficult problems. In our opinion the system in this form could prove helpful in searching for information in technical documentation in the developer's workflow. However, a series of surveys would help support this statement.

That seems promising and we aim to continue our research. In future work, we see two main ideas for improving our solution. The first one revolves around the usage of more complex Information Retrieval techniques. As we have seen in Sects. 5 and 6, the retriever seems to be the bottleneck of the described system. The usage of more sophisticated algorithms and/or technologies should improve both metrics and the general usability of the system. Because of that, in the future we will change the way of storing documents, as the used library allows us to use different search engines (Elasticsearch, as in [16] and OpenSearch) or different types of databases (SQL, NoSQL, Vector - for a list of providers see: https://docs.haystack.deepset.ai/docs/document_store). This would be crucial in deploying our solution as those ways of storing data are way more scalable and efficient.

The second idea is to use LLMs in our system. This would change the form of our approach to generative question answering. It seems more user-friendly, and if the right data is provided to the node responsible for generating answers, this approach could fix the problem of reducing answers to one character in the Reader node (for reference see Sect. 7, especially Table 6). Those two branches along with surveys conducted among the developers are our next goals.

# References

1. Aurpa, T.T., Rifat, R.K., Ahmed, M.S., Anwar, M.M., Ali, A.B.M.S.: Reading comprehension based question answering system in Bangla language with transformer-based learning. Heliyon **8**(10), e11052 (2022)
2. Brown, T.B., et al.: Language models are few-shot learners (2020)
3. Chen, D., Fisch, A., Weston, J., Bordes, A.: Reading Wikipedia to answer open-domain questions. arXiv preprint arXiv:1704.00051 (2017)
4. Deriu, J., et al.: Survey on evaluation methods for dialogue systems. Artif. Intell. Rev. **54**(1), 755–810 (2021)
5. Devlin, J., Chang, M.W., Lee, K., Toutanova, K.: Bert: pre-training of deep bidirectional transformers for language understanding (2019)
6. Ed-Douibi, H., Daniel, G., Cabot, J.: OpenAPI bot: a chatbot to help you understand REST APIs. In: Bielikova, M., Mikkonen, T., Pautasso, C. (eds.) ICWE 2020. LNCS, vol. 12128, pp. 538–542. Springer, Cham (2020). https://doi.org/10.1007/978-3-030-50578-3_40
7. Es, S., James, J., Espinosa-Anke, L., Schockaert, S.: Ragas: automated evaluation of retrieval augmented generation (2023)
8. He, Y., Zhu, Z., Zhang, Y., Chen, Q., Caverlee, J.: Infusing disease knowledge into BERT for health question answering, medical inference and disease name recognition. In: Webber, B., Cohn, T., He, Y., Liu, Y. (eds.) Proceedings of the 2020 Conference on Empirical Methods in Natural Language Processing (EMNLP), pp. 4604–4614. Association for Computational Linguistics (2020). https://doi.org/10.18653/v1/2020.emnlp-main.372
9. Lee, H.J.O., Kim, H.J., Jang, M.G.: Descriptive question answering in encyclopedia. In: Nagata, M., Pedersen, T. (eds.) Proceedings of the ACL Interactive Poster

and Demonstration Sessions, pp. 21–24. Association for Computational Linguistics, Ann Arbor (2005). https://doi.org/10.3115/1225753.1225759

10. Liu, Y., et al.: Roberta: a robustly optimized bert pretraining approach (2019)
11. Muwanei, S., Ravana, S.D., Hoo, W.L., Kunda, D.: Prediction of the high-cost normalised discounted cumulative gain (NDCG) measure in information retrieval evaluation. Inf. Res. **27**(2) (2022). https://doi.org/10.47989/IRPAPER928
12. Qu, C., Yang, L., Qiu, M., Croft, W.B., Zhang, Y., Iyyer, M.: Bert with history answer embedding for conversational question answering. In: Proceedings of the 42nd International ACM SIGIR Conference on Research and Development in Information Retrieval (SIGIR 2019). ACM (2019). https://doi.org/10.1145/3331184.3331341
13. Reddy, S., Chen, D., Manning, C.D.: COQA: a conversational question answering challenge (2019)
14. Robertson, S., Zaragoza, H.: The probabilistic relevance framework: Bm25 and beyond. Found. Trends® Inf. Retriev. **3**(4), 333–389 (2009). https://doi.org/10.1561/1500000019
15. Wang, D., Nyberg, E.: A long short-term memory model for answer sentence selection in question answering. In: Zong, C., Strube, M. (eds.) Proceedings of the 53rd Annual Meeting of the Association for Computational Linguistics and the 7th International Joint Conference on Natural Language Processing (Volume 2: Short Papers), pp. 707–712. Association for Computational Linguistics, Beijing (2015). https://doi.org/10.3115/v1/P15-2116
16. Wang, Z., Ng, P., Ma, X., Nallapati, R., Xiang, B.: Multi-passage BERT: a globally normalized BERT model for open-domain question answering. arXiv preprint arXiv:1908.08167 (2019)
17. Yang, S., Zou, L., Wang, Z., Yan, J., Wen, J.R.: Efficiently answering technical questions - a knowledge graph approach. Proc. AAAI Conf. Artif. Intell. **31**(1) (2017). https://doi.org/10.1609/aaai.v31i1.10956
18. Yang, W., et al.: End-to-end open-domain question answering with bertserini. arXiv preprint arXiv:1902.01718 (2019)
19. Yang, Y., Yih, W.T., Meek, C.: WikiQA: a challenge dataset for open-domain question answering. In: Màrquez, L., Callison-Burch, C., Su, J. (eds.) Proceedings of the 2015 Conference on Empirical Methods in Natural Language Processing, pp. 2013–2018. Association for Computational Linguistics, Lisbon (2015). https://doi.org/10.18653/v1/D15-1237

# Interpretable Dense Embedding for Large-Scale Textual Data via Fast Fuzzy Clustering

Olzhas Kozbagarov[1,2], Rustam Mussabayev[1,2]($\boxtimes$) (iD),
Alexander Krassovitskiy[1,2] (iD), and Nursultan Kuldeyev[1]

[1] Satbayev University, 22a Satpaev Street, 050013 Almaty, Kazakhstan
[2] Institute of Information and Computational Technologies, 28 Shevchenko Street, 050010 Almaty, Kazakhstan
rustam@iict.kz

**Abstract.** Efficient and interpretable analysis of large-scale textual data is a major challenge in big data. This paper introduces a novel approach for creating interpretable dense embeddings from extensive text corpora. Our method uses FlexiClust, a fast fuzzy clustering algorithm, combined with word co-occurrence analysis to generate semantically rich text embeddings. This approach balances interpretability, simplicity, and efficiency by merging precise word co-occurrence analysis with advanced fuzzy clustering, forming dense, interpretable vector representations. The proposed technique addresses limitations of traditional sparse vectors and complexities of neural network models, offering improvements in text vectorization. It is particularly beneficial for applications such as news aggregation, content recommendation, semantic search, topic modeling, and text classification in large datasets.

**Keywords:** Interpretable text embedding · Vector representation · Big data · Fuzzy Clustering · Topic modeling · Word co-occurrence · Large text corpora

## 1 Introduction

In the field of big data providing the interpretability of text analysis methods is also important as their efficiency especially in the context of processing large-scale textual data [1]. The big amount of text data circulated today presents both opportunities and challenges in deriving meaningful insights. Efficient processing of these large datasets is possible using advanced algorithms and increased computational power, but the clarity and understandability/interpretability of the results remain insufficient [2]. This is especially significant in fields like healthcare, finance, and public policy, where decision-making should heavily rely on the

This research was funded by the Science Committee of the Ministry of Science and Higher Education of the Republic of Kazakhstan (grant no. BR21882268).

transparency and explainability of data analysis [3]. Combining interpretability with efficiency is a complex task. Available highly accurate models are often not transparent obscuring the logic behind their predictions. At the same time more simple models may struggle with the scale and complexity of large datasets. In this paper we propose an approach that addresses this challenge by enhancing the interpretability of text analysis results without compromising their efficiency in processing large volumes of data.

Traditional approaches in text analysis such as statistical methods and vector-space models are fundamental in this field [4]. These methods are instrumental for solving tasks of text categorization, sentiment analysis and topic modeling. However, as the size and complexity of textual datasets have grown, these traditional methods have encountered limitations related with their scalability and interpretability.

This research addresses the challenge of developing a text analysis method that is scalable for large datasets while also ensuring the interpretability of its results. Current methods typically offer either scalability or interpretability but rarely both. This paper proposes a new method designed to efficiently handle large text volumes while also providing clear and interpretable outcomes. The proposed method holds potential for various applications such as semantic search and topic modeling, offering a balance between interpretability, simplicity, and efficiency [5].

## 2 Related Works

Textual information can be represented in vectors primarily in two forms: sparse and dense. The length of sparse vectors matches the dictionary size, and their values correspond to word frequency based information. To optimize sparse vector representation, the key strategy is to manage the content and size of the dictionary [6]. This includes restricting the dictionary to specific parts of speech, setting word frequency limits, and eliminating less informative words. Usually sparse vectors are very large, matching the size of the dictionary, which can make them computationally difficult to handle. Additionally, if two texts cover the same topic but use different synonyms or related terms, their vector representations might vary greatly, even though they are topically similar.

Text representation has evolved with various models, each bringing a unique approach. Doc2Vec creates a single embedding for a whole text, capturing its overall meaning [7]. Word2Vec, on the other hand, focuses on individual words, highlighting how their meanings change based on context [8].

Google's BERT model is a significant advancement, especially in understanding the context of words in sentences [9]. It looks at surrounding words to get a deeper sense of meaning. GloVe works differently, using statistics on how often words appear together to find shared meanings [10]. ELMo goes deeper, considering the entire sentence, both before and after a word, to represent its meaning [11].

Newer models like GPT and its updates (GPT-2, GPT-3) are known for generating text that's both relevant and flows well [12]. FastText, from Facebook, improves on Word2Vec by breaking down words into smaller parts, helping it understand shorter words and those with similar roots [13].

Neural network text embeddings typically have 100–1000 dimensions (with 300 being common). Also they have lack of interpretability and require significant setup and resources that making them complex for Big Data applications.

Each model has its strengths making them useful for various language processing tasks. This variety shows the ongoing growth and change in the ways of text vectorization.

# 3  Description of the Proposed Text Vectorization Method

In this section the newly developed approach is presented, which provides a method for obtaining interpretable text embeddings of textual information. This approach involves several steps.

## 3.1  Forming the Target Dictionary

The initial task is to create a dictionary from the corpus texts by consisting of nouns and named entities and excluding stopwords and very rare words to reduce computational complexity and focus on thematically significant units.

## 3.2  Grouping Words According to Their Semantic and Thematic Relatedness

At this stage, all words in the dictionary are grouped by topic, where a 'topic' is defined by words frequently co-occurring in texts. This grouping is crucial for creating interpretable text embeddings and representing similar themes comparably that helps identify thematic patterns across texts.

To group words by topic a co-occurrence matrix $X$ is formed with dimensions defined by text count and dictionary size. Each $X$ element counts word occurrences in texts, assuming words co-occur if in the same text and ignoring their order (i.e. using "bag of words" approach). Word vector representations are generated by:

$$V = X^T \cdot X \tag{1}$$

As a result, a symmetric matrix $V$ sized by the dictionary is obtained. It represents word vectors showing word contextual relatedness via co-occurrence. Similar vectors should indicate higher cosine similarity that suggests their topic relevance. Then the matrix $D$ calculates pairwise cosine distances between word vectors in $V$:

$$D = 1 - \cos(V_1, V_2) = 1 - \frac{V_1 \cdot V_2}{|V_1| \cdot |V_2|} \tag{2}$$

The calculation of both matrices (1) and (2) can be efficiently optimized using modern software and hardware tools for high-performance computing, such as SIMD-accelerated vector mathematics or GPU-based calculations.

The obtained distance matrix $D$ can serve as an input for a fuzzy clustering algorithm to create topic-related word clusters. Fuzzy clustering allows for word vectors to belong to multiple clusters due to their potential relevance to various topics, unlike 'crisp' clustering where each word is only assigned to a single cluster.

### 3.3    Fuzzy Clustering of Word Vectors

Let's describes the fuzzy clustering algorithm (Algorithm 1) that was used for clustering of word vectors. The input data for this algorithm is the pairwise distance matrix $D$. This algorithm has three parameters: $d$ - the radius of neighborhood to generate the primitive first-order clusters; $q$ - similarity threshold for combining them into complex second-order clusters and $s_{min}$ - minimal size of resulting clusters. The fuzzy clustering algorithm works as follows:

1. Initially, the algorithm forms $n$ primitive clusters from each data point in the dataset, selecting neighbors within radius $d$ around each point to create a cluster based on the Greedy Algorithm, when a locally optimal solution is chosen at each stage to obtain a final solution close to the optimal solution. The primitive cluster starts with two points: the $i$-th point and its nearest neighbor. Points are then added from the neighborhood set if they are the closest to the cluster and all pairwise distances within the cluster are $\leq d$, until no more points meet this criterion. Primitive clusters are formed independently from each neighborhood of a point that allowing points to belong to multiple clusters and introduces fuzziness in the later stages of algorithm. A higher $d$ value increases algorithm computation time due to a larger neighborhood radius which expands potential element combinations for cluster formation.
2. In the second step unique primitive clusters are selected from the initial set. Due to the high number of these clusters merging those with overlapping words is crucial for denser vector representation.
3. At the third step, the matrix $\mathbf{J}$ of pairwise similarities is calculated between all unique primitive clusters. To calculate the similarity $\mathbf{J}_{ij}$ between two unique primitive clusters $\mathcal{U}_i$ and $\mathcal{U}_j$, the modified Jaccard measure is used as follows:

$$\mathbf{J}_{ij} = \frac{|\mathcal{U}_i \cap \mathcal{U}_j|}{\min(|\mathcal{U}_i|, |\mathcal{U}_j|)} \tag{3}$$

where $|.|$ denotes the size of a set, and $U_i$ and $U_j$ are the sets of all point indices constituting the $i$-th and $j$-th unique primitive clusters, respectively.
4. In the fourth step, we aggregate primitive clusters into second-order clusters by identifying explicit and implicit connections, resulting in more complex cluster forms. Aggregation merges $i$-th and $j$-th primitive clusters with Jaccard similarity $\mathbf{J}_{ij} \geq q$ or those connected either explicitly (pairwise ratio $\geq q$) or implicitly via common conjoint clusters. For example, with three clusters

$A$, $B$, and $C$: if $A$ and $B$, and $B$ and $C$ have pairwise similarities above $q$, but $A$ and $C$ do not, we merge $A$, $B$, and $C$ into one complex cluster, despite the lower $A$-$C$ similarity. Setting the threshold $q$ adjusts the final cluster count: higher $q$ values yield fewer clusters.

5. In the final step, second-order clusters smaller than $s_{min}$ are removed, retaining only clusters meeting this minimum size, defined as the third parameter of fuzzy clustering algorithm.

The fuzzy clustering algorithm uses few parameters and constructing clusters based on pairwise distances without predefining cluster count, which varies with dataset semantics. Its simplicity and design enable parallelization for ensuring fast execution on large-scale datasets.

Algorithm 1 shows the general version of the proposed FlexiClust algorithm. The source code of FlexiClust in Python is available at https://github.com/R-Mussabayev/flexiclust/.

Parameters of algorithm influence the cluster count, ranging from zero to thousands. Merged clusters form thematic categories setting text embedding dimensions equal to their number. The output of the fuzzy clustering process of word vectors provides a stand-alone, autonomous topic model that can be applied in other various NLP applications.

## 3.4   The Process of Text Embedding Construction

Text embeddings are constructed by determining the degree of relatedness $R(T, C_i)$ between each thematic category $C_i$, obtained in the previous stage, and the text $T$ for which we are creating its embedding representation:

$$\text{Emb(T)} = [R(\text{T}, \text{C}_1), R(\text{T}, \text{C}_2), \ldots, R(\text{T}, \text{C}_N)] \qquad (4)$$

where

$$R(\text{T}, \text{C}_i) = \frac{|T \cap C_i|}{\min(|T|, |C_i|)} \qquad (5)$$

In equations above $T$ and $C_i$ represent the sets of words in the text and the $i$-th thematic category, respectively, $Emb(T)$ represents the embedding of the text, $R(T, C_i)$ is the degree of relatedness of the text $T$ to the $i$-th thematic category $C_i$.

The value of the text embedding's component indicates the level of relatedness between the publication and the corresponding thematic category; the higher the value, the stronger the relatedness of the publication to the topic. This indicates the interpretability of the resulting text embeddings.

**Input:** $\mathbf{D}$ - distance matrix, $d$ - distance threshold, $q$ - linkage threshold,
　　　　$s$ - minimum cluster size;
**Function** FlexiClust$(\mathbf{D}, d, q, s)$:

 $n \leftarrow$ number of rows in $\mathbf{D}$ ;
 $P \leftarrow$ initialize collection of $n$ empty sets ;
 **for** $i \leftarrow 0$ **to** $n$ **do**
  Indices$_{\leq d} \leftarrow [j \mid D_{ij} \leq d]$ ;
  **if** *Indices$_{\leq d}$ not empty* **then**
   $start \leftarrow$ position of $i$ in Indices$_{\leq d}$ ;
   $p \leftarrow \{start\}$ ;
   $\mathbf{D}_{sub} \leftarrow$ submatrix of $\mathbf{D}$ consisting of rows and columns both with indices Indices$_{\leq d}$ ;
   $m \leftarrow$ number of rows in $\mathbf{D}_{sub}$ ;
   $C \leftarrow \{0, 1, \ldots, m-1\} \setminus \{start\}$ ;
   **while** *C not empty* **do**
    $C \leftarrow \{j \in C \mid \sum_{k \in p}(\mathbf{D}_{sub}(k,j) \leq d) = |p|\}$ ;
    **if** *C not empty* **then**
     $best \leftarrow$ argmin over $j \in C$ of $\sum_{k \in p} \mathbf{D}_{sub}(k,j)$ ;
     $p \leftarrow p \cup \{C[best]\}$ ;
     $C \leftarrow C \setminus \{C[best]\}$
    **end**
   **end**
   $P[i] \leftarrow \{$Indices$_{\leq d}[k]$ for each $k$ in $p\}$ ;
  **end**
 **end**
 $\mathcal{U} \leftarrow$ identify unique non-empty sets in $P$ with size at least $s$;
 $n_u \leftarrow$ number of elements in $\mathcal{U}$ ;
 Initialize matrix $\mathbf{J}$ of size $n_u \times n_u$ with diagonal elements set to 1 ;
 **for** $i \leftarrow 0$ **to** $n_u - 1$ **do**
  **for** $j \leftarrow i + 1$ **to** $n_u$ **do**
   $\mathbf{J}_{ij} \leftarrow \frac{|\mathcal{U}_i \cap \mathcal{U}_j|}{\min(|\mathcal{U}_i|, |\mathcal{U}_j|)}$ ;
   $\mathbf{J}_{ji} \leftarrow \mathbf{J}_{ij}$ ;
  **end**
 **end**
 $\mathcal{S} \leftarrow$ empty set of second-order clusters ;
 **for** $i \leftarrow 0$ **to** $n_u - 1$ **do**
  **for** $j \leftarrow i + 1$ **to** $n_u$ **do**
   **if** $\mathbf{J}_{ij} \geq q$ **then**
    $\mathcal{S} \leftarrow \mathcal{S} \cup \{\mathcal{U}_i \cup \mathcal{U}_j\}$ ;
   **end**
  **end**
 **end**
 $\mathcal{S}_{final} \leftarrow$ filter $\mathcal{S}$ by size $\geq s_{\min}$;
 **return** $\mathcal{S}_{final}$ ;

**Algorithm 1:** FlexiClust Algorithm

**Result:** Labels for each base and Merged Bases
**Input:** Set of bases $\mathcal{B} = \{B_1, B_2, ..., B_n\}$, Pairwise similarity matrix $S$,
  Threshold $q$
**Output:** Labels $\mathcal{L}$ for each base, Merged bases $\mathcal{M}$

Construct a graph $G(V, E)$ where each vertex $v_i$ corresponds to a base $B_i \in \mathcal{B}$.
  Add an edge $(v_i, v_j) \in E$ if $S[i, j] \geq q$ for $i > j$ (considering the lower triangle
  of $S$);

Find connected components in $G$, assigning each vertex a label $l_i$ indicating its
  component. Store these labels in $\mathcal{L}$;

Group vertices by their labels to identify sets of bases that will be merged
  based on connectivity;

Initialize $\mathcal{M} = \emptyset$;
**foreach** *unique label l in* $\mathcal{L}$ **do**
  | Find all bases $\{B_j | \mathcal{L}[j] = l\}$;
  | Merge these bases to form a new base $M_l = \bigcup_{B_j | \mathcal{L}[j]=l} B_j$;
  | Add $M_l$ to $\mathcal{M}$, associating it with label $l$;
**end**

return $\mathcal{L}, \mathcal{M}$;
  **Algorithm 2:** Improved Merging of Bases Based on Connectivity

## 4    Experimental Setup

### 4.1    Description of the Text Corpus

The approach was evaluated on a corpus [14] of 186,183 news articles from 2008
to 2020, split 70% (130,477) for training and 30% (55,706) for testing. For fuzzy
clustering and dictionary construction was used only the training set.

### 4.2    First Step: Text Preprocessing and Dictionary Construction

Word sequences for each text in the corpus were obtained through the tokenization procedure using the NLTK library in Python. Punctuation was removed,
words were lemmatized, and a morphological analysis was conducted. As a
result, all nouns including named entities were identified. High-frequency and
low-frequency words were excluded from consideration, i.e., words that appear
in fewer than 10 publications or in more than 40,000 publications. The remaining
words formed the dictionary, which consisted of 15,710 noun words.

### 4.3    Second Step: Grouping of the Dictionary Words According
to Their Thematic Relatedness

Initially, matrix $X$ of dimensions $130,477 \times 15,710$ (publications x vocabulary
size) was computed recording word occurrence frequencies. Word vectors were
derived using formula (1), producing a symmetric $15,710 \times 15,710$ matrix $V$,
with each row representing a sparse vector of some word.

Then, the matrix $D$ of pairwise cosine distances between these vector representations, with the same dimensions, was calculated using the formula (2). Afterwards, the newly proposed fuzzy clustering algorithm was applied using the matrix $D$ as the input data.

### 4.4 Third Step: Merging of the Clusters to Obtain Thematic Categories

The implemented procedure for constructing thematic categories operates relatively quickly. For instance, in the case under consideration, to perform topic modeling on a CPU with 8 cores took about one minute when using the parameter $d = 0.37$. In contrast, utilizing the GPU implementation of the fuzzy clustering algorithm under NVIDIA CUDA reduced the time to 10 s. Thus, in a relatively short period, this approach enables the obtaining of fuzzy clustering results at various values of $d$, allowing for their comparison to achieve the desired outcomes. Examples of thematic categories obtained when $d = 0.37$ are presented in Table 1. The first category consists of words related to US foreign policy; the second category is related to art, namely opera and ballet; the third one to traffic accidents; the fourth to weather; and so on.

**Table 1.** Examples of obtained thematic categories

| Thematic Category |
| --- |
| USA, dispatch, Rex, Washington, John, state department, Obama, ally, Hilary, ... |
| Placido, philharmonic society, prima ballerina, opera, theater, alt, ... |
| Jeep, side, scorcher, ditch, cuvette, passenger, turn, damage, movement, ... |
| Weather service, cooling, calm, air, weather station, thermometer, ... |
| Infant, placenta, obstetrician, pregnancy, feeding, mother, ... |
| Kabul, Talib, Taliban, Hamid, Afghan, Afghanistan, ... |
| City, highway, line, branch, ... |
| Penalty shot, hockey player, hockey throw, burn-out, overtime, puck, ... |
| Matchmaking, groom, marriage, dowry, ... |
| Officer, platoon, battalion, company, commander, ... |

The different values of parameters provide different number of thematic categories. The resulting number of the thematic categories are given in Table 2.

### 4.5 Fourth Step: Construction of Text Embeddings

The categories obtained at the third stage were used to obtain text embeddings of news publications by calculating the level of relatedness between texts and each of categories using the formula (5). The text embedding matrices $T_{training}$

**Table 2.** The number of categories obtained using different combinations of parameter values

| Radius of neighborhood $d$ | Similarity threshold $q$ | Minimal size of resulting clusters $s_{min}$ | Resulting number of categories |
|---|---|---|---|
| 0.43 | 0.8 | 5 | 1,266 |
| 0.43 | 0.5 | 10 | 55 |
| 0.31 | 0.5 | 2 | 923 |
| 0.34 | 0.8 | 2 | 3,966 |
| 0.35 | 0.9 | 2 | 5,514 |

and $T_{test}$ were calculated for both train and test sets correspondingly. Each row of these matrices is the text embedding for the corresponding publication having the number of elements equal to the number of used thematic categories.

### 4.6 Assessing the Quality of the Proposed Embeddings in Solving the Task of Thematic Categorization of Publications

The matrix $D$ representing the pairwise cosine distances (2) between the text embeddings of the test and training sets was calculated having dimensions of 55,706 × 130,477. Then, the k-smallest values were determined in each row of this matrix: the corresponding $k$ column indices indicate the publications from the training set that are similar to the given publications in the test set according to their.

Afterwards, a comparison between the topics of the publications in the test set and the topics of their corresponding k-nearest neighbors' publications from the training set was performed to assess quality of obtained embeddings. For this purpose, the topic tag names that were assigned to each article by the news agency itself were used. Table 3 presents example tag names assigned to the publications in the corpus.

**Table 3.** Tags assigned by the news agency to their publications

| Publication fragments | Tags assigned by the news agency |
|---|---|
| In the night of March 5/6, the Champions League 1/8 finals were played in two pairs - Real Madrid took home Dutch Ajax, Dortmund Borussia played at home with English Tottenham ... | Champions League, Real Madrid, Football |
| The popular online cinema Netflix has acquired the rights to film adaptation of the novel of the Colombian writer and Nobel laureate Gabriel Garcia Marquez "One Hundred Years of Solitude" - one of the most famous books of the 20th century ... | Literature Series |

The total number of unique tags was 3,032. The frequencies of some tags, for example, is shown in Table 4.

**Table 4.** Frequencies of tags assigned to publications

| Tag names | Number of Occurrences | Percentage |
|-----------|-----------------------|------------|
| Law enforcement | 11,067 | 5.9% |
| Justice | 8,237 | 4.4% |
| Children | 8,054 | 4.3% |
| Traffic accident | 5,789 | 3.1% |
| Murder | 4,916 | 2.6% |

A 2.5 tags were assigned to an average publication (minimum = 0, maximum = 13). Statistics on the number of assigned tags are shown in Table 5. Most often, two or three tags were assigned to average publication.

**Table 5.** Distribution of assigned tag numbers among publications

| Tags number | Number of publications | Percentage |
|-------------|------------------------|------------|
| 2 | 73,502 | 39.5% |
| 3 | 51,215 | 27.5% |
| 1 | 31,700 | 17.0% |
| 4 | 20,155 | 10.8% |

If a publication and its corresponding k-nearest neighbors have a complete coincidence of tags, then it is assumed that the result of the text embeddings approach is successful in identifying thematically similar texts. The higher the overall level of tag matching, the better. The level of tag correspondence between the $i$-th publication's tags and the tags of its k-nearest neighbors was determined through the following formula:

$$J(i, k\text{-nn}) = \frac{|T_i \cap T_{k\text{-nn}}|}{\max(|T_i|, |T_{k\text{-nn}}|)} \tag{6}$$

where $T_i$ denotes the set of tags associated with publication $i$, $T_{k-nn}$ denotes the set of tags associated with the k-nearest neighbors of publication $i$.

To obtain the total assessment of the quality of vector representations, all values referenced in (6) are summed for all publications in the test set:

$$Q = \sum_i J(i, k\text{-nn}) \tag{7}$$

where $J(i, k\text{-nn})$ is the $i$-th publication's estimate of (6).

# 5    Experimental Results

## 5.1    Random Simulation of $k$ Neighbors

For the purpose of comparative analysis of the results received using the developed approach, the estimate $Q$ (7) was calculated for the case where each publication from the test set is uniformly randomly assigned to $k$ publications from the training set. Using this approach, we can obtained measures for the worse-case scenario, i.e., for random assignment of k-nearest neighbors. The "Random Assignment" row of the Table 6 shows the estimates obtained for $k = 1, 2, 3, 5, 10$ for the test set (which includes the 55,706 publications).

## 5.2    Sparse Vector Representation Based on the Word Frequencies

For the purpose of comparative analysis of the results obtained using the developed approach, the vector representation of publications in the form of sparse vectors. In this case, each publication was represented by a vector which has the size of the dictionary. Here, each vector element represents the TF-IDF measure of the word in the given publication. The results of the assessment of $Q$ are shown in "Sparse Vector" row of Table 6.

## 5.3    Text Embeddings Based on Neural Networks Model

For the purpose of comparative analysis of the results obtained using the developed approach, the vector representation of publications constructed using word2vec model was considered. The word2vec model was trained on the training set using its standard parameter values. Words embeddings for each word, with a dimension equal to 300, were obtained. Text embeddings were calculated as weighted average of all word embeddings. The maximum value of $Q$ was achieved with the following parameter values: $window = 3$, $sg = 1$, $epochs = 100$. The results are shown in "Word2Vec Embeddings" row of Table 6.

## 5.4    Proposed Text Embeddings

The $Q$ values obtained using the proposed text embeddings are shown in rows 4–6 of Table 6. These results were obtained after performing parametric identification of the parameters $(d, q, s_{min})$ with the goal of maximizing the $Q$ value, subject to the constraint that dimensions do not exceed 5,000. The results demonstrate that in most cases a decrease in the number of considered thematic categories leads to decrease in the $Q$ value obtained. Furthermore, it is shown that it is possible to build text embeddings using proposed approach with dimensions ranging from 300 to 2,000 without a strong decrease in the $Q$ values.

**Table 6.** Estimates of $Q$ for various methods and k-nearest neighbors

| Method | k = 1 | k = 2 | k = 3 | k = 5 | k = 10 |
|---|---|---|---|---|---|
| Random Assignment | 702 | 773 | 766 | 740 | 728 |
| | 1.3% | 1.4% | 1.4% | 1.3% | 1.3% |
| Sparse Vector (TF-IDF) | 22,433 | 24,342 | 23,374 | 21,904 | 20,053 |
| | 40.2% | 43.6% | 41.9% | 39.2% | 35.9% |
| Word2Vec Embeddings | 20,491 | 22,224 | 21,344 | 20,044 | 18,377 |
| | 37% | 40% | 38% | 36% | 33% |
| Proposed Embeddings with dimension equal to 3,966 ($d = 0.34$, $q = 0.8$, $s_{min} = 2$) | 17,369 | 18,760 | 17,886 | 16,609 | 14,976 |
| | 31% | 34% | 32% | 30% | 27% |
| Proposed Embeddings with dimension equal to 1,298 ($d = 0.42$, $q = 0.8$, $s_{min} = 5$) | 16,801 | 18,108 | 17,336 | 16,171 | 14,709 |
| | 30% | 33% | 31% | 29% | 26% |
| Proposed Embeddings with dimension equal to 650 ($d = 0.42$, $q = 0.8$, $s_{min} = 10$) | 15,262 | 16,521 | 15,836 | 14,736 | 13,449 |
| | 27% | 30% | 28% | 26% | 24% |

## 5.5   Conclusions

The proposed approach offers several advantages over sparse vector representations and neural network-based embeddings: it is relatively simple and straightforward, resulting in fast computational speed. The main advantage of the developed approach for calculating text embeddings is that it provides explicitly interpretable results with each value of the vector's components indicating the level
of relatedness of text to a given topic.

The obtained experimental results presented in rows 4–6 of Table 6 indicate
that using the proposed approach, it is possible to generate dense and interpretable embeddings that provide quality comparable to that obtained using
sparse vector representations, which are uninterpretable but with dimensions
several times smaller than those of sparse vector representations. Moreover,
although the proposed approach for obtaining embeddings is slightly inferior
to complex neural network methods, its simplicity and interpretability of the
resulting embeddings may offer a more significant advantage for some crucial
applications (such as healthcare) than the insignificant loss in resulting quality.

In the future, it is planned to study the issue of applying this model to
other NLP tasks and to compare the results of topic modeling produced by the
developed approach with other popular topic models, such as LDA, BigARTM
and others.

# References

1. Jain, S., Wallace, B.C.: Attention is not explanation. In: Proceedings of the 2019 Conference of the North American Chapter of the Association for Computational Linguistics: Human Language Technologies, vol. 1 (Long and Short Papers), pp. 3543–3556 (2019). https://doi.org/10.18653/v1/N19-1357
2. Sinoara, R.A., Camacho-Collados, J., Rossi, R.G., Navigli, R., Rezende, S.O.: Knowledge-enhanced document embeddings for text classification. Knowl.-Based Syst. **163**, 955–971 (2019). https://doi.org/10.1016/j.knosys.2018.10.026
3. Sha, Y., Wang, M.D.: Interpretable predictions of clinical outcomes with an attention-based recurrent neural network. In: ACM-BCB' 2017: Proceedings of the 8th ACM International Conference on Bioinformatics, Computational Biology, and Health Informatics, pp. 233–240. ACM SIGBIO; Association for Computing Machinery (2017). https://doi.org/10.1145/3107411.3107445, 8th ACM International Conference on Bioinformatics, Computational Biology,and Health Informatics (ACM-BCB), Boston, MA, 20–23 August 2017
4. Kim, D., Seo, D., Cho, S., Kang, P.: Multi-co-training for document classification using various document representations: TF-IDF, LDA, and Doc2Vec. Inf. Sci. **477**, 15–29 (2019). https://doi.org/10.1016/j.ins.2018.10.006
5. Linh, N.V., Anh, N.K., Than, K., Dang, C.N.: An effective and interpretable method for document classification. Knowl. Inf. Syst. **50**(3), 763–793 (2017). https://doi.org/10.1007/s10115-016-0956-6
6. Unnikrishnan, P., Govindan, V.K., Kumar, S.D.M.: Enhanced sparse representation classifier for text classification. Expert Syst. Appl. **129**, 260–272 (2019). https://doi.org/10.1016/j.eswa.2019.04.003
7. Lau, J.H., Baldwin, T.: An empirical evaluation of Doc2Vec with practical insights into document embedding generation. In: Proceedings of the 1st Workshop on Representation Learning for NLP, pp. 78–86 (2016). https://doi.org/10.18653/v1/W16-1609
8. Ji, S., Satish, N., Li, S., Dubey, P.K.: Parallelizing Word2Vec in shared and distributed memory. IEEE Trans. Parallel Distrib. Syst. **30**(9), 2090–2100 (2019). https://doi.org/10.1109/TPDS.2019.2904058
9. Zhang, H., Shafiq, M.O.: Survey of transformers and towards ensemble learning using transformers for natural language processing. J. Big Data **11**(1), 25 (2024). https://doi.org/10.1186/s40537-023-00842-0
10. Sakketou, F., Ampazis, N.: A constrained optimization algorithm for learning glove embeddings with semantic lexicons. Knowl.-Based Syst. **195**, 105628 (2020). https://doi.org/10.1016/j.knosys.2020.105628
11. Peters, M.E., et al.: Deep contextualized word representations. In: Proceedings of the 2018 Conference of the North American Chapter of the Association for Computational Linguistics: Human Language Technologies, vol. 1 (Long Papers), pp. 2227–2237 (2018). https://doi.org/10.18653/v1/N18-1202
12. Brown, T.B., Mann, B., Ryder, N., Subbiah, M., et al.: Language models are few-shot learners. In: Proceedings of the 34th International Conference on Neural Information Processing Systems, NIPS 2020. Curran Associates Inc., Red Hook, NY, USA (2020)
13. Choi, J., Lee, S.W.: Improving FastText with inverse document frequency of subwords. Pattern Recogn. Lett. **133**, 165–172 (2020). https://doi.org/10.1016/j.patrec.2020.03.003
14. Yakunin, K., et al.: KazNewsDataset: single country overall digital mass media publication corpus. Data **6**(3), 31 (2021). https://doi.org/10.3390/data6030031

# M2DS: Multilingual Dataset for Multi-document Summarisation

Kushan Hewapathirana[1,2(✉)] ⓘ, Nisansa de Silva[1] ⓘ, and C.D. Athuraliya[2] ⓘ

[1] Department of Computer Science & Engineering, University of Moratuwa,
Moratuwa, Sri Lanka
{kushan.22,nisansa}@cse.mrt.ac.lk
[2] ConscientAI, Battaramulla, Sri Lanka
{kushan,cd}@conscient.ai

**Abstract.** In the rapidly evolving digital era, there is an increasing demand for concise information as individuals seek to distil key insights from various sources. Recent attention from researchers on Multi-document Summarisation (MDS) has resulted in diverse datasets covering customer reviews, academic papers, medical and legal documents, and news articles. However, the English-centric nature of these datasets has created a conspicuous void for multilingual datasets in today's globalised digital landscape, where linguistic diversity is celebrated. Media platforms such as British Broadcasting Corporation (BBC) have disseminated news in 20+ languages for decades. With only 380 million people speaking English natively as their first language, accounting for less than 5% of the global population, the vast majority primarily relies on other languages. These facts underscore the need for inclusivity in MDS research, utilising resources from diverse languages. Recognising this gap, we present the Multilingual Dataset for Multi-document Summarisation (M2DS), which, to the best of our knowledge, is the first dataset of its kind. It includes document-summary pairs in five languages from BBC articles published during the 2010–2023 period. This paper introduces M2DS, emphasising its unique multilingual aspect, and includes baseline scores from state-of-the-art MDS models evaluated on our dataset.

**Keywords:** Multi-document Summarisation · Multilingual · Natural Language Processing

## 1 Introduction

The art of document summarisation relies on intricate language skills: the ability to navigate through extensive texts, extract important information, and distil it into concise summaries. In recent years, the surge in deep learning within Natural Language Processing (NLP) has sparked significant interest among researchers in this particular task [1,2,28]. Summarisation stands as a significant challenge in

N.-T. Nguyen et al. (Eds.): ICCCI 2024, CCIS 2165, pp. 219–231, 2024.
https://doi.org/10.1007/978-3-031-70248-8_17

NLP, gaining paramount importance as the demand for easily digestible content continues to soar [9,21].

The field of multi-document summarization (MDS) faces a shortage of comprehensive datasets, unlike the advancements in single-document summarization (SDS). While SDS datasets have expanded to include multilingual summarization, MDS is still relatively new but shows promise. Recent MDS research has explored various domains, such as customer reviews, academic papers, medical and legal documents, and news articles, with a predominant focus on the English language [1,2,21,28]. Despite the availability of extensive SDS datasets like CNN/Daily Mail [15], Gigaword Corpus [31], Newsroom corpus [13], and New York Times [35], there is a scarcity of datasets specifically designed for versatile MDS applications, though MDS datasets like DUC[1], TAC[2], and Multi-News [10] exist which predominantly serve the news domain and limited to the English language.

However, in a world boasting over 7,000 languages, the crucial requirement for multilingual approaches in MDS has become evident. According to the 26th edition of Ethnologue published in 2023, only 380 million people speak English natively as their first language, which accounts for less than 5% of the global population, and the total English-speaking population (i.e. as the first language and second language) is 20% by 2023, which means that the vast majority of the global population is primarily dependent on other languages [8]. This underscores the importance of an inclusive approach in MDS research, where multilingual models cater for diverse languages.

To address this, the research introduces the Multilingual Dataset for Multi-document Summarisation (M2DS). This dataset aims to facilitate the development of robust MDS models across diverse languages, including low-resource languages, for real-world applications. Covering languages such as English, Japanese, Korean, Tamil, and Sinhala, M2DS is considered a pioneering effort in multilingual MDS, complementing existing single-document summarisation datasets in the multilingual domain.

## 2     Related Work

This section aims to delve into the MDS landscape, exploring existing datasets, multilingual text summarisation datasets, and current state-of-the-art models. This provides insights into the diverse facets and recent advancements in MDS.

### 2.1     Major MDS Datasets Across Diverse Domains

Despite being essential for various applications, MDS datasets are relatively scarce compared to SDS datasets. However, the following key datasets have significantly influenced summarisation research. DUC and TAC datasets had set

---

[1] https://duc.nist.gov.
[2] https://tac.nist.gov.

early benchmarks in the news domain [2,28]. The Multi-News [10] dataset offers substantial size and traceability in the news domain. WikiSum [24] leverages Wikipedia and search engine results for abstractive summarisation challenges. Multi-XScience [27] blends arXiv[3] papers and Microsoft Academic Graph [37] (MAG) for scientific writing challenges. BigSurvey [25] and MS$\hat{2}$ [6] contribute to scientific writing, focusing on comprehensive summaries and consolidating conflicting evidence, respectively.

Domain-specific datasets like Rotten Tomatoes [19] and WikiHow [17] diversify summarisation research into movie reviews and knowledge base articles. In customer reviews, Opinosis [11] and OPOSUM [3] are significant, with Opinosis providing professional-written golden summaries for model training and evaluation, and OPOSUM including domain and polarity information across six product categories.

## 2.2  Existing MDS Models

Transformer architecture-based models, particularly those pre-trained on large datasets, have gained attention for their ability to capture inter-document relationships and generate informative summaries. Examples include BERT-SUM [26], using a hierarchical encoder, BART [20], designed as a denoising auto-encoder, PEGASUS [42], leveraging self-supervised learning, and T5 [33], a text-to-text transformer.

In the MDS domain, PRIMERA [41], based on the LongFormer Encoder-Decoder (LED) [4] architecture, stands out, surpassing previous models with a synthetic summary generation strategy during pre-training. DAMEN [30], tailored for the medical domain, combines BERT models with discriminative methods. CGSUM [5] introduces a citation-guided summarization approach for scientific papers.

Despite these advancements, challenges persist in accurately reflecting conflicting information, especially in multi-document scenarios [7]. In multilingual MDS, progress is limited, often relying on linear programming models, and summaries are often in English rather than the original languages, limiting language coverage [29].

## 2.3  Prior Work on Multilingual MDS

The Workshop on Multilingual Summarisation (MultiLing)[4] within the ACL anthology has been a crucial focal point in multilingual summarisation research and the 2013 workshop specifically focused on Multilingual MDS [12]. During this event, a Multilingual MDS corpus was constructed, featuring languages like Arabic, English, Greek, Chinese, Romanian, Czech, Hebrew, and Spanish. The corpus creation involved selecting English texts and employing a sentence-by-sentence translation approach for the featured languages [9,21].

---

[3] https://arxiv.org.

[4] https://aclanthology.org/venues/multiling/.

In terms of model concepts, Marina et al.(2013) [29] introduced a novel text representation model extending the classic Vector Space Model [34] to Hyperplane and Half-spaces. They reformulated the extractive summarisation problem as an optimisation task using linear programming, addressing the challenge of representing a large number of extracts without explicit computation. The optimal solution was found by minimising a distance function in polynomial time. While an evaluation was not conducted, the authors suggested potential assessments using Recall-Oriented Understudy for Gisting Evaluation (ROUGE) scores [23]. This nuanced approach to multilingual MDS, focusing on innovative text representation models and optimisation strategies, has laid a foundation for further exploration and evaluation in subsequent research endeavours [29].

### 2.4    Existing Multilingual Text Summarisation Datasets

In recent years, there has been a notable increase in research exploring the benefits of summarising across diverse languages, particularly in bilingual settings [39]. Our focus is directed towards datasets relevant to multilingual summarising, covering SDS, MDS, and Cross-Lingual Summarization (CLS).

While multilingual SDS has seen significant progress, research in multilingual MDS is limited. CLS, involving generating summaries in one language for documents in another language, has gained momentum with the development of multilingual SDS. Many multilingual SDS efforts have transitioned to include CLS components in their datasets [9, 14, 21, 39].

Key multilingual SDS datasets include MLSUM [36], featuring 1.5 million news articles across multiple languages; XL-Sum [14], a diverse dataset containing 1.35 million articles in 44 languages; WikiLingua [18], one of the largest parallel multilingual summarization datasets; MLGSum [40], drawing from various news providers; and M3LS, comprising over 1.11 million multilingual multimodal instances across 20 languages. Despite their contributions for SDS, the exploration of multilingual MDS is limited due to the absence of a dedicated high-quality dataset.

## 3    M2DS Dataset

This section provides an overview of the data sources, collection, and preprocessing procedures employed in this study. The dataset, named M2DS, consists of news articles in five languages, each paired with professionally written summaries sourced from the BBC. The summaries, crafted by editors, include links to the original articles for reference. The study emphasises transparency and reproducibility, with a commitment to providing links and scripts for replicating the dataset from the specified sources.

### 3.1    Dataset Development

In the rapidly changing digital landscape, the significant increase in online news articles has led to a growing demand for concise and informative content. To

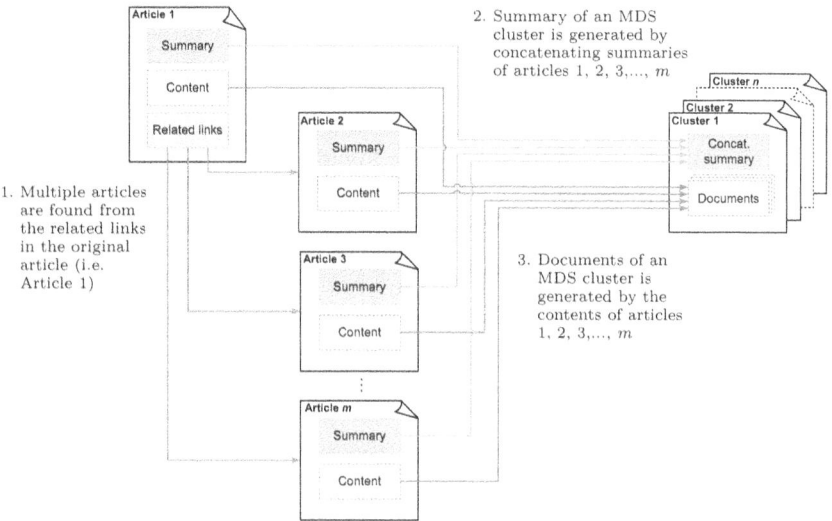

**Fig. 1.** Process of dataset development. The golden summary for each original article was generated by logically combining its own summary and summaries of its related articles, whereas the original article and the related articles served as the collection of multi-documents. These pairs formed multi-document clusters.

**Table 1.** MDS datsaset statistics. The sources are as follows: [*]Xiao et al.(2022) [41], [†]DeYoung et al.(2021) [6], [‡]DeYoung et al.(2023) [7], [•]Fabbri et al.(2019) [10], [°]Lu et al.(2020) [27], [‖]Li et al.(2022) [22], [ᶦ]Liu et al.(2018) [24], [ᶜ]Liu et al.(2023) [25], [†]Leon et al.(2020) [19].

| Dataset | No. of documents | No. of clusters | Avg. no. of documents per cluster | Domain |
|---|---|---|---|---|
| Multi-News[•] | 56.0k[*] | 16.0k | 3.5[*] | News articles[•] |
| Multi-Xscience[°] | 40.0k[*] | 14.0k | 2.8[*] | Related work section in scientific articles[°] |
| Wikisum[ᶦ] | 1.5M[*] | 37.5k | 40.0[*] | Wikipedia articles[ᶦ] |
| BigSurvey-MDS[ᶜ] | 430.0k[*] | 7.0k | 61.4[*] | Human-written survey papers on various domains[ᶜ] |
| PEERSUM[‖] | 11.9k[‖] | 1.5k | 7.8[‖] | Peer reviews of scientific publications |
| MS^2[†] | 470.0k[†] | 20.0k | 23.5[†] | Reviews of scientific publications in medical domain[†] |
| Rotten Tomato Dataset[†] | 244.0k[‡] | 9.0k | 26.8[‡] | Movie reviews[‡] |
| M2DS | 180.0k | 51.5k | 3.5 | News articles |
| – English | 67.0k | 17.0k | 3.9 | |
| – Tamil | 32.0k | 10.0k | 3.2 | |
| – Japanese | 29.0k | 11.0k | 2.6 | |
| – Korean | 27.0k | 8.0k | 3.4 | |
| – Sinhala | 23.5k | 5.5k | 4.2 | |

address this need, the Multilingual Multi-document Summarisation Dataset (M2DS) has been introduced. Emphasising language inclusivity, the dataset focuses on linguistic diversity and uses BBC News as the primary source due to its global coverage and articles available in multiple languages.

We utilised the M3LS dataset to extract links of parsed articles in each language. This dataset served as a valuable foundation for creating our dataset by providing corresponding Twitter page links for each BBC news article. To ensure the reliability of the M3LS dataset [39], the authors conducted a manual assessment of article and summary quality, evaluating factors such as informativeness, length, and the ability to capture essential information. This assessment involved a meticulous review of 100 articles in four languages from their dataset. For each article, the authors carefully read the text and assigned a score between 1–5 to the golden summary, with 5 representing the best possible summary that captures most of the crucial information from the given article and vice versa. Notably, more than 70 articles across the evaluated languages received a score of over 4 out of 5 in their analysis [39].

Assuming uniformity in the quality of articles published by BBC across various domains, the authors extrapolated that this high-quality standard holds true for every language in their dataset [39]. This verification process ensured the overall quality and reliability of the BBC articles and their summaries derived from the M3LS dataset for our study.

Once we completed the first stage, we had a SDS dataset similar to the M3LS dataset, consisting of (1) BBC articles extracted from the links included in the M3LS dataset, (2) the corresponding summaries, and (3) links to related articles which are listed in the original articles. The transformation from an SDS dataset to a MDS dataset involved extracting article-summary pairs from related links, which is illustrated by Fig. 1. To ensure the quality of these summaries and the relatedness of articles in each cluster, we manually verified a sample of 10 clusters, containing 2–10 articles per cluster for English, Tamil, and Sinhala languages.

Our dataset spans from 2010 to 2021, incorporating articles sourced from the M3LS dataset dated between 2010 and 2021. To expand temporal coverage, we collected links from the front page of the BBC News site, focusing on articles from December 2021 to December 2023, and ensured non-duplication. To handle duplicates, we meticulously removed repeated links, guaranteeing the uniqueness of each document cluster. Consequently, the dataset contains a diverse range of articles, free from duplication. The dataset is structured in the Hugging Face `DatasetDict` format, offering ease of access[5].

### 3.2   Dataset Composition

The M2DS dataset encompasses articles in Sinhala, English, Japanese, Korean, and Tamil languages. Our aspiration is for the M2DS dataset to serve as a

---

[5] The dataset can be found at https://huggingface.co/datasets/KushanH/m2ds and https://osf.io/7gjtm/files/osfstorage. The code and pre-trained models are available at https://github.com/KushanMH/m2ds.

**Table 2.** ROUGE scores of selected models on different domain datasets. Note: This study utilises multiple datasets from various domains.(Results obtained from Hewapathirana et al.(2023) [16]). The Multi-News dataset [10] consists of news articles, Multi-XScience [27] focuses on scientific papers, WikiSum [24] provides Wikipedia summaries, and Rotten Tomatoes [19] covers movie reviews. These diverse datasets offer valuable resources for training and evaluating summarization models. The sources for the results are as follows: *Xiao et al.(2022) [41] and •DeYoung et al.(2023) [7]

| Dataset | | PRIMERA | PEGASUS | LED |
|---------|-----|---------|---------|-----|
| Multi-News | R-1 | **42.0**[*] | 32.0[*] | 17.3[*] |
| | R-2 | **13.6**[*] | 10.1[*] | 3.7[*] |
| | R-L | **20.8**[*] | 16.7[*] | 10.4[*] |
| Multi-Xscience | R-1 | **29.1**[*] | 27.6[*] | 14.6[*] |
| | R-2 | **4.6**[*] | 4.6[*] | 1.9[*] |
| | R-L | **15.7**[*] | 15.3[*] | 9.9[*] |
| WikiSum | R-1 | **28.0**[*] | 24.6[*] | 10.5[*] |
| | R-2 | **8.0**[*] | 5.5[*] | 2.4[*] |
| | R-L | **18.0**[*] | 15.0[*] | 8.6[*] |
| Rotten Tomatoes | R-1 | 25.4[•] | **27.4**[•] | 25.6[•] |
| | R-2 | 8.4[•] | **9.5**[•] | 8.0[•] |
| | R-L | 19.8[•] | **21.1**[•] | 19.6[•] |

catalyst, sparking research interest in languages that have received less exploration. Each language-specific cluster within the dataset comprises two to ten documents.

The M2DS dataset comprises 180,000 documents organized into 51,500 clusters, with an average of 3.5 documents per cluster. English-language news articles contribute the highest number of documents at 67,000, whereas Sinhala has the lowest count at 23,500. The average documents per cluster vary across languages, with Japanese news having the lowest at 2.6 and Sinhala having the highest at 4.2. (See Table 1).

### 3.3  Dataset Comparison

As the M2DS dataset is the first of its kind, we decided to compare it with existing MDS datasets. We conducted a comprehensive comparison with existing MDS datasets across various domains, considering factors such as the total number of documents, number of clusters, and the number of documents per cluster. This approach provides valuable insights into the positioning of our dataset within the landscape of English-centric MDS datasets.

To offer a holistic view, we present both aggregated numbers and language-specific statistics. This multi-faceted analysis allows for a nuanced understanding of the dataset's characteristics in comparison to established datasets as shown in Table 1. Comparatively, when assessing M2DS against other MDS datasets, our dataset stands out with a significant overall number of documents. However, on a language-wise comparison, it exhibits a relatively lower count per language, emphasising the importance of considering linguistic variations in dataset analysis.

It is important to note that certain statistical metrics, such as average sentence length, average token count, and average word count per article and per cluster, were not included in the comparison. The rationale behind this omission is the inherent linguistic differences across languages. For instance, languages like Japanese and Korean may convey the same meaning with a lesser number of words, or in some cases, they might encapsulate an entire sentence with a single character. Consequently, direct comparisons based on these metrics could be misleading due to the diverse linguistic structures and expressions employed by different languages.

## 4   Experiments

In this section, we outline the experiments conducted, taking the dataset size relative to existing English-centric MDS datasets into consideration. The dataset was partitioned into training, testing, and validation sets, following a 90-5-5 split for languages other than English [39]. For English, we adopted an 80-10-10 split, aligning with the practices of previous researchers in MDS dataset creation [17,24,25]. A meticulous evaluation of various MDS models was carried out to establish robust baselines. Additionally, we explored the efficacy of open-source large language models, aiming to set a strong baseline for future research.

### 4.1   Pre-trained Model Selection

In the context of Multilingual MDS, there is a noticeable gap in the literature concerning the absence of transformer-based models. Recognising the robustness of such models, the approach in this study involved evaluating pre-trained models to establish baselines. Model selection was guided by an extensive literature review, considering factors like model performance, ROUGE scores, publication year, and venue.

The evaluation focused on three summarising models: PRIMERA, PEGASUS, and LED. PRIMERA demonstrated superior performance in previous studies, while PEGASUS showed superior sentiment understanding, particularly on the Rotten Tomatoes dataset. LED, a widely used pre-trained model, served as a baseline in the existing literature (See Table 2).

Among the models under consideration, PRIMERA emerged as a promising choice due to its distinctive approach to MDS [16]. Seeking to minimize dependency on dataset-specific modeling, it consolidates multiple documents into a single extended sequence, employing the LED architecture known for its computational efficiency. PRIMERA incorporates a sparse "local+global" attention mechanism in the encoder and introduces special document separator tokens

(`<doc-sep>` ) to indicate document boundaries. Inspired by models like PEGA-SUS, PRIMERA adopts a unique masking strategy based on the Entity Pyramid framework, to address the limitations in selecting representative information for summarisation [4, 20, 41].

**Table 3.** Comparison of performance across fine-tuned models on the M2DS dataset

| Language | | Models | | | | | | | |
|---|---|---|---|---|---|---|---|---|---|
| | | LEAD-3 | RANDOM | CENTROID | PRIMERA | PEGASUS | LED | Llama 2 |
| Sinhala | R-1 | 0.06 | 5.7 | 4.5 | 5.7 | 4.1 | 3.6 | **20.2** |
| | R-2 | 0.0 | 0.05 | 0.1 | 2.2 | 2.1 | 1.9 | **6.5** |
| | R-L | 0.06 | 5.1 | 3.9 | 3.2 | 2.8 | 2.9 | **17.3** |
| Japanese | R-1 | 3.5 | 2.3 | 1.9 | 6.3 | 5.7 | 5.9 | **7.7** |
| | R-2 | 0.0 | 0.01 | 0.05 | **3.2** | 1.3 | 1.4 | 0.8 |
| | R-L | 3.5 | 1.9 | 1.7 | 4.1 | 3.3 | 2.7 | **6.8** |
| Korean | R-1 | 2.4 | 1.4 | 1.3 | 5.4 | 5.5 | 4.6 | **8.5** |
| | R-2 | 0.4 | 0.02 | 0.03 | 1.1 | **1.4** | 0.8 | 1.0 |
| | R-L | 2.3 | 1.3 | 1.3 | 2.3 | 2.9 | 1.9 | **8.1** |
| Tamil | R-1 | 6.8 | 1.6 | 2.2 | 4.4 | 3.8 | 3.7 | **10.2** |
| | R-2 | 0.9 | 0.0 | 0.06 | 1.1 | 0.7 | 0.4 | **3.1** |
| | R-L | 6.2 | 1.6 | 1.9 | 2.2 | 1.7 | 1.3 | **9.8** |
| English | R-1 | 1.2 | 6.4 | 7.6 | **28.7** | 22.5 | 20.5 | 20.8 |
| | R-2 | 0.0 | 0.05 | 3.8 | 12.3 | 9.9 | 10.1 | **13.5** |
| | R-L | 1.1 | 5.7 | 7.6 | 17.1 | 14.7 | 15.2 | **19.2** |

## 4.2   Baselines

For our baseline models, we explore simpler extractive approaches and statistical methods alongside pretrained models. In the extractive category, we employ LEAD-3 and RANDOM [39]. LEAD-3 extracts the first three sentences from the source text as the final summary, while RANDOM recursively selects words randomly from the source text until the threshold summary length is reached. These approaches serve as unbiased reference points for understanding and comparing more complex models.

In the statistical approach, we experiment with CENTROID, inspired by [32]. CENTROID ranks sentences based on centrality scores derived from the words within each sentence, utilising TF-IDF scores to measure word similarity. We extract top sentences from each ranking until the threshold summary length is achieved.

Moving to pre-trained models, we select PRIMERA, PEGASUS, and LED, training them on each language's respective training set. For tokenization, we use a space-based tokenizer for Sinhala and Tamil, the original tokenizer for

PRIMERA and LED in other languages, and a space-based tokenizer for PEGA-SUS in all languages except English. For English, we report results both with and without fine-tuning.

Additionally, we present baseline scores for Llama 2, which is an open Large Language Model (LLM) [38]. Llama 2, an updated version of Llama 1 and a formidable 7 billion-parameter causal decoder-only model, is introduced by Meta AI[6]. We limit ourselves to using open LLMs to ensure reproducibility within the research community.

**Table 4.** Comparison of performance across models originally trained on English datasets, on English articles of the M2DS dataset

| Language | | Models | | | | | |
|---|---|---|---|---|---|---|---|
| | | PRIMERA | PRIMERA (fine-tuned) | PEGASUS | PEGASUS (fine-tuned) | LED | LED (fine-tuned) |
| English | R-1 | 23.6 | **28.7** | 18.6 | 22.5 | 17.1 | 20.5 |
| | R-2 | 8.8 | **12.3** | 9.1 | 9.9 | 7.1 | 10.1 |
| | R-L | 13.6 | **17.1** | 12.4 | 14.7 | 13.2 | 15.2 |

## 5  Analysis and Discussion

In our baseline evaluations, Llama 2 7B outperforms all other models, showcasing its robust performance. Notably, PRIMERA excels slightly better in the English language, indicating its effectiveness in capturing linguistic nuances specific to that language. However, when assessing the state-of-the-art MDS models fine-tuned on our dataset, we observed a discernible drop in performance compared to their previous performance under English-centric news domain datasets as depicted in Table 2. This phenomenon could stem from the models struggling to capture language-specific information unique to each language in our multilingual dataset (See Table 3).

A noteworthy observation is the lower scores in LEAD-3, which extracts only the first three sentences as the summary. This suggests that our dataset exhibits better quality, addressing the issues found in TAC/DUC datasets, where the first three sentences often serve as summaries, leading models to learn biased patterns.

Contrary to the trend of using LLMs for MDS, our findings suggest that simpler models, such as PRIMERA specifically designed for MDS tasks, may be more effective. This is evident from PRIMERA's superior performance in English when compared to Llama 2. Designing task-specific models like PRIMERA which perform well without extensive fine-tuning, could be a more effective approach with respect to resource constraints. Additionally, it is crucial to note that Llama 2, without fine-tuning, achieves competitive results, highlighting its potential

---

[6] https://ai.meta.com.

for zero-shot learning and its effectiveness across diverse datasets compared to models specifically trained on individual datasets.

Furthermore, it is essential to emphasise the scalability of models like Llama 2, indicating their potential for handling larger datasets and their adaptability across various domains. Additionally, future research should explore Transfer Learning techniques to enhance the performance of MDS models across different languages, minimising the observed drop in performance. Finally, understanding the impact of dataset quality on model evaluation is crucial, and our dataset's higher quality, as reflected in low LEAD-3 scores, underscores the significance of curating datasets that truly represent the summarisation task.

Additionally, we conducted a comparison of PRIMERA's and other models' performance with and without fine-tuning on the English language subset of our dataset (See Table 4). Although PRIMERA excels with a zero-shot approach surpassing other models, its scores are slightly lower when compared to other MDS models trained on news domain datasets. This suggests that our dataset presents challenges for models, underscoring its quality. Furthermore, all the models have improved their performance when they are fine-tuned. For instance, PRIMERA's score increased from 23.6 to 28.7, exhibiting the highest improvement among other models.

## 6 Conclusion and Future Directions

The study introduces the M2DS dataset to fill the gap in multilingual MDS datasets. While existing MDS datasets have made strides in various domains, they mostly focus on English, leaving a void in multilingual representation. M2DS, with document-summary pairs in five languages, stands out as the pioneering multilingual MDS dataset.

The evaluation of M2DS against existing datasets demonstrates its potential and unique contribution to the field. Baseline scores from state-of-the-art MDS techniques provide a benchmark for future research in multilingual settings. Llama 2 7B outperforms other models, showcasing robust performance. PRIMERA excels slightly better in English, indicating effectiveness in capturing language-specific nuances.

The introduction of M2DS opens avenues for future research, enabling researchers to enhance the robustness of MDS models across diverse linguistic contexts. Possible directions include language-specific model tuning, exploring multilingual model development, and extending M2DS into diverse domains beyond news articles for broader applications.

## References

1. Abid, A.M.: Multi-document text summarization using deep belief network. Int. J. Adv. Sci. Res. Eng. (2022)
2. Afsharizadeh, M., Ebrahimpour-Komleh, H., et al.: A survey on multi-document summarization and domain-oriented approaches. J. Inf. Syst. Telecommun. **1**(37), 68 (2022)

3. Angelidis, S., Lapata, M.: Summarizing opinions: Aspect extraction meets sentiment prediction and they are both weakly supervised. In: EMNLP, pp. 3675–3686 (2018)
4. Beltagy, I., Peters, M.E., et al.: Longformer: the long-document transformer. arXiv preprint arXiv:2004.05150 (2020)
5. Chen, J., Cai, C., Jiang, X., Chen, K.: Comparative graph-based summarization of scientific papers guided by comparative citations. In: Proceedings of the 29th International Conference on Computational Linguistics, pp. 5978–5988 (2022)
6. DeYoung, J., Beltagy, I., van Zuylen, M., Kuehl, B., Wang, L.L.: MS$\hat{2}$: multi-document summarization of medical studies. In: EMNLP, pp. 7494–7513 (2021)
7. DeYoung, J., Martinez, S.C., Marshall, I.J., Wallace, B.C.: Do multi-document summarization models synthesize? arXiv preprint arXiv:2301.13844 (2023)
8. Eberhard, D.M., G.F.S., Fennig, C.D.: Ethnologue: languages of the Americas and the pacific (2023)
9. Elhadad, M., Miranda-Jiménez, S., Steinberger, J., Giannakopoulos, G.: Multi-document multilingual summarization corpus preparation, Part 2: Czech, Hebrew. In: Proceedings of the MultiLing 2013 Workshop on Multilingual Multi-document Summarization, pp. 13–19 (2013)
10. Fabbri, A.R., Li, I., She, T., Li, S., Radev, D.: Multi-news: a large-scale multi-document summarization dataset and abstractive hierarchical model. In: ACL, pp. 1074–1084 (2019)
11. Ganesan, K., Zhai, C., Han, J.: Opinosis: a graph based approach to abstractive summarization of highly redundant opinions. In: Coling 2010, pp. 340–348 (2010)
12. Giannakopoulos, G.: Multi-document multilingual summarization and evaluation tracks in ACL 2013 multiling workshop. In: Proceedings of the Multiling 2013 Workshop on Multilingual Multi-document Summarization, pp. 20–28 (2013)
13. Grusky, M., Naaman, M., Artzi, Y.: Newsroom: a dataset of 1.3 million summaries with diverse extractive strategies. In: ACL, pp. 708–719 (2018)
14. Hasan, T., Bhattacharjee, et al.: Xl-sum: large-scale multilingual abstractive summarization for 44 languages. In: ACL-IJCNLP 2021, pp. 4693–4703 (2021)
15. Hermann, K.M., et al.: Teaching machines to read and comprehend. Adv. Neural Inf. Process. Syst. **28** (2015)
16. Hewapathirana, K., De Silva, N., Athuraliya, C.D.: Multi-document summarization: a comparative evaluation. In: 2023 IEEE 17th International Conference on Industrial and Information Systems (ICIIS), pp. 19–24. IEEE (2023)
17. Koupaee, M., Wang, W.Y.: Wikihow: a large scale text summarization dataset. arXiv preprint arXiv:1810.09305 (2018)
18. Ladhak, F., Durmus, E., Cardie, C., Mckeown, K.: Wikilingua: a new benchmark dataset for cross-lingual abstractive summarization. In: EMNLP 2020, pp. 4034–4048 (2020)
19. Leon, S.: Rotten tomatoes movies and critic reviews dataset (2020). https://bit.ly/RTdataset. Accessed 24 June 2023
20. Lewis, M., Liu, Y., et al.: Bart: denoising sequence-to-sequence pre-training for natural language generation, translation, and comprehension. In: ACL, pp. 7871–7880 (2020)
21. Li, L., Forăscu, C., El-Haj, M., Giannakopoulos, G.: Multi-document multilingual summarization corpus preparation, part 1: Arabic, English, Greek, Chinese, Romanian. In: Proceedings of the Multiling 2013 Workshop on Multilingual Multi-document Summarization, pp. 1–12 (2013)
22. Li, M., Qi, J., Lau, J.H.: Peersum: a peer review dataset for abstractive multi-document summarization. arXiv preprint arXiv:2203.01769 (2022)

23. Lin, C.Y.: Rouge: a package for automatic evaluation of summaries. In: Text Summarization Branches Out, pp. 74–81 (2004)
24. Liu, P.J., Saleh, M., et al.: Generating wikipedia by summarizing long sequences. arXiv preprint arXiv:1801.10198 (2018)
25. Liu, S., Cao, J., Yang, R., Wen, Z.: Generating a structured summary of numerous academic papers: dataset and method. arXiv preprint arXiv:2302.04580 (2023)
26. Liu, Y., Lapata, M.: Text summarization with pretrained encoders. In: EMNLP-IJCNLP, pp. 3730–3740 (2019)
27. Lu, Y., Dong, Y., Charlin, L.: Multi-xscience: a large-scale dataset for extreme multi-document summarization of scientific articles. In: EMNLP, pp. 8068–8074 (2020)
28. Ma, C., Zhang, W.E., et al.: Multi-document summarization via deep learning techniques: a survey. ACM Comput. Surv. (2020)
29. Marina, L., Natalia, V.: Multilingual multi-document summarization with poly. In: Proceedings of the MultiLing 2013 Workshop on Multilingual Multi-document Summarization (2013)
30. Moro, G., Ragazzi, L., Valgimigli, L., Freddi, D.: Discriminative marginalized probabilistic neural method for multi-document summarization of medical literature. In: ACL, pp. 180–189 (2022)
31. Napoles, C., Gormley, M.R., Van Durme, B.: Annotated gigaword. In: Proceedings of the Joint Workshop on Automatic Knowledge Base Construction and Web-scale Knowledge Extraction (AKBC-WEKEX), pp. 95–100 (2012)
32. Radev, D., Jing, H., Budzikowska, M.: Centroid-based summarization of multiple documents: sentence extraction, utility-based evaluation, and user studies. In: NAACL-ANLP 2000 Workshop: Automatic Summarization (2000)
33. Raffel, C., Shazeer, N., et al.: Exploring the limits of transfer learning with a unified text-to-text transformer. J. Mach. Learn. Res. **21**(1), 5485–5551 (2020)
34. Salton, G.: A vector space model for information retrieval. J. ASIS 613–620 (1975)
35. Sandhaus, E.: The New York Times Annotated Corpus (2008). https://catalog.ldc.upenn.edu/LDC2008T19
36. Scialom, T.D., et al.: Mlsum: the multilingual summarization corpus. In: EMNLP, pp. 8051–8067 (2020)
37. Sinha, A., Shen, Z., et al.: An overview of Microsoft Academic Service (MAS) and applications. In: WWW, pp. 243–246 (2015)
38. Touvron, H., Martin, et al.: Llama 2: open foundation and fine-tuned chat models. arXiv preprint arXiv:2307.09288 (2023)
39. Verma, Y., Jangra, A., Verma, R., Saha, S.: Large scale multi-lingual multi-modal summarization dataset. In: ACL, pp. 3602–3614 (2023)
40. Wang, D., Chen, J., Zhou, H., Qiu, X., Li, L.: Contrastive aligned joint learning for multilingual summarization. In: ACL-IJCNLP 2021, pp. 2739–2750 (2021)
41. Xiao, W., Beltagy, I., Carenini, G., Cohan, A.: Primera: pyramid-based masked sentence pre-training for multi-document summarization. In: ACL, pp. 5245–5263 (2022)
42. Zhang, J., Zhao, Y., et al.: Pegasus: pre-training with extracted gap-sentences for abstractive summarization. In: ICML, pp. 11328–11339. PMLR (2020)

# uMentor: LLM-Powered Chatbot
# for Harnessing Technology Books
# in Digital Library

Lan T. K. Nguyen[1] , Long D. Pham[2] , and Hoa N. Nguyen[2](✉)

[1] VNU University of Social Sciences and Humanities, Hanoi, Vietnam
`lanntk@vnu.edu.vn`
[2] VNU University of Engineering and Technology, Hanoi, Vietnam
{`20021388,hoa.nguyen`}`@vnu.edu.vn`

**Abstract.** Large language models (LLMs) are currently attracting considerable interest and being extensively utilised across all disciplines. Chatbots such as ChatGPT are used by many people and are also employed as internal personal assistants in numerous firms. However, in education, ChatGPT is currently limited in providing comprehensive knowledge and resources related to technology curriculum, with a most focus on English material. To some extent, the access, exploitation and comprehension of the content and information in technological fields in terms of materials and knowledge are relatively limited for our local Vietnamese students due to individual skills and linguistic obstacles. The study presents uMentor, a method that uses the pretrained LLaMA-2 model trained on Vietnamese language datasets to help students enhance learning quality in technological fields with the aid of a virtual mentor. For uMentor, we firstly propose a process to collect, normalize, and improve a specific instruction dataset on technology themes in Vietnamese. We then perform the fine-tuning the pretrained LLaMA-2 model with our instruction dataset. Consequently, we build a virtual mentor prototype to assist students with academic issues and offer assistance on technology book discovery, study road-maps, schedules, and technology-related resources. The preliminary results indicate a notable enhancement in the learning and teaching environment because of uMentor's efficacy in resolving knowledge deficiencies and increasing academic engagement.

**Keywords:** Large Language Models · Virtual Mentor · Chatbot · Book Understanding · Digital Library

## 1 Introduction

There is an increasing need for jobs in technological sectors to support the digital economy and societal requirements. The number of students enrolling in technology-related courses is increasing rapidly. In 2023, at Vietnam National

N.-T. Nguyen et al. (Eds.): ICCCI 2024, CCIS 2165, pp. 232–244, 2024.
https://doi.org/10.1007/978-3-031-70248-8_18

University Hanoi (VNU), there was a 27% rise in students choosing technology and other AI-related disciplines compared to 2022. Several prominent institutions in Vietnam, such as VNU, have introduced a range of short training courses, majors, and diplomas to attract students interested in emerging subjects. Postgraduates have also seen a significant increase in the number of students enrolling to enhance and perfect their talents, not just for undergraduate education. Learning resources for these technical sectors are currently limited, and there is a lack of good tools or software systems to help students utilise them efficiently [19]. Technology students require tools with extensive knowledge, such as chatGPT, to provide information and knowledge related to the curriculum and learning materials of technology subjects, based on assessments, evaluations, and teaching experience. The software tools should function as guides and virtual learning mentors for students.

There are various methods available to assist students in their virtual learning paths. There has been a significant increase in the online learning community over the years, allowing students to independently search for and select solutions. Learning management systems (LMS) have been developed over the course of decades, with each institution having its own LMS to facilitate internal assistance, guidance, communication, and information exchange [16]. Many businesses seek to develop a learning advising system supported by AI models that can offer timely solutions tailored to individual users' preferences, behaviours, and learning goals over the next 5 years. Students might benefit from receiving help on numerous issues such as establishing programmes, training roadmaps, schedules, learning materials, and academic consultations.

Large Language Models (LLMs) have been researched, developed, and widely applied. Typical examples include ChatGPT (10/2022), Perplexity (8/2022), ClaudeAI (7/2023), Gemini (12/2023). LLMs have undoubtedly advanced the creation of chatbots capable of engaging with consumers to enable timely question-answer interactions across multiple knowledge fields. ChatGPT, rightnow, may not be able to assist students in exploring and answering in-depth inquiries in technical sectors. Additionally, it has limitations in terms of languages available for Vietnamese students. These limitations can be entirely overcome with further research to explore open-source LLM models, such as Meta's LLaMA-2 (short for Large Language Model Meta AI 2). This is a pre-trained LLM models (using 7, 13 and 70 billion parameters) trained on a large amount of data to generate logical and natural answers [23].

The study, therefore, aims to utilise LLMs technology to create a virtual mentor for technology students at VNU-UET to support their learning paths. Our research problem is established with the following challenges:

1. Chatbots like chatGPT are unable to provide comprehensive knowledge and content related to technology curricula and materials.
2. There is no specialized corpus that gathers comprehensive curricula and materials to serve students in technology fields.
3. Accessing, exploring, and understanding the content and knowledge within technology curricula and materials is relatively limited for technology stu-

dents at VNU due to limitations in both knowledge and language. Therefore, systems and services like chatbots are essential to ensure the role of a virtual mentor for them.

From the research objective, the main results we obtained in this study include:

1. Proposing a method to support Vietnamese students in enhancing the quality of learning in technology subjects through virtual mentorship based on LLMs.
2. Collecting, normalizing, and enriching a specialized instruction dataset for technology subjects in Vietnamese. Subsequently, fine-tuning the pretrained LLaMA-2 model for Vietnamese. Consequently, we have developed a virtual mentor prototype to assist students in answering questions, providing advice on study programs, schedules, study materials, and answering academic questions in the field of technology.

The remainder of this paper is organized as follows. Section 2 introduces some fundamental theory and related works in knowledge integration and management. Section 3 presents our proposed method. In Sect. 4, we discuss and evaluate our proposed work. Finally, Sect. 5 concludes our contributions and describes some future works.

## 2   Background and Related Works

Organizations are increasingly embracing chatbot technology, placing particular emphasis on consumer chatbots for internal use cases and business-to-business (B2B) interactions. Practical implementations of these solutions have increased the productivity [21] and efficiency of an expanding array of industries [11], including, education [1,22], climate change [24], medical pharmacy [7] financial services [6], retail [10,13], oilfield [26], and other B2B interactions automation. Depending on the requirements of various organisations, bots with industry-specific commercial applications are developed. In [27], the author suggested a system within a Chinese telecommunications company in which conversational AI-powered voice navigation in the customer service hotline would increase the initial customer service's efficacy in addressing client complaints and inquiries. Additional examples include instances where voice navigation can comprehend the travel needs of passengers, thereby enabling the driver to concentrate on the road ahead [3,30]. The domain of chatbot business applications has effectively mitigated the issue of inequitable distribution of educational resources by enabling the dissemination and advancement of personalized learning. In this context, user demands enhance the development of educational products that are more diverse and personalized, thereby enhancing adaptability for users across all levels. The following table represent the summary of literature reviews on how chatbots have transformed businesses' activities and operations in diverse fields.

In education [18,19], the authors have concentrated on improving personalized learning and rapid feedback to boost engagement among teachers and students. This is achieved through the implementation of intelligent tutoring systems, automated grading systems, and chatbot-led interviews for data collecting. The immediate assistant provides administrative help and academic guidance to information users promptly. Some of the similar examples could be found in the fields of academic libraries where have established a 24/7 reference services to enhance the efficiency of library operation, offering not only personalized experiences but also aiding the library acquisition, collection, dissemination, discovery as well as cataloging and metadata management [2,14,15]. Lastly, the challenging sector involves the enhancement of citizen engagement, service delivery, efficiency, and accessibility through the implementation of chatbots in government administrations. This includes effectively managing feedback, insights, inquiries, and public services for local citizens [4,25].

In brief, the implementation of chatbots has revolutionized the capabilities of technologies in diverse sectors, emphasizing their pivotal functions in augmenting service industry provision, enhancing workforce productivity, and fostering user participation through the provision of interactive, responsive, and personalized assistance to a multitude of stakeholders through the utilization of NLP and LLMs to not only revolutionize conventional procedures but also generate more inventive resolutions. The astute observations have unveiled the wide-ranging utility of substantial advantages offered by conversational AI. Furthermore, they have demonstrated how chatbots are transforming operational workflows, interactions, and customer experiences across diverse sectors. Certain instances have highlighted the transformative potential of AI's implications not only for industries but also for academic institutions and academic libraries. These implications pertain to the establishment of new library services and user engagements, with an emphasis on a collaborative approach in which AI systems supplement the expertise of library and school professionals. By doing so, academic libraries can provide services that are more effective, informed, and user centric.

## 3   Proposed Method

In this section, the proposed method to develop such specialized virtual learning mentor is described, supporting technology students and leveraging LLMs.

### 3.1   Approach Direction

With the aim of forming a virtual learning mentor to support Vietnamese students in the field of technology, our approach is formed on the following main ideas:

1. Collect, clean, standardize, and process data to create a specialized dataset gathering course materials, study resources in the field of technology.
2. Supplement and enrich the dataset with a set of common questions related to technology subjects that students often encounter to form a training dataset.

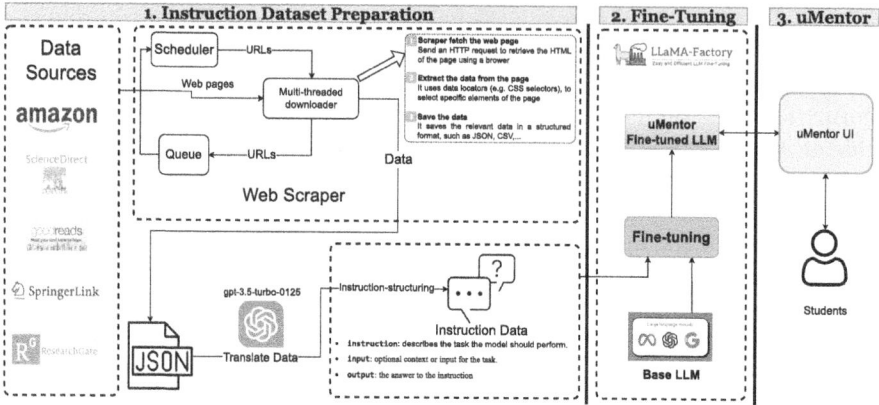

**Fig. 1.** Overview of uMentor

3. Evaluate and select the pretrained LLMs in Vietnamese to perform fine-tuning based on the collected training dataset.
4. Develop a chatbot acting as a virtual mentor to assist students in searching, looking up, exchanging, and synthesizing key content, knowledge within technology course materials and study resources.

The installation of uMentor has been carried out gradually, addressing the problems identified. The initial specialised dataset is established by data collecting and processing, primarily extracted from metadata in libraries and leading book publishing companies like Amazon, O'Reilly, Packt, Wiley, Apress, and Pearson. When crawling books from websites, it's important to remember that only the metadata description is being collected, not the actual text, to prevent copyright infringement. The second stage is refining all the metadata records we have gathered using Vistral, a LLMs previously trained with Vietnamese text. The last stage involves creating a chatbot to act as a virtual mentor for VNU-UET students, assisting them with information retrieval and knowledge sharing related to technology literature.

In this section, we outline the fine-tuning approach and data generation process for training the Vistral [5] model to suit the field of book understanding and personalized recommendation of the VNU Digital Library, called uMentor, which is being piloted at the University of Engineering and Technology (VNU-UET). The ultimate goal is to improve the accuracy of book recommendation systems, enhance user experience, and expand the application scope of the model across the entire VNU library network, creating an advanced and effective personalized book advising system.

## 3.2   Instruction Dataset Preparation

To accomplish the task of book understanding and personalized recommendation, it is necessary to construct a set of instructions for books. This process is divided into 3 stages Fig. 1.

The first step in building textbook guidelines involves meticulous data collection. Data is carefully selected from reputable websites such as Goodreads and Amazon, ensuring accuracy and currency. Collected data encompasses detailed information about each book, including title, description, publisher, as illustrated in Fig. 2. The data collection scope focuses on books on artificial intelligence and machine learning, a rapidly developing field that aligns with the learning and research needs of students at VNU-UET.

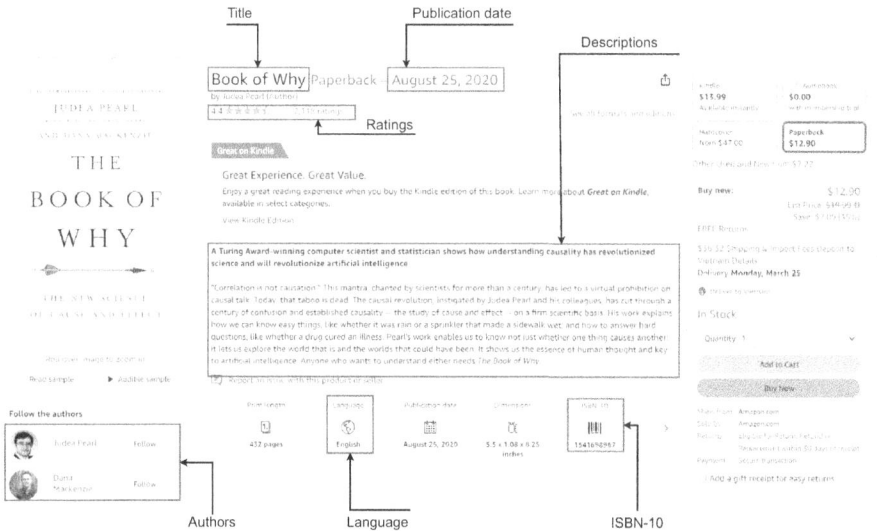

**Fig. 2.** Book Metadata

Specifically, we collect book metadata. This process involves collecting information such as title, authors, publisher, and more. As the dataset was in English, we used the GPT-3.5-Turbo-0125 model to translate the data into Vietnamese in order to build a high-quality Vietnamese dataset. For book descriptions, which are often lengthy, we split them into sentences and group sentences into paragraphs with a maximum of 1024 tokens per paragraph. After translation, the paragraphs are recombined to form the original description. This process of splitting and translating in segments ensures the logical flow and meaning of the content.

Finally, from the Vietnamese dataset, we constructed an instructions dataset to facilitate the fine-tuning of the model. The instructions dataset will encompass a collection of question-answer pairs that users are interested in when seeking information about a book, such as inquiries about the author and the content.

### 3.3   LLM Fine-Tuning Process

Currently, there are a number of LLMs dedicated to Vietnamese, such as ViG-PTQA, BloomZ, PhoGPT4B, and Vistral. In this work, we leverage the Vistral [5] model as the backbone for fine-tuning, a multi-turn conversational large language model for Vietnamese. Vistral is extended from the Mistral 7B [12] model using diverse data for continual pre-training and instruction tuning. Vistral achieves the highest benchmark scores among open-source Vietnamese models on the VMLU benchmark (A Vietnamese Multitask Language Understanding Benchmark Suite for Large Language Models), second only to GPT-4. However, fine-tuning the 7B parameter model requires a large GPU memory budget. To overcome this challenge, we utilize the LLaMA-Factory [8] framework and apply the LoRA [9] architecture for fine-tuning the model.

### 3.4   AI-Powered Academic Mentor

The goal of our research is to develop a virtual mentor called uMentor to support students in learning technology and engineering subjects. With uMentor, students are empowered to take charge of their learning. They can tap into a wealth of knowledge and resources through interactive Q&A sessions, tailored learning pathways, and curated book recommendations, all designed to support their academic success. After fine-tuning the Vistral model, we will build a functional interface that utilizes the model to create uMentor. The interface will be similar to other popular LLM-based chatbots: students ask questions, and uMentor receives, analyzes, and processes the questions using the fine-tuned LLM before providing an answer.

## 4   Experiments and Evaluation

In order to prove the ability and efficiency of our proposed method, we carry out a prototype of LLM-powered virtual mentor aimed to provide comprehensive technological textbook knowledge at VNU-UET to respond to the following research questions:

- RQ1: Can uMentor answer questions, provide advice on formulation programs, training road-maps, schedules, learning materials in technology field?
- RQ2: In what way the uMentor can overcome the language barriers and information skills at Vietnamese students in technology?

The following sections will describe our experimental results and evaluation.

### 4.1   Experiment Environment

**Finetuning:** The entire process was conducted on a single A100-80GB GPU. The training configuration included a maximum input sequence length of 512, a LoRA rank (k) of 8, a total batch size of 64, and a learning rate of 5e–5. The trainable LoRA parameters amounted to approximately 3.4M and were fine-tuned for 3 epochs, taking about 6 h to complete.

**Deployment:** All of our experiments deploy on an NVIDIA Tesla T4 (16 GB) GPU, 2 × Xeon Platinum 8160, and 256 GB of RAM. Besides that, we implement our method using the PyTorch framework, which robustly supports deep learning.

## 4.2   Experiment Results

Our proposed solution outlined in 3 involved developing uMentor and integrating it into our digital library system VNU-LIC to assist students in technology-related subjects. In the following subsections, we will analyze in detail the experimental results to address all the research questions posed.

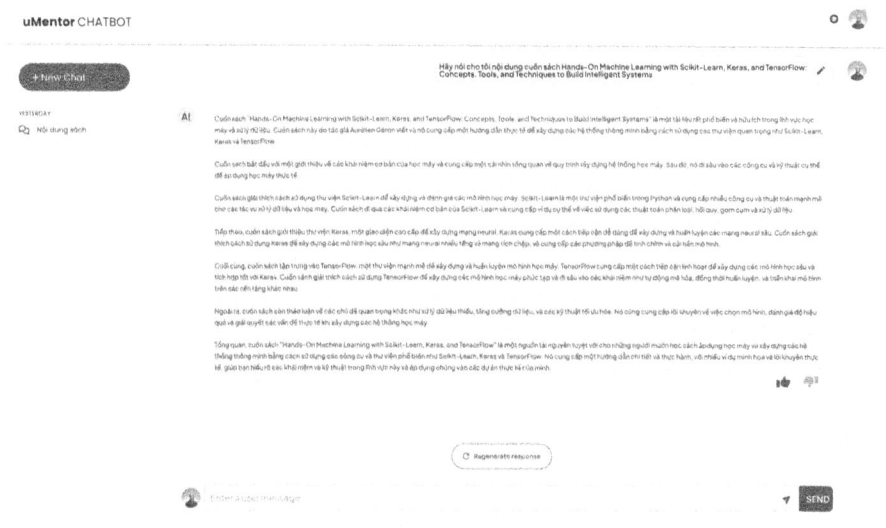

**Fig. 3.** uMentor User Inteface

**Instruction Dataset:** It plays a crucial role in fine-tuning LLMs [28], providing them with detailed instructions on how to perform tasks. This leads to increased efficiency and accuracy, as well as improved control and flexibility over the fine-tuning process. Consequently, the use of instruction datasets helps to optimize performance and accelerate the development of LLMs.

The instruction dataset was constructed to train large language models capable of supporting learning in Vietnamese. After the collection and processing process, we have created a dataset with nearly 18,000 high-quality instructions for science textbooks. The dataset contains instructions related to textbooks in the domains of Artificial Intelligence and Machine Learning, encompassing

6 areas: *Computer Vision - Pattern Recognition, Artificial Intelligence Expert Systems, Artificial Intelligence - Semantics, Machine Theory, Natural Language Processing, Computer Neural Networks*

**Fine-Tuning LLM:** To enhance its performance as a book mentor, we fine-tuned the pretrained LLM on the Instruction Dataset collected above. Through this process, the model becomes more familiar with the knowledge from the books. After fine-tuning, the new LLM is capable of providing advice and support to users with higher accuracy and richer information. It can understand the user's intent and provide relevant answers, meeting their search and learning needs.

**uMentor: Enhancing Technology Learning for Students:** To deploy uMentor on the web environment, we use the llama.cpp library. This library allows us to enable LLM inference with minimal setup and state-of-the-art performance on a wide variety of hardware - locally and in the cloud. An example of the uMentor deployment on the web environment can be seen in Fig. 3.

**Table 1.** Evaluation of uMentor's LLM Performance

|  | ROUGE-1 | ROUGE-2 | BLEU | $P_{BERT}$ | $R_{BERT}$ | $F_{BERT}$ |
|---|---|---|---|---|---|---|
| GPT-3.5-Turbo | 0.5029 | 0.2854 | 0.1395 | **0.7252** | 0.6783 | 0.7008 |
| BaseModel | 0.5847 | 0.3248 | 0.1444 | 0.7119 | 0.6861 | 0.6985 |
| **uMentor** (our) | **0.6252** | **0.3399** | **0.1675** | 0.7145 | **0.6914** | **0.7025** |

## 4.3   Evaluation and Discussion

In the first research question RQ1, the establishment of uMentor will act as a virtual assistant to respond to questions from UET's students. The process of fine-tuning for a certain subject involves selecting technology books to assist both the chatbot and users in providing tailored replies related to technology resources, with a particular emphasis on technology books. For the second question RQ2, designing uMentor with a Vietnamese context will familiarise the model with Vietnamese vocabulary, expressions, leading to improved responses and support in Vietnamese scenarios. This is the inaugural academic Vietnamese chatbot in this environment supporting accessing a vast dataset of Vietnamese and English technology literature which will enable students to focus on practical applications and professional abilities.

To evaluate the performance of uMentor, we focused on assessing the model's interpretation capability by using book summary recommendation task. The goal was to examine the model's ability to understand books through their summaries, which in turn could help to stimulate the reader's interest during interaction.

As there is no benchmark available for this task in Vietnamese, we constructed an evaluation dataset by collecting book summaries from the Amazon website. These summaries were written either by the authors themselves or by Amazon's book editing experts. The evaluation dataset consists of 30 most popular books across 6 categories of AI and Machine Learning.

To assess models' performances, we utilize automated metrics (ROUGE-1, ROUGE-2, BLEU-1, and BERTScore). ROUGE (Recall-Oriented Understudy for Gisting Evaluation) [17] and BLEU (Bilingual Evaluation Understudy) [20] are used to assess the fluency and overlap of the summary text, while BERTScore [29] evaluates the ability of the summary to convey meaning and naturalness.

### 4.4   LLM Ablation Study and Analysis

Table 1 shows the evaluation results on the book summarization task, and the uMentor model significantly outperformed both GPT-3.5-Turbo and BaseModel. Specifically, uMentor model achieved a ROUGE-1 score of 0.6252, which is 24.3% higher than GPT-3.5-Turbo's 0.5029 and 6.9% higher than BaseModel's 0.5847. uMentor also had the highest ROUGE-2 score of 0.3399, surpassing GPT-3.5-Turbo by 19.1% and BaseModel by 4.6%.

In terms of BLEU score, which evaluates the quality of translation and text generation based on n-gram overlap between the model output and the reference set, uMentor model again leads with a score of 0.1675, 20.1% higher than GPT-3.5-Turbo and 15.9% higher than BaseModel. This demonstrates that uMentor model has the best ability to generate text that closely mimics real-world data.

Regarding the BERTScore metrics, $P_{\text{BERT}}$, $R_{\text{BERT}}$, and $F_{\text{BERT}}$, uMentor model still shows superiority over the other two models. uMentor model $P_{\text{BERT}}$ is 0.7145, compared to 0.7252 of GPT-3.5-Turbo and 0.7119 of BaseModel, indicating that although it slightly lags behind in terms of precision, uMentor model still has competitive performance. For $R_{\text{BERT}}$, uMentor model achieves a score of 0.6914, surpassing BaseModel by 0.7% and GPT-3.5-Turbo by 1.9%, indicating a better recall ability. Finally, $F_{\text{BERT}}$, the harmonic mean of precision and recall, is at its highest for uMentor model at 0.7025, compared to 0.7008 of GPT-3.5-Turbo and 0.6985 of BaseModel.

Overall, the uMentor-model demonstrated a significant advantage over the other two models in the task of summarizing technical books. In terms of ROUGE, BLEU, and BERT metrics, uMentor achieved scores 24.3%, 20.1%, and 0.2% higher than GPT-3.5-Turbo, respectively, along with superior performance on other metrics. This demonstrates the potential of the uMentor model fine-tuning approach and emphasizes the need for incorporating instructions in book summarization to achieve the best results in the future.

## 5   Conclusions

The collaboration between humans and machines can be challenging. uMentor, the first to utilize LLMs as a virtual mentor, has the potential to be a powerful technology in assisting students at VNU-UET by leveraging the technology books of VNU-LIC. It supports them in their current workflows and tasks, enabling collaboration not only with machines but also with individuals within the same institutions. One of the approaches suggests creating high-quality Vietnamese datasets tailored for various technological fields related to books.

Challenges remain in obtaining technology books in Vietnamese and accessing English content. To enhance uMentor, more stakeholders need to be involved. uMentor could serve as a advisor, guiding students to develop more skills. uMentor could be trained to incorporate expertise into its algorithms, acting as a personalized coach and teammate. This suggests a future where humans and machines work together in real-time based on collective intelligence. uMentor's conversational AI has future implications. Deploying chatbots that go beyond basic conversation capabilities can be time-consuming and costly, especially when considering issues related to books and copyrights. Various challenges are arising within the interconnected departments of VNU, particularly regarding the deployment of software. These challenges include flexibility, scalability, security, and assuring non-copyright infringement. Currently, the trial version does not need to access or store user profile information. In the future, we aim to incorporate the initial trial version of uMentor into our present digital library to provide digital library services for all VNU departments, schools, and majors. The chatbot may need to access user profile information and credentials to provide useful functions and personalise services once it becomes official. These days, organisations must prioritise investing more resources in growing AI technology, especially in crucial industries like education, to align with their strategic visions and long-term goals.

# References

1. Atlas, S.: Chatgpt for higher education and professional development: a guide to conversational AI (2023)
2. Bagchi, M.: Conceptualising a library chatbot using open source conversational artificial intelligence. DESIDOC J. Libr. Inf. Technol. **40**(6) (2020)
3. Balog, K.: Conversational AI from an information retrieval perspective: remaining challenges and a case for user simulation (2021)
4. Chen, T., Gascó-Hernandez, M., Esteve, M.: The adoption and implementation of artificial intelligence chatbots in public organizations: evidence from US state governments. Am. Rev. Public Administrat. 02750740231200522 (2023)D
5. Chien, V.N., et al.: Vistral-7b-chat - towards a state-of-the-art large language model for Vietnamese (2023)
6. George, A.S., George, A.H.: A review of chatgpt AI's impact on several business sectors. Partners Univ. Int. Innov. J. **1**(1), 9–23 (2023)
7. Han, T., et al.: Medalpaca–an open-source collection of medical conversational AI models and training data. arXiv preprint arXiv:2304.08247 (2023)
8. hiyouga: Llama factory (2023). https://github.com/hiyouga/LLaMA-Factory
9. Hu, E.J., et al.: Lora: low-rank adaptation of large language models. CoRR arxiv:2106.09685 (2021)

10. Jan, I.U., Ji, S., Kim, C.: What (de) motivates customers to use AI-powered conversational agents for shopping? the extended behavioral reasoning perspective. J. Retail. Consum. Serv. **75**, 103440 (2023)
11. Ji, H., Han, I., Ko, Y.: A systematic review of conversational AI in language education: focusing on the collaboration with human teachers. J. Res. Technol. Educ. **55**(1), 48–63 (2023)
12. Jiang, A.Q., et al.: Mistral 7b (2023)
13. Kamoonpuri, S.Z., Sengar, A.: Hi, may AI help you? an analysis of the barriers impeding the implementation and use of artificial intelligence-enabled virtual assistants in retail. J. Retail. Consum. Serv. **72**, 103258 (2023)
14. Kaushal, V., Yadav, R.: The role of chatbots in academic libraries: an experience-based perspective. J. Aust. Libr. Inf. Assoc. **71**(3), 215–232 (2022)
15. Khan, R., Gupta, N., Sinhababu, A., Chakravarty, R.: Impact of conversational and generative AI systems on libraries: a use case large language model (llm). Sci. Technol. Libr. 1–15 (2023)
16. Le, H.V., Phung, O.V., Nguyen, H.N.: Information security risk management by a holistic approach: a case study for Vietnamese e-government (2020). https://doi.org/10.1007/978-3-662-58611-2_5. http://ijcsns.org/07_book/html/202006/202006008.html
17. Lin, C.Y.: ROUGE: a package for automatic evaluation of summaries. In: Text Summarization Branches Out, pp. 74–81. Association for Computational Linguistics, Barcelona (2004). https://aclanthology.org/W04-1013
18. Meshram, S., Naik, N., Megha, V., More, T., Kharche, S.: Conversational AI: chatbots. In: 2021 International Conference on Intelligent Technologies (CONIT), pp. 1–6. IEEE (2021)
19. Nguyen, L.T., Nguyen, H.N., Nguyen, S.H.: From fragmented data to collective intelligence: a data fabric approach for university knowledge management. In: Nguyen, N.T., et al. (eds.) ICCCI 2023. LNCS, vol. 14162, pp. 16–28. Springer, Heidelberg (2023). https://doi.org/10.1007/978-3-031-41456-5_2
20. Papineni, K., Roukos, S., Ward, T., Zhu, W.J.: Bleu: a method for automatic evaluation of machine translation. In: Annual Meeting of the Association for Computational Linguistics (2002). https://api.semanticscholar.org/CorpusID:11080756
21. Saka, A.B., Oyedele, L.O., Akanbi, L.A., Ganiyu, S.A., Chan, D.W., Bello, S.A.: Conversational artificial intelligence in the AEC industry: a review of present status, challenges and opportunities. Adv. Eng. Inf. **55**, 101869 (2023)
22. Son, N.H., et al.: Vnu-lic digital knowledge center model: transforming big data into knowledge. In: Tseng, Y.H., Katsurai, M., Nguyen, H.N. (eds.) ICADL 2022. LNCS, vol. 13636, pp. 516–522. Springer, Heidelberg (2022). https://doi.org/10.1007/978-3-031-21756-2_44
23. Touvron, H., et al.: Llama 2: open foundation and fine-tuned chat models (2023)
24. Vaghefi, S.A., et al.: Chatclimate: grounding conversational AI in climate science. Commun. Earth Environ. **4**(1), 480 (2023)
25. Vassilakopoulou, P., Haug, A., Salvesen, L.M., Pappas, I.O.: Developing human/AI interactions for chat-based customer services: lessons learned from the Norwegian government. Eur. J. Inf. Syst. **32**(1), 10–22 (2023)
26. Wang, F.Y., Miao, Q., Li, X., Wang, X., Lin, Y.: What does chatgpt say: the dao from algorithmic intelligence to linguistic intelligence. IEEE/CAA J. Automatica Sinica **10**(3), 575–579 (2023)
27. Zhang, L.: Improvement of voice navigation system based on customer service. In: 2023 IEEE 3rd International Conference on Information Technology, Big Data and Artificial Intelligence (ICIBA), vol. 3, pp. 535–538. IEEE (2023)

28. Zhang, S., et al.: Instruction tuning for large language models: a survey. arXiv preprint arXiv:2308.10792 (2023)
29. Zhang, T., Kishore, V., Wu, F., Weinberger, K.Q., Artzi, Y.: Bertscore: evaluating text generation with bert (2020)
30. Zhou, X., Zheng, Y.: Research on personality traits of in-vehicle intelligent voice assistants to enhance driving experience. In: Kromker, H. (ed.) HCII 2023. LNCS, vol. 14048, pp. 236–244. Springer, Heidelberg (2023). https://doi.org/10.1007/978-3-031-35678-0_15

# Advancements in Text Subjectivity Analysis: From Simple Approaches to BERT-Based Models and Generalization Assessments

Margit Antal[1]([⊠]) ⓘ, Krisztian Buza[1,2] ⓘ, and Szilárd Nemes[1,2]

[1] Department of Mathematics-Informatics,
Sapientia Hungarian University of Transylvania, Targu Mures, Romania
`manyi@ms.sapientia.ro`
[2] Faculty of Finance and Accountancy, Budapest Business School,
Budapest, Hungary
`buza@biointelligence.hu, nemesszili@gmail.com`

**Abstract.** Text subjectivity is an important research topic due to its applications in various domains such as sentiment analysis, opinion mining, social media monitoring, clinical research and patient feedback analysis. While rule-based approaches dominated this field at the beginning of the 21st century, contemporary works rely on transformers, a specific neural network architecture designed for language modeling. This paper explores the performance of various BERT-based models, including our fine-tuned BERT (Bidirectional Encoder Representations from Transformer) model, and compares them with pre-built models. To assess the generalization abilities of the models, we evaluated the models on benchmark datasets. Additionally, the models underwent evaluation on two synthetic datasets created using large language models. To ensure reproducibility, we have made our implementation publicly available at https://github.com/margitantal68/TextSubjectivity.

**Keywords:** text subjectivity · transformers · BERT · synthetic datasets

## 1 Introduction

Text subjectivity classification is crucial for understanding and analyzing human opinions and sentiments. It helps to identify whether a piece of text expresses personal opinions, beliefs, or feelings, or whether it presents facts or neutral information. Subjectivity classification plays a vital role in various applications, including sentiment analysis, opinion mining, and summarization. It enables us to filter out objective information and focus on extracting and analyzing subjective expressions, which are essential for understanding human perspectives and opinions. The ability to accurately classify text as subjective or objective is essential for building intelligent systems that can effectively interact with humans.

Early models for subjectivity classification relied on rule-based approaches, which involved hand-crafted rules and features to capture subjective expressions. While these models were effective for simple cases, they were often limited in their ability to generalize to new domains.

Machine learning transformed text subjectivity classification, improving predictions with advanced statistical models. However, challenges remain in capturing long-range dependencies and context-specific details.

The rise of transformer-based models marked a paradigm shift in text subjectivity classification. Transformers are neural network architectures that are particularly well-suited for capturing long-range dependencies and contextual information. By enabling models to learn semantic relationships between words and phrases across the entire text, transformers have significantly improved the accuracy of subjectivity classification.

Language models such as BERT [3] and GPT (Generative Pretrained Transformers) have recently achieved remarkably success in several fields of AI (Artificial Intelligence). While BERT is an open source model that can be fine-tuned by anyone, in case of GPT models, not even the pre-training data is public, let alone the details of the fine-tuning.

This study aims to compare the performance of differently fine-tuned BERT models with GPT 3.5 for text subjectivity classification. Specifically, we will investigate their accuracy, precision, recall, and F1-score on benchmark subjectivity datasets. To further assess the generalization capabilities of differently fine-tuned models, we utilized synthetic datasets generated by Google's Bard[1] and the Jurassic 2 model of AI21[2]. These datasets provided a controlled environment for evaluating model performance across various linguistic nuances and task requirements.

The subsequent sections of the paper are structured as follows: We begin with a review of the key studies in text subjectivity classification. Subsequently, we introduce our pipeline and the pre-built models we considered in this study. Section 4 covers details of datasets, implementation, and the obtained results. Section 6 concludes the study by outlining key findings and observations.

## 2   Related Work

While there is a rich literature on text classification, see e.g. [16] and the references therein, only a few studies focus on the subjectivity of texts. Next, we will point out these studies as they are most closely related to our work.

Early studies employed a rule-based approach, identifying linguistic patterns to classify text segments as either subjective or objective [17]. In 2004, Pang and Lee [10] created an English dataset from movie reviews from Rotten Tomatoes and official movie descriptions of IMDb (Internet Movie Database) with the assumption that reviews are subjective and descriptions are objective. This dataset became one of the leading benchmarks in text subjectivity analysis.

---

[1] https://bard.google.com/chat.

[2] https://www.ai21.com/.

Machine learning, especially deep learning approaches were introduced for text subjectivity detection in more recent studies. Zhao et al. [20] introduced a self-adaptive hierarchical sentence model that achieved superior classification performance on state-of-the-art datasets compared to the utilization of convolutional [5] and recurrent neural networks [6].

More recently, large-scale pre-trained language models based on transformers, have emerged as the state of the art of text modeling. Most prominent models include the autoregressive GPT [15] and the auto-encoding BERT [3].

Chen et al. recently achieved remarkable performance on various benchmark datasets by employing contrastive learning during the fine-tuning phase of pre-trained BERT language model [2].

Another recent study reported subjectivity classification results on cross-lingual datasets using variants of BERT models [13].

Antici et al. introduced a novel dataset for sentence-level subjectivity detection in news articles [1]. The study presents results on text subjectivity classification using logistic regression, support vector machine, and fine-tuned BERT models on their dataset.

Nandi et al. ran an empirical evaluation of various word embedding models for subjectivity analysis tasks [9]. Their results shows that BERT embeddings performs extremly well for text subjectivity classification compared to Word2Vec [7], GloVe [11] and ELMo [12].

Although GPTs have been used to various natural language processing tasks, see e.g. [4] and the references therein, none of the aforementioned works used GPTs for the recognition of text subjectivity.

## 3    Models

In this section, we introduce the models employed for the classification of texts according to subjectivity. We start with presenting our pipeline, which facilitates the incorporation of various embedding layers, and subsequently, we introduce the pre-built models.

### 3.1    Our Pipeline

In our proposed pipeline, see Fig. 1, we use two types of embeddings for text representation: (i) the 'EmbeddingBag' from PyTorch[3], (ii) BERT embeddings with trainable and also with frozen parameters.

Embeddings are dense vector representations that capture the semantic relationships and contextual information of words or entities. The PyTorch EmbeddingBag layer is a versatile tool for efficiently computing embeddings in natural language processing tasks, allowing users to specify vocabulary size and embedding dimension parameters. This layer provides a way to compute embeddings for a variable-length sequence of text and aggregate them in a memory-efficient

---

[3] https://pytorch.org.

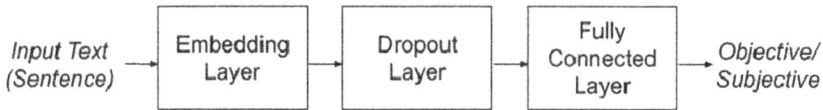

**Fig. 1.** Text classification pipeline

manner. An EmbeddingBag layer is useful for simple natural language classification tasks, such as sentence classification. The benefit of employing an EmbeddingBag layer is that padding is unnecessary, as each sentence, regardless of its length, is represented by a single embedding vector. We designate our model, built upon EmbeddingBag, as *embedding-bag*.

Text embeddings can be computed using a BERT language model [3]. This model is pre-trained on a dataset consisting of BookCorpus [21] (11,038 unpublished books) and English Wikipedia. Because of its training approach, BERT has learned latent representations of words in context. From a memory efficiency perspective, BERT embeddings need more memory since they capture contextual information for each word in a sentence.

Regarding the training of the embedding we note that in case of EmbeddingBag, we train the embeddings at the same time as the rest of the model. But in case of BERT, it's a bit different. BERT is already pretrained, and we have two choices: we can use it as it is (with frozen weights), or we can fine-tune it with the rest of the model. We utilized both options, referred to as *bert-finetuned* and *bert-frozen*.

In both models, we employed the BERT tokenizer, a WordPiece tokenizer [18] that divides words into smaller subword units. This facilitates the handling of words that may not be included in the vocabulary of the model.

**Table 1.** Models and their properties. Note that the final two models, namely *bert-finetuned-wiki* and *gpt3.5turbo*, did not undergo training. Consequently, the count of trainable parameters is unavailable.

| Model | Vocabulary size | #embedding | #parameters | #trainable parameters |
|---|---|---|---|---|
| embedding-bag | 30K | 64 | 1.3M | 1.3M |
| bert-finetuned | 30K | 768 | 110M | 110M |
| bert-frozen | 30K | 768 | 110M | 1,538 |
| bert-finetuned-wiki | 30K | 768 | 110M | – |
| gpt3.5turbo | – | 384 | 20B | – |

## 3.2  Pre-built Models

The first pre-built model was taken from HuggingFace[4] [14]. This model is based on the *bert-base-uncased* model [3], which has been fine-tuned on the Wiki Neutrality Corpus consisting of 180,000 biased and neutralized sentence pairs. The model can be used to classify text as subjectively biased vs. neutrally toned. We will denote this model as *bert-finetuned-wiki*.

The second pre-built model is the widely recognized GPT-3.5-turbo model. Henceforth, it will be denoted as *gpt3.5turbo*. This model was used through its public API.

The models employed and their key features are outlined in Table 1.

## 4  Experimental Results

### 4.1  Datasets

Our experiments were conducted on several datasets.

SUBJ [10] is a movie review dataset containing 5000 subjective sentences from Rotten Tomatoes reviews and 5000 objective sentences extracted from IMDb (Internet Movie Database) summaries. This dataset was split into 3 balanced parts, 8100 sentences for training, 900 for validation and 1000 for evaluation/testing.

TASKSOURCE, as presented in [1], comprises a novel dataset containing 1049 sentences (411 subjective and 638 objective). These sentences are extracted from continuous coverage of political affairs in online news outlets. Given its small size, the dataset served solely for evaluation purposes. Specifically, we utilized the test segment of the dataset, comprising 219 sentences (106 objective and 113 subjective), which has nearly balanced distribution between the two categories.

The SUBJ_BARD dataset, generated by Bard[5] (a large language model from Google AI), comprises sentences covering a broad array of topics including food, travel, nature, music, art, and love. This dataset comprises **100** sentences (50 subjective and 50 objective).

Jurassic-2 Ultra[6] is AI21's most powerful multi-language model, delivering outstanding quality. Among the supported use-cases are question answering, summarization, advanced information extraction, and generating ideas for tasks that require reasoning and logic. We used this model (j2-ultra) through its public API with a 0.8 temperature parameter, and asked to generate 50 subjective and 50 objective sentences (denoted as the SUBJ-AI21 dataset).

Table 2 displays the essential characteristics of the utilized datasets.

---

[4] https://huggingface.co/cffl/bert-base-styleclassification-subjective-neutral.

[5] https://bard.google.com/chat.

[6] https://docs.ai21.com/docs/jurassic-2-models.

**Table 2.** Datasets

| Dataset | Year | Synthetic | Train | Validation | Test | Total |
|---|---|---|---|---|---|---|
| SUBJ | 2004 | No | 8100 | 900 | 1000 | 10000 |
| TASKSOURCE | 2023 | No | 731 | 99 | 219 | 1049 |
| SUBJ-BARD | 2024 | Yes | | | 100 | 100 |
| SUBJ-AI21 | 2024 | Yes | | | 100 | 100 |

## 4.2    Implementation Details

The evaluations were conducted on a Gigabyte Technology Co., Ltd. Z490M computer equipped with Intel Core i9-10900K CPU@ 3.7 GHz x 20, 32 GB RAM, and NVIDIA GeForce RTX 2080 graphics card. The system ran on Ubuntu 22 operating system. The experiments were executed using Python 3.10 and PyTorch 2.1.2.

In the case of *embedding-bag* model, we conducted training for 10 epochs employing the AdamW optimizer with a learning rate of $10^{-3}$. Our model has a vocabulary size of 30M, and we opted for an embedding size of 64. Although alternative embedding sizes like 128 and 256 were tested, the choice of 64 yielded the highest classification accuracy on the test set.

While fine-tuned models show strong empirical performance, it's essential to note that the fine-tuning process itself is inherently unstable [8]. While the creators of BERT models [3] propose 3 epochs for fine-tuning, other work show that allocating more training time can stabilize fine-tuning [19]. Our BERT model *bert-finetuned* underwent training for 30 epochs utilizing the AdamW optimizer, learning rate of $10^{-5}$ and incorporating a weight decay of 0.01.

## 4.3    Experimental Results

The main results are presented in Table 3. It can be seen that fine tuning the BERT model using our training data produced the best results regardless of evaluation metrics. We observe that using the BERT model without fine tuning is simply not working. Therefore, the *bert-frozen* model will be excluded from subsequent measurements. Remarkably good results are generated by the simplest model, *embedding-bag*, which learned its parameters through training from scratch.

Concerning the pre-built models, their performance, particularly in terms of precision, is relatively poor. Precision focuses on the accuracy of positive predictions and is particularly useful when the cost of false positives is high. It provides insights into the model's ability to avoid falsely classifying negative instances as positive. Higher precision indicates fewer false positives relative to true positives.

The TASKSOURCE dataset is inadequate for fine-tuning due to its small size. Our models trained on the SUBJ dataset and the two pre-built models were evaluted on the testing part of the TASKSOURCE dataset. The results are

**Table 3.** Evaluation results of selected models on the testing subset (consisting of 1000 sentences) of the SUBJ dataset.

| Model | Accuracy | Precision | Recall | F1 |
|---|---|---|---|---|
| embedding-bag | 0.928 | 0.929 | 0.926 | 0.927 |
| bert-finetuned | **0.973** | **0.976** | **0.970** | **0.973** |
| bert-frozen | 0.504 | 0.611 | 0.022 | 0.042 |
| bert-finetuned-wiki | 0.708 | 0.664 | 0.842 | 0.743 |
| gpt3.5turbo | 0.659 | 0.601 | 0.944 | 0.734 |

shown in Table 4. The aim of this experiment was to observe the behavior of the models on a novel dataset. All four models failed to achieve acceptable performance, obtaining consistently low precision and generating numerous false positives (objective sentences classified as positives). Although our models exhibited very poor recall, the pre-built models achieved a recall exceeding 0.9, indicating a more effective capture of positive instances, particularly the subjective ones in our case. This difference may be attributed to their fine-tuning processes.

**Table 4.** Evaluation results of selected models on the testing subset (consisting of 219 sentences) of the TASKSOURCE dataset.

| Model | Accuracy | Precision | Recall | F1 |
|---|---|---|---|---|
| embedding-bag | 0.579 | 0.596 | 0.575 | 0.585 |
| bert-finetuned | 0.611 | 0.725 | 0.398 | 0.514 |
| bert-finetuned-wiki | 0.648 | 0.607 | 0.902 | 0.725 |
| gpt3.5turbo | 0.659 | 0.601 | 0.944 | 0.734 |

The next experiment involved assessing the models on synthetic datasets. The outcomes for the SUBJ-BARD datasets are presented in Table 5, while the results for the SUBJ-AI21 dataset are summarized in Table 6. Considering both datasets, the model named *embedding-bag* demonstrated the least favorable performance on average, whereas the *gpt3.5turbo* model showcased the best performance.

Benchmark datasets allow to compare our findings with those of other researchers. In Table 7, we showcase the performance of our fine-tuned model in comparison to previous results achieved on the SUBJ dataset. It is crucial to note that these results were obtained under slightly varied evaluation conditions.

To determine the required volume of training data for achieving satisfactory test accuracy, we conducted an additional experiment. We partitioned the SUBJ dataset in a manner consistent with the primary experiment: 81% for training, 9% for validation, and 10% for testing. Subsequently, we utilized varying percentages of the 81% training data for the training phase, ranging from 10% to

**Table 5.** Evaluation results of selected models on SUBJ-BARD (consisting of 100 sentences) dataset.

| Model | Accuracy | Precision | Recall | F1 |
|---|---|---|---|---|
| embedding-bag | 0.680 | 0.636 | 0.840 | 0.724 |
| bert-finetuned | 0.830 | 1.000 | 0.660 | 0.795 |
| bert-finetuned-wiki | 0.680 | 0.660 | 0.740 | 0.698 |
| gpt3.5turbo | 0.910 | 0.847 | 1.000 | 0.917 |

**Table 6.** Evaluation results of selected models on AI21 (consisting of 100 sentences) dataset.

| Model | Accuracy | Precision | Recall | F1 |
|---|---|---|---|---|
| embedding-bag | 0.580 | 0.587 | 0.540 | 0.562 |
| bert-finetuned | 0.720 | 1.000 | 0.440 | 0.611 |
| bert-finetuned-wiki | 0.860 | 0.909 | 0.800 | 0.851 |
| gpt3.5turbo | 0.979 | 1.000 | 0.960 | 0.979 |

100%, with increments of 10% at each iteration. This process was repeated ten times as part of the primary experiment, employing a limited amount of training data. The resulting models were then assessed on the identical test set, comprising 1000 examples. The accuracy and F1 scores for the *embeddingbag* model (depicted on the left) and the *bert-finetuned* model (depicted on the right) for varying amounts of training data are illustrated in Fig. 2.

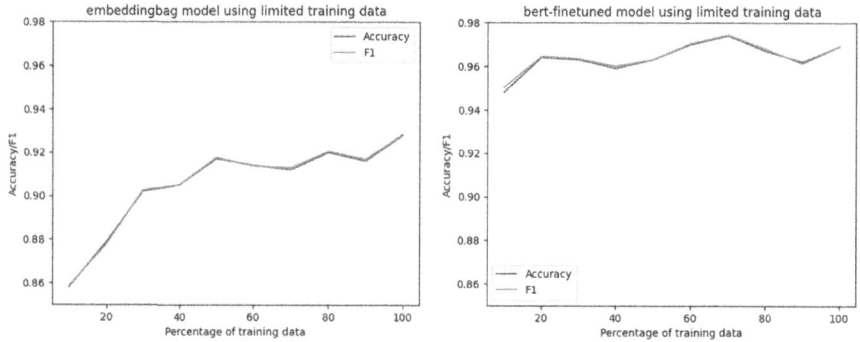

**Fig. 2.** Accuracy and F1 for different sizes of the training dataset.

**Table 7.** Evaluation results on SUBJ benchmark dataset.

| Model | Accuracy | Precision | Recall | F1 |
|---|---|---|---|---|
| AdaSent [20] | 0.955 | – | – | – |
| BERT-base + LSTM [9] | 0.966 | 0.964 | 0.965 | 0.965 |
| RoBERTa + DualCL [2] | 0.973 | 0.972 | 0.982 | 0.977 |
| Our bert-finetuned | 0.973 | 0.976 | 0.970 | 0.973 |

## 5   Discussion

Initially, we examine the outcomes derived from a real-world context, utilizing evaluation data sourced from the SUBJ dataset.

The data shown in Table 3 indicate that models fine-tuned for particular tasks surpass the performance of broad-scope large language models like *gpt3.5turbo*. This includes not only the *bert-finetuned* model but also the ssimpler *embedding-bag* model, both achieving high levels of accuracy.

To assess our model's effectiveness, we benchmarked it against other outcomes from the SUBJ dataset (refer to Table 7). The accuracy of our optimized model (*bert-finetuned*) reached 0.973, positioning it alongside leading models like RoBERTa [2].

Experiments on varying training data sizes (Fig. 2) reveal that the performance of both the *embedding-bag* and *bert-finetuned* models improves with increased training data volume, but even smaller amounts of training data are sufficient for achieving satisfactory test accuracy.

Next, we analyze the outcomes from our two synthetic datasets SUBJ-BARD and SUBJ-AI21. As shown in Tables 5 and 6 the *gpt3.5turbo* model outperforms other models on both datasets. The model achieved the highest F1 scores (0.917 on SUBJ-BARD and 0.979 on SUBJ-AI21), highlighting its superior ability to handle tasks such as subjectivity classification of synthetic text.

## 6   Conclusion

In this paper we studied text subjectivity classification. Inline with recent findings, we fine-tuned a BERT model and compared its performance against a baseline model and two pre-built models. The performance evaluation indicates that having sufficient data for fine-tuning a pretrained model yields the most optimal results. Not only is this approach performance-oriented, but it also proves to be cost-effective and secure. Surprisingly, even the simplest model, *the embedding-bag*, outperformed *gpt3.5turbo* on the SUBJ benchmark dataset. On the other hand, when lacking sufficient data for fine-tuning, one might consider utilizing publicly available and well-documented pre-built models like *bert-finetuned-wiki*. Evaluations on synthetic datasets demonstrate that the *gpt3.5turbo* model excels in detecting subjectivity in synthetic texts.

## 7     Appendix

We provide the input prompt utilized for the text classification task in the GPT-3.5-turbo model.

prompt = f""""You are an expert linguist. You should decide about a sentence whether is a subjective or an objective sentence. A sentence is subjective if its content is based on or influenced by personal feelings, tastes, or opinions. Otherwise, the sentence is objective. More precisely, a sentence is subjective if one or more of the following conditions apply: 1. expresses an explicit personal opinion from the author (e.g., speculations to draw conclusions); 2. includes sarcastic or ironic expressions; 3. gives exhortations of personal auspices; 4. contains discriminating or downgrading expressions; 5. contains rhetorical figures that convey the author's opinion. Please classify the following text text in one of the two classes: (SUBJECTIVE, OBJECTIVE). Please, answer with a single word: OBJECTIVE or SUBJECTIVE""""

The instruction utilized for generating sentences is as follows:

prompt = f""""You are an expert linguist. Please, generate 10 objective sentences. A sentence is considered objective when it presents information in a factual and unbiased manner, without expressing personal opinions, emotions, or interpretations. Otherwise, the sentence is subjective.""""

## References

1. Antici, F., et al.: A corpus for sentence-level subjectivity detection on English news articles (2023)
2. Chen, Q., Zhang, R., Zheng, Y., Mao, Y.: Dual contrastive learning: text classification via label-aware data augmentation (2022)
3. Devlin, J., Chang, M.W., Lee, K., Toutanova, K.: BERT: pre-training of deep bidirectional transformers for language understanding. In: Burstein, J., Doran, C., Solorio, T. (eds.) Proceedings of the 2019 Conference of the North American Chapter of the Association for Computational Linguistics: Human Language Technologies, volume 1 (Long and Short Papers), pp. 4171–4186. Association for Computational Linguistics, Minneapolis (2019). https://doi.org/10.18653/v1/N19-1423
4. Forman, N., Udvaros, J., Avornicului, M.S.: Chatgpt: a new study tool shaping the future for high school students. Int. J. Adv. Nat. Sci. Eng. Res. **7**(4), 95–102 (2023)
5. Kim, Y.: Convolutional neural networks for sentence classification. In: Moschitti, A., Pang, B., Daelemans, W. (eds.) Proceedings of the 2014 Conference on Empirical Methods in Natural Language Processing (EMNLP), pp. 1746–1751. Association for Computational Linguistics, Doha (2014). https://doi.org/10.3115/v1/D14-1181
6. Lai, S., Xu, L., Liu, K., Zhao, J.: Recurrent convolutional neural networks for text classification. Proc. AAAI Conf. Artif. Intell. **29**(1), 2267–2273 (2015). https://doi.org/10.1609/aaai.v29i1.9513
7. Mikolov, T., Sutskever, I., Chen, K., Corrado, G.S., Dean, J.: Distributed representations of words and phrases and their compositionality. In: Burges, C., Bottou, L., Welling, M., Ghahramani, Z., Weinberger, K. (eds.) Advances in Neural Information Processing Systems. vol. 26. Curran Associates, Inc. (2013)

8. Mosbach, M., Andriushchenko, M., Klakow, D.: On the stability of fine-tuning BERT: misconceptions, explanations, and strong baselines. In: 9th International Conference on Learning Representations, ICLR 2021, Virtual Event, 3–7 May 2021. OpenReview.net (2021). https://openreview.net/forum?id=nzpLWnVAyah

9. Nandi, R., Maiya, G., Kamath, P., Shekhar, S.: An empirical evaluation of word embedding models for subjectivity analysis tasks. In: 2021 International Conference on Advances in Electrical, Computing, Communication and Sustainable Technologies (ICAECT), pp. 1–5 (2021). https://doi.org/10.1109/ICAECT49130.2021.9392437

10. Pang, B., Lee, L.: A sentimental education: sentiment analysis using subjectivity. In: Proceedings of Annual Meeting of the Association for Computational Linguistics (ACL), pp. 271–278 (2004)

11. Pennington, J., Socher, R., Manning, C.: GloVe: global vectors for word representation. In: Moschitti, A., Pang, B., Daelemans, W. (eds.) Proceedings of the 2014 Conference on Empirical Methods in Natural Language Processing (EMNLP), pp. 1532–1543. Association for Computational Linguistics, Doha (2014). https://doi.org/10.3115/v1/D14-1162

12. Peters, M.E., et al.: Deep contextualized word representations. In: Walker, M., Ji, H., Stent, A. (eds.) Proceedings of the 2018 Conference of the North American Chapter of the Association for Computational Linguistics: Human Language Technologies, Volume 1 (Long Papers), pp. 2227–2237. Association for Computational Linguistics, New Orleans (2018). https://doi.org/10.18653/v1/N18-1202

13. Přibáň, P., Steinberger, J.: Czech dataset for cross-lingual subjectivity classification. In: Calzolari, N., et al. (eds.) Proceedings of the Thirteenth Language Resources and Evaluation Conference, pp. 1381–1391. European Language Resources Association, Marseille (2022). https://aclanthology.org/2022.lrec-1.148

14. Pryzant, R., Martinez, R.D., Dass, N., Kurohashi, S., Jurafsky, D., Yang, D.: Automatically neutralizing subjective bias in text (2019)

15. Radford, A., Narasimhan, K.: Improving language understanding by generative pre-training (2018). https://api.semanticscholar.org/CorpusID:49313245

16. Revina, A., Buza, K., Meister, V.G.: It ticket classification: the simpler, the better. IEEE Access **8**, 193380–193395 (2020)

17. Riloff, E., Wiebe, J.: Learning extraction patterns for subjective expressions. In: Proceedings of the 2003 Conference on Empirical Methods in Natural Language Processing, pp. 105–112 (2003). https://aclanthology.org/W03-1014

18. Wu, Y., et al.: Google's neural machine translation system: bridging the gap between human and machine translation (2016)

19. Zhang, T., Wu, F., Katiyar, A., Weinberger, K.Q., Artzi, Y.: Revisiting few-sample BERT fine-tuning. In: 9th International Conference on Learning Representations, ICLR 2021, Virtual Event, 3–7 May 2021. OpenReview.net (2021). https://openreview.net/forum?id=cO1IH43yUF

20. Zhao, H., Lu, Z., Poupart, P.: Self-adaptive hierarchical sentence model. In: Proceedings of the 24th International Conference on Artificial Intelligence (IJCAI 2015), pp. 4069–4076. AAAI Press (2015)

21. Zhu, Y., et al.: Aligning books and movies: towards story-like visual explanations by watching movies and reading books. In: 2015 IEEE International Conference on Computer Vision (ICCV), pp. 19–27 (2015). https://doi.org/10.1109/ICCV.2015.11

# Hybrid Approach Text Generation
# for Low-Resource Language

Diana Rakhimova[1,2](✉) [iD], Eşref Adali[3] [iD], and Aidana Karibayeva[1,2] [iD]

[1] Al-Farabi Kazakh National University, Almaty 050040, Kazakhstan
di.diva@mail.ru
[2] Institute of Information and Computational Technologies, Almaty 050010, Kazakhstan
[3] Istanbul Technical University, 34485 Maslak, Sarıyer, Istanbul, Turkey

**Abstract.** Text generation is an important tool used by many companies in various fields such as chatbots, search engines, and question and answer systems, and is a hot trend in artificial intelligence. Generating texts and sentences can be used for both educational and entertainment purposes. Generating texts and sentences for children in natural language processing plays an important role in children's development. This helps them improve their reading, comprehension and communication skills in the language. Currently, many languages of the world belong to the class with the low resources. The field of text generation for low-resource languages is still at an early stage of development and there are many problems that need to be solved. One of the main problems is the lack of big data and linguistic resources in the public domain, which makes it difficult to effectively apply modern machine learning methods. As well as the lack of modern methods and tools for analyzing the processing of these languages. This article presents a hybrid approach to text generation on the example of the Turkish and Kazakh languages. These languages belong to a large group of Turkic languages along with Kyrgyz, Tatar, Uzbek and other languages. An approach based on neural learning using the LSTM model is proposed and implemented, considering the structural and semantic properties of the language. Training and testing are carried out on the assembled corpus (for various types of text genres). The quality of text generation was assessed based on the BLEU metric.

**Keywords:** Text generation · low recourse language · Kazakh language · Turkish languages · TF-IDF · RNN · LSTM

## 1 Introduction

The languages of the world divided into six categories based on available resources in terms of labeled and unlabeled data [1]. More than 88% of the world's languages are in the least resource class, and only 25 languages belong to the two high resource classes. The term "low-resource languages" (LRL) refers to natural languages with some (or all) of the following properties: limited distribution on the Internet, lack of electronic resources for language and speech processing, including monolingual corpora, bilingual electronic dictionaries, spelling and phonetic transcriptions of speech, pronunciation dictionaries, etc.

The Turkic language is one of the largest language families, with over 160 million speakers in the world [2]. Turkic languages are related languages. As a result of the complex long history of the Turkic-speaking peoples, all Turkic languages have acquired, to one degree or another, changes at the morphemic, lexical, and syntactic levels. At the same time, the Turkic languages retained common features.

Turkic languages have strict systems of word formation and rules for affix conjunction. Turkish, Kazakh, Tatar, Uzbek, and Kyrgyz like other Turkic languages, are grammatically similar in terms of the types of affixes. For example, the plural is formed using a suffix that has 6 phonetic variants in the Kazakh language (-лар, -лер, -дар, -дер, -тар, -тер), in Kyrgyz there are 12 variants (-лар, -лер, -лор, -лөр, -дар, -дер, -дор, -дөр, -тар, -тер, -тор, -төр), and for Turkish 2 option (lar ler).

For example, the Kazakh word 'балалар [balalar] (бала + лар)', the Turkish word 'çocuklar (çocuk + lar)', and the Kyrgyz word 'балдар (бал + дар)', which translated as 'children' had similar plural affixes. The main common features of the Turkic languages in the field of morphology are found mainly in the bases of nouns, verbs, numerals, adjectives, personal and demonstrative pronouns. There is also its own basic order and word organization rules with the help of many affixes. In [3] based on the analysis done, an assembly of endings into a complete set of endings for several Turkic languages was developed. From the union of all derivative language endings into a single list of language endings. A complete set of language endings, combined with a variety of language bases, determines the morphological and logical model of a given language. Based on a four-step process, the full set of estimated endings includes: 4679 Kazakh endings, 4768 Kyrgyz endings, and 747 Uzbek endings. For example, the longest word in the Kazakh language "қанағаттандырылмағандықтарыңыздан" consists of 33 letters and is translated as "because of your dissatisfaction" when respectfully addressed to some people. Where the stem consists of 7 characters and the rest is a set of suffixes: " қанағат + тан + дырыл + ма + ған + дық + тар + ыңыз + дан ".

Having a complex morphological structure and rich semantic context, unfortunately, at the present time, the Turkic group of languages belongs to a low-resource group. And many countries and peoples do not have access to high-quality electronic resources and information technologies in their native language. This article will consider the tasks of text synthesis for a low-resource language using the example of the Kazakh language, which is part of the Turkic group.

Generating text in the Kazakh language is essential for language preservation and modernization. The importance of Kazakh text production lies in its ability to facilitate communication, education, and cultural conservation. Historically, the language was marginalized, and the development of text-generation tools can help to increase its use and increase its growth. In addition, Kazakh text production can be used for various purposes such as language learning, speech recognition, and machine translation [4].

Modernization of Kazakh language through text generation is crucial to preserving national identity.

## 2 Related Works

In recent years, artificial intelligence products have been developed using text generation models. The well-known models for text generation based on neural networks: GPT-3 (Generative Pre-trained Transformer 3) is a neural network model that generates texts of different topics and lengths [5], BERT (Bidirectional Encoding Representations from Transformers) is a model of a neural network that can process large amounts of text information and make contextual conclusions, GROVER (Generating Review of Different Emotions and Ratings) is a neural network model that generates texts with different emotions and ratings, XLNet is a neural network model that does not only take into account the words in the previous text, but also the following ones to process and generate text in a wide context, GPT-2 (Generative Pretrained Transformer 2) is a neural network model developed by OpenAI that generates highly precise natural texts [5, 6]. Also, on the Internet there are various text generation products for various applications such as Jasper AI, GrowthBar, Frase, Copysmith, Hypotenuse AI, Copy AI, Writer [6].

These models have been applied to many resource-rich languages such as English, Spanish, German, Russian, and others [5–7]. Unfortunately, Kazakh, like many other Turkic languages, is resource-poor, which currently hinders achieving excellent results using neural network-based approaches.

The study attempted to create summary/original content on a specific topic using Wikipedia TR for Turkish and a data pool generated from hundreds of thousands of scientific publications [8]. In implementing the model, HNM and LSTM approaches were used. The comparison results show not a bad result. The syntactic structures of the Turkish language were also considered. Different numbers of words and texts created with different functions were used for the dataset. They were compared with the survey method, and it was proven that LSTM-RNN can produce more successful natural language texts for Turkish.

There are several methods that can be employed for text generation in the Kazakh language. One of the most common approaches is a rule-based approach, which involves segmenting words into inflectional classes and utilizing grammatical rules to generate new text [9]. This approach is often used for constructing simple sentences and is based on the principles of Kazakh grammar and syntax.

The authors investigate the task of correcting generated synthetic text for languages with limited resources [10]. In most cases, such synthetic text contains numerous errors that need to be carefully examined and corrected using additional tools. These errors must be automatically rectified to avoid performance degradation of the system. The approach to automatic error correction is based on employing finite-state automata to propose candidates for correcting typos in words.

Another method used for text generation in the Kazakh language is the neural network approach. This approach involves training artificial neural networks to learn patterns and structures of the Kazakh language for machine translation systems [11]. Neural networks can be trained on large datasets of Kazakh text, enabling them to generate new text that is grammatically correct and semantically meaningful. It has been shown that this approach is particularly effective for generating complex sentence structures and has led to significant improvements in machine translation systems [12].

The authors successfully developed a question-answering system for the Kazakh language using the classic Seq2Seq model [13]. This approach has been tested in a specific domain and follows a question-and-summary answer scheme.

In article [14], the authors present text generation in the Kazakh language using Bag-of-Words (BoW) models for consumer sentiment analysis in social media. The proposed BoW model is applied to analyze the sentiment of generated texts from social networks. This approach has been tested within a limited domain and with incomplete structural and semantic language forms.

A pattern of the POS sequence with appropriate endings was presented in [15]. This type of text generation was used to increase the volume of the low-resource corpus of the Kazakh language.

The complete system of endings (CSE) for the Kazakh language and its application in various tasks of natural language processing are presented in works [16–18]. The CSE-model is used in stemming, morphological segmentation, and morphological analysis of Kazakh texts. The mentioned model can also be applied to the sentence generation problem.

The above approaches and studies have been aimed at addressing specific NLP tasks for the Kazakh language. Many of the developed solutions are based on rule-based approaches due to the limited resources available for the Kazakh language. Despite the challenges, these works contribute significantly to the development of this scientific field. Unfortunately, now, there is a lack of technological tools for robust text generation in the Kazakh language, like existing systems.

## 3   Problems of Text Generation in the Turkish and Kazakh Languages

The Kazakh language has several characteristics that can make text generation a challenging task. Some of these problems include:

- Grammatical complexity: The Kazakh language has a rich system of grammatical forms and inflections. For example, verbs can change depending on tense, person, number, and mood, while nouns can have different forms depending on case and number. This complexity can make generating syntactically correct sentences a more challenging task.
- Phonetic and graphical features: The Kazakh language uses its own alphabet, which is different from Latin or Cyrillic. Additionally, the Kazakh language has many sounds that may be unfamiliar to those not familiar with the language.
- Lexical peculiarities: The Kazakh language contains many unique words and expressions that may be unfamiliar to non-speakers. This can make contextual text generation a more challenging task.
- Ambiguity: The Kazakh language contains many words that have multiple meanings depending on the context. This can lead to the text generation program incorrectly interpreting the meaning of a word and using it in inappropriate contexts.

One of the main challenges in generating text in the Kazakh language is the limited availability of data in Kazakh. This can make it difficult for natural language processing

(NLP) systems to accurately analyze and generate text in Kazakh. However, efforts are being made to address this issue. For example, software for statistical machine translation of texts into Kazakh has been developed, which can help improve the quality of Kazakh language data. Additionally, algorithms for synthesis and analysis of word forms in the Kazakh language have been proposed, which can help improve the accuracy of Kazakh text generation. Further research and development in this field can help address the problem of limited data in the Kazakh language and improve the quality of text generation in Kazakh.

Today, the frequent use of chat bots is growing every day. However, the GPT-4 produces unpredictable texts that are not clear in content when generated for languages with low resources. The results of generating texts for the Kazakh, Kyrgyz and Tatar languages, the text does not have the correct data, the content of the text is meaningless. In the case of generating the Tatar text, due to the scarcity of the Tatar language, GPT-4 produced the text in Turkish. And for the Turkish language the text was not bad, with correct data. As a result, we can conclude that for the generation of low-resource languages, artificial intelligence models have not yet been developed, we need big data for low-resource languages.

## 4  Description of the Hybrid Text Generation Approach in the Low Resource Language

To solve the given tasks, it is necessary to first determine the quantity and format of the input and output data. Based on the problem statement, it is much easier to determine the format of the output data.

Next, an approach to generating texts in the Kazakh language will be presented. This hybrid approach consists of several blocks:

1. Data Collection and Processing Block.
2. Structural and Semantic Analysis of Text Block.
3. Machine Learning using a Neural Network model.

The output data should consist of a narrative-interactive sequence of text blocks that are logically and semantically connected to each other. Now, let's assume that the number of iterations (user choices), the number of choice options at each iteration, and the size of text blocks should be determined at the neural network level, depending on related input parameters. A brief description of the operation of each block will be provided.

### 4.1  Data Collection and Processing

In the initial tasks and challenges, the lack of a sufficient amount of well-processed electronic resources posed a difficulty. At the first stage, two tasks are solved: word preprocessing; division of the text into separate words and phrases. The first task is language-dependent; therefore, the morphological feature of the Kazakh language is taken into account here. To solve this problem, the system of full endings of the language is involved. And for the second, a simple approach was used - the tokenization procedure, with the help of which the entire text is divided into separate words. The lexicon-free stemming algorithm based on the CSE (Complete Set of Endings) morphology model [19] was applied for text processing and normalization.

The Kazakh language belongs to a language with few resources. Collecting a high-quality, large-scale parallel corpus for the Kazakh language is one of the urgent problems in natural language processing. Collection and processing of the corpus for the Kazakh language is considered [20–23].

Parallel corpora were collected and crawled from the official websites of the Republic of Kazakhstan. All assembled parallel corpus aligned, cleaned, normalized, tokenized.

To conduct the experiment, various text corpora were collected and processed for the Kazakh language. Corpora set 1 consists of 130 056 sentences in scientific literature genre, corpora set2 consists of 107 125 sentences in publicist style genre.

The linguistic corpus for the Kazakh language was collected using crawling from various portals. For the genre of scientific literature, sources of scientific journals and abstracts from portals were used. Scientific articles and reports of republican publications are Bulletin of al-Farabi KazNU, Bulletin of Satbayev University. Additionally, text resources from the Kazakh language corpus on the portal https://github.com/NLP-KazNU were also utilized. The collected data was processed and prepared manually to train the model.

## 4.2 Structural and Semantic Properties of Text in Kazakh Language

Errors in the structure of the sentence make it difficult to understand the text. The structure of a sentence depends not only on grammar, but also on style and sequence. The texts use sentences of various lengths and structures, and simple and complex sentences have their own translation errors. The availability of large datasets for popular languages (in terms of research) makes it possible to correct grammatical and structural errors based on modern technologies (such as machine learning, hybrid approaches, etc.). Research for low-resource languages in most cases suffers from a lack of well-structured and well-formed large datasets.

Structural properties of the sentence of the Kazakh language. In linguistics, there is a system of typological classification of languages based on the order of words in a sentence. This system is based on the basic order in which the subject (according to the English subject), the predicate (according to the English verb) and the direct object (according to the English object) appear in the sentence. The Kazakh language, like many Turkic languages, in its structural and syntactic structure, belongs to the group SOV - Subject Object Verb.

Formal models of the syntax of simple sentences of the Kazakh language are focused on the allocation of three types of phrasal structures: subjective phrasal structure (SP), verbal phrasal structure (VP) and object phrasal structure (OP). The basis of the subjective phrasal structure is the subject, the basis of the verbal phrasal structure is the verb, and the basis of the object phrasal structure is the object of action.

Using the Backus notation [24], the formal model of the structure of the sentence syntax of the Kazakh language will have the following form:

S::=<SP> <OP> <VP>|<OP> <SP> <VP>|<OP> <SP> <VP>|

(This rule represents all possible variants of structures at the level of the entered phrase structures).

$<SP>::=<N>|<Adj><SP>|<Num><SP>|<N><SP>$
$<SP>::=<SP><Conn><SP>$
$<VP>::=<V>|<Aux><VP>|<Adv><VP>$
$<OP>::=<N>|<Adj><OP>|<OP><Conn><OP>$

Here, $<Adj>$ is an adjective, $<Num>$ is a numeral, $<Conn>$ are conjunctions, $<Aux>$ are auxiliary verbs, $<Adv>$ is an adverb.

The analysis of the structural and semantic meanings of simple sentences is very clear and does not have much difficulty since it has a strict sequence and rules for education.

Next, an analysis and model of complex sentence structures will be presented using the Kazakh language as an example. Also, in the Kazakh language there are complex sentences that differ in their properties and rules. For example, in practice, two main types of complex sentences in the Kazakh language are used: compound sentences (salalas qurmalas) and complex sentences (sabaqtas qurmalas). Each type of complex sentence is further divided into 6 subtypes. And each subtype is formed by its own specific morphological rules and auxiliary words.

Table 1 below shows Model of the order of internal sentences of a complex sentence.

**Table 1.** Model of the order of internal sentences of a complex sentence of Kazakh language.

| Sentence type in Kazakh | Order of internal sentences in Kazakh |
| --- | --- |
| *SALALAS QURMALAS (жалғаулықты) compound sentences* | |
| Ynggailas salalas | [MS1] және [MS2] |
| | [MS1 __ әрі __], [әрі MS2] |
| Talgauly salalas | [MS1], әйтпесе (әлде, немесе) [MS2] |
| Kezektes salalas | [MS1... бірде...], бірде [MS2] |
| | Біресе [MS1], біресе [MS2] |
| Sebep-saldar salalas | [MS1], сондықтан [MS2] |
| *SABAQTAS QURMALAS complex sentences* | |
| Shartty bagynyngqy-sabaqtas | (DS... v1), [MS] |
| Qarsylyqty bagynyngqy-sabaqtas | (DS... v2), [MS] |
| Mezgil bagynyngqy-sabaqtas | (DS... v3), [MS] |
| Sebep bagynyngqy-sabaqtas | (DS... v4), [MS] |
| Qimyl-syn bagynyngqy-sabaqtas | (DS... v5), [MS] |
| Maqsat bagynyngqy-sabaqtas | (DS... v6), [MS] |

Here MS- is the main sentence; DS -is the dependent sentence; vn is a group of verbs with different endings. A complete description of the model for the formation of structures of complex sentences for the Kazakh language is presented in the work [12].

To identify semantic properties, the morphological analysis of the Kazakh language and the TF-IDF (Term Frequency - Inverse Document Frequency) algorithm [25] were used to determine the frequency and keywords in the sentence and text. Further, for each found candidate for keywords, signs are distinguished, according to which the degree of its importance and semantic links are estimated. The main objective of keyword detection algorithms is to find appropriate candidates, identify attributes, and rank them.

Based on the algorithm for determining keywords, properties, and linguistic resources of the Kazakh language, an algorithm for extracting semantic properties has been developed. In [26] the paper presents a complete description of the algorithm for the semantic analysis of the Kazakh language.

For the algorithm to identify semantic connections, the following steps were performed:

- text preprocessing: normalization and morphological analysis of the text;
- definition of structure and type of sentences;
- identification of keywords using the TF-IDF method;
- building semantic associations based on the syntactic rules of the language and keywords of the text.

This block begins with a statistical analysis of the text, counting the number of words, stop words, characters, and morphological analysis. Then, the TF-IDF algorithm is used to find semantically important words in the text. Annotating this text with semantic analysis tools allows for improved quality of text generation. Based on the obtained SSP-structural and semantic properties and a set of rules, it is possible to improve the quality of text generation.

# 5  Practical Results

After marking up the text and encoded characters, we proceed to the construction of training sequences. Remember that RNN layers learn sequences of tokens. Our goal here is to pass sequences and offset sequences to train a text generation model.

Then we take and upload training sequences as an example. Since we are training the network with sequences, we will create an input sequence and then a target one. When using RNN, the target sequence will be the input sequence shifted by one character.

Thus, when we train the network, the input symbol must always match the expected output symbol. An example of what this looks like is shown below.

Various inquiries were conducted to test the model. For instance, the word "ғылым"-"science" was inputted into the model (see Fig. 1 and Fig. 2).

After testing the model, the word "Ғылым" ("science") was inputted, resulting in a generated text that has a good semantic understanding and correct structural forms for the Kazakh language (see Fig. 3).

LSTM networks are performance intensive and may take time to train on CPU servers. Therefore, several parameters have been defined below to determine the number of training periods.

During the training process, a model was prepared with a total of seven layers and 4,070,247 parameters,

```
start = time.time()
states = None
next_char = tf.constant(['Ғылым'])
result = [next_char]

for n in range(1000):
  next_char, states = one_step_model.generate_one_step(next_char, states=states)
  result.append(next_char)

result = tf.strings.join(result)
end = time.time()
print(result[0].numpy().decode('utf-8'), '\n\n' + '_'*80)
print('\nRun time:', end - start)
```

Ғылымы жеткен, солына табылмайды.
Мал, мақтан, оған ре емес пе?
Осы екеуі маған қалай табадада саламыз ешнәрсе шықпы, жермеген жерде құмар қылып,
Осының бір фәрдәдән жүниеді қанірет - бәрінің де білін құрметтейін десең, жатқан т
Құдай тағала әрне жаратты, бір түрлі пайдалы хиямәқ үшін кісі шығадам берей қалып,
Осы күнде қазақ ішінде «ісі барамын жемістіген нәрсені хайуандарды асырайтуғын жан
Бұрынғы қазақ жайын шықты.
Еске болы жоқ» деген - өмертің назаза өзі әуелі мал табу к рек, малға мінге қалған
Оп - алла тағаланың фиғыл ғазимләрінің аттары, олардың мағынасын біл һәм сегіз сиф
Кедей болса, ұрлық хик

Run time: 3.2470293045843945

**Fig. 1.** An example of text generation using the input word in Kazakh "Ғылым" ("science").

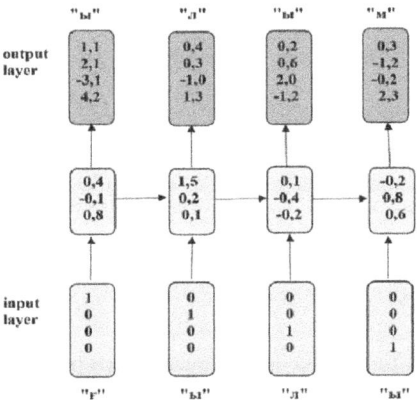

**Fig. 2.** An RNN training process built for the word "Ғылым" ("science").

When the exercise began, the model did not know how to write Kazakh words and did not even know that words are text units. The structure of the output is like that of a game—blocks of text usually begin with uppercase letters, as in the dataset. The model is trained on small batches of text (100 characters each) and can still generate longer sequences of text with a coherent structure.

Next, training was carried out for the classical LSTM neural network model, and LSTM trained on the basis of SSP. The length of the generated texts is 60 and 90 words. Table 2 shows examples of queries and results in the Kazakh language of the trained model based on the hybrid approach. An English translation is also provided (*translated using https://translate.google.com/). From the results obtained, it can be seen that the words and structures of the sentence are correct. The selected sentences are related to the subject of the request (Table 2).

**Table 2.** Table captions should be placed above the tables.

| Input data | Қазақстан республикасы- Republic of Kazakhstan |
|---|---|
| Output data (generation result) | Қазақстан Республикасы, Шығыс Еуразияда орналасқан елдің өзіндік республикалық жүйесімен басқарылатын мемлекет. Қазақстан Республикасының ең кіші мемлекеттік бірлігін құрастырушы мекеме Парламент, Президент және Ұлттық сәбиекорлық суды. Қазақстан, Қазақстан жерінің шекарасы, Тянь-Шань тасмағының ортақтамасы және Орталық Азияда орналасқаны арқылы бұрынғы шамамен Гүней Еуразияда орналасқан биік алқабылардың бірі |
| | The Republic of Kazakhstan is a country in Eastern Eurasia governed by its own republican system. The constituent institutions of the smallest state unit of the Republic of Kazakhstan are the Parliament, the President and the National Child Protection Court. Kazakhstan is one of the high valleys located in South Eurasia, due to the border of the land of Kazakhstan, the border of the Tien-Shan belt and its location in Central Asia |

The following data samples are used for training and testing neural networks: An LSTM neural network trained based on SSP was tested on data samples of the collected corpora. Testing was carried out on the test samples described earlier. 175 generated examples were used for testing. The quality assessment was carried out based on the BLEU metric [27]. When assessed by the BLEU metric uses sentence-level smoothing (see Fig. 3):

THE EVALUATION OF THE OBTAINED RESULTS
OF THE TEXT GENERATION MODEL TESTING IN
THE KAZAKH LANGUAGE

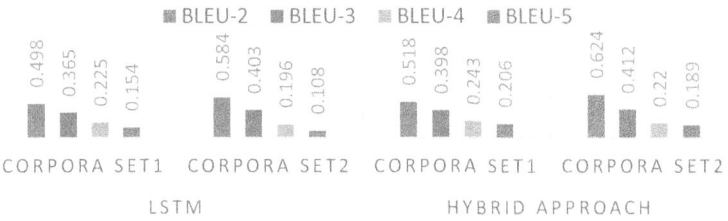

**Fig. 3.** The evaluation indicators for the BLEU metric, the developed approach.

As shown in Fig. 3 on the graph, the evaluation indicators for the BLEU metric, the developed approach (LSTM + SSP) produces the best indicators, this is due to the identified structural and semantic associations of the language. Also on the test, the results for the Turkish language are higher than for the Kazakh language, this is due to the volume of corpora specified for training the model.

The experiments were conducted on the workstation with the following specifications: Core i7 4790K, 32 GB RAM, 1 TB SSD, and NVIDIA GeForce RTX 2070 Super. From Fig. 3, the quality of the examples generated by hybrid approach the LSTM neural network trained based on the SSP approach is inferior to the examples from the LSTM training model by the BLEU metric.

## 6    Conclusion and Future Work

The developed hybrid approach for generating texts for low-resource languages is presented on the example Kazakh and Turkish languages. Overall, generating text and sentences for children in natural language processing has great potential for developing children's language skills and can be a useful tool in the educational process. Using presented technology can create exciting and educational texts and sentences that capture children's attention and improve their language skills. Such materials can help children expand their vocabulary and better understand the grammatical rules of the Kazakh and Turkish languages. To implement this approach, the following tasks were resolved: collection and processing of resources and datasets; revealing structural and semantic properties (SSP) of a language based on formal grammar; building an LSTM neural network model trained based on SSP rules. Practical training and test samples were carried out. During the experiment, the lowest results were observed in the generation of texts of the literary and scientific genres. This was due to the specifics of the subject matter and the structural forms of the texts themselves. For the scientific genre, only scientific articles were considered, which limited the learning process of the model. To solve the problem and improve the quality of the model, we plan to increase the number and quality of cases in the future.

The results obtained were evaluated by the BLEU metric. The developed model for generating texts in the Kazakh language based on the hybrid gave fairly good results. According to the results obtained, it was revealed that the hybrid approach gives a better value than the conventional machine learning method for low-resource languages.

The approach presented in this article may be applicable to other languages with limited resources, it is also necessary to emphasize the importance of understanding the characteristics of the languages being studied.

**Acknowledgments.** This study was funded by the grant Project (grant number IRN AP 19577833, AP09259556) of Ministry of Science and Higher Education of the Republic of Kazakhstan.

**Disclosure of Interests.** The authors have no competing interests to declare that are relevant to the content of this article.

# References

1. Joshi, P., Santy, S., Budhiraja, A., Bali, K., Choudhury, M.: The state and fate of linguistic diversity and inclusion in the NLP world. In: Proceedings of the 58th Annual Meeting of the Association for Computational Linguistics, pp. 6282–6293 (2020)
2. Savelyev, A., Robbeets, M.: Bayesian phylolinguistics infers the internal structure and the time-depth of the Turkic language family. J. Lang. Evol. **5**(1), 39–53 (2020)
3. Tukeyev, U.A.: A new computational model for Turkic languages morphology and processing. J. Prob. Comput. Sci. Inf. Technol. **1**(1), 47–54 (2023)
4. Dotton, Z.; Wagner, J.D.: A Grammar of Kazakh. https://slaviccenters.duke.edu/sites/slavic centers.duke.edu/files/file-attachments/kazakh-grammar.pdf. Accessed 09 Jan 2024
5. York, A.: 10 Best AI Text & Content Generator Tools in 2023. https://clickup.com/blog/ait ext-generator/. Accessed 08 Jan 2024
6. Devlin, J., Chang, M.W., Lee, K., Toutanova, K.: BERT: pre-training of deep bidirectional transformers for language understanding, pp. 1–16 (2019)
7. Birkett, A.: The 8 best AI text generators to 10X content production. Experimentation Leader & Content Agen. https://www.alexbirkett.com/ai-text-generator/. Accessed 09 Jan 2024
8. Dogan, E.B., Kaya, B., Mungen, A.: Generation of original text with text mining and deep learning methods for Turkish and other languages. In: 2018 International Conference on Artificial Intelligence and Data Processing (IDAP), Malatya, Turkey, pp. 1–9 (2018)
9. Makhambetov, O., Makazhanov, A., Sabyrgaliyev, I., Yessenbayev, Zh.: Data-driven morphological analysis and disambiguation for Kazakh. In: International Conference on Intelligent Text Processing and Computational Linguistics, pp. 151–163 (2015)
10. Kartbaev, A., Mamyrbayev, O., Khairova, N., Ybytaeva, G., Abilkair, N., Mussayeva, D.: Correction of Kazakh synthetic text using finite state. J. Theor. Appl. Inf. Technol. **99**(22) (2021)
11. Karyukin, V., Rakhimova, D., Karibayeva, A., Turganbayeva, A., Turarbek, A.: The neural machine translation models for the low-resource Kazakh-English language pair. PeerJ Comput. Sci. **9**(e1224), 1–20 (2023)
12. Turganbayeva, A., Rakhimova, D., Karyukin, V., Karibayeva, A., Turarbek, A.: Semantic connections in complex sentences for post-editing machine translation in the Kazakh language. Information **13**(411), 1–13 (2022)
13. Rakhimova, D., Suleimenov, Y., Akhmet, G.: The task of synthesizing the Kazakh language based on the seq2seq approach for a question-answer system. In: The 7th International Conference on Computer Science and Engineering, Dyarbakir, Turkey, pp. 289–293 (2022)
14. Nurlybayeva, A., Abd Almisreb, A., Norzeli, M.S., Musab, A.M.Ali.: Kazakh text generation using neural bag-of-words model for sentiment analysis. Southeast Europe J. Soft Comput. **11**(2), 29–39 (2022)
15. Tukeyev, U., Karibayeva, A., Abduali, B.: Neural machine translation system based on synthetic corpora. In: III International Conference of Computational Methods in Engineering Science, Poland, pp. 1–5 (2018)
16. Tukeyev, U. A., Turganbaeva, A.: Lexicon - free stemming for the Kazakh language. In: International Scientific Conference "Computer Science and Applied mathematics", Almaty, pp. 84–88 (2016)
17. Tukeyev, U., Karibayeva, A., Zhumanov, Zh.: Morphological segmentation method for Turkic language neural machine translation. Cogent Eng. 1–15 (2020)
18. Tukeyev, U.; Karibayeva, A.: Inferring the complete set of Kazakh endings as a language resource. In: ICCCI 2020: Advances in Computational Collective Intelligence, Da Nang, Vietnam, pp. 741–751 (2020)

19. NLP-KazNU. https://github.com/NLP-KazNU. Accessed 03 Jan 2024
20. Tukeyev, U., Zhumanov, Zh., Rakhimova, D., Karibayeva, A., Amirova, D.: Complex technology of machine translation resources extension for the Kazakh language. Varia Informatica **1**, 1–14 (2017)
21. Zhumanov, Zh., Madiyeva, A., Rakhimova, D.: New Kazakh parallel text corpora with online access. In: International Conference on Computational Collective Intelligence, pp. 501–508 (2017)
22. Rakhimova, D., Zhumanov Zh.: Complex technology of machine translation resources extension for the Kazakh language. In: Advanced Topics in Intelligent Information and Database Systems, pp. 297–307 (2017)
23. Zhumanov, Zh., Tukeyev, U.: integrated technology for creating quality parallel corpora. In: Proceedings of ICCCI 2021, Greece, pp. 511–524 (2021)
24. Backus–Naur form. https://en.wikipedia.org/wiki/Backus%E2%80%93Naur_form. Accessed 08 Dec 2023
25. TF-IDF, https://en.wikipedia.org/wiki/Tf%E2%80%93idf. Accessed 05 Jan 2024
26. Rakhimova, D., Turarbek, A., Kopbosyn, L.: Hybrid approach for the semantic analysis of texts in the Kazakh language. In: Hong, T.-P., Wojtkiewicz, K., Chawuthai, R., Sitek, P. (eds.) ACIIDS 2021. CCIS, vol. 1371, pp. 134–145. Springer, Singapore (2021). https://doi.org/10.1007/978-981-16-1685-3_12
27. Papineni, K., Roukos, S., Ward, T., Zhu, W.-J.: BLEU: a method for automatic evaluation of machine translation. In: Proceedings of the 40th Annual Meeting of the Association for Computational Linguistics, Philadelphia, pp. 311–318 (2002)

# Data Mining and Machine Learning

# ROCKET with Dynamic Convolution for Time Series Classification

Krisztian Buza[1(✉)][iD] and Margit Antal[2][iD]

[1] Faculty of Finance and Accountancy, Budapest Business School,
Budapest, Hungary
buza@biointelligence.hu
[2] Department of Mathematics-Informatics, Sapientia Hungarian University of
Transylvania, Targu Mures, Romania
manyi@ms.sapientia.ro

**Abstract.** Time series classification is an important research topic due
to its prominent applications in industry, medicine, and finance. While
in the early 2000s, techniques based on dynamic time warping (DTW)
dominated this field, many recent works are based on Random Convolutional Kernel Transform (ROCKET). In this paper, we aim at combining the advantages of DTW and ROCKET. In particular, we incorporate dynamic convolution into ROCKET, thus we call the resulting
approach *DynamicROCKET*. We perform experiments on 10 publicly
available real-world time-series datasets and demonstrate that our approach, DynamicROCKET, may lead to statistically significant improvement in terms of classification accuracy. In order to promote the use of
DynamicROCKET, we made our implementation publicly available in
our github repository at https://github.com/kr7/DynamicROCKET.

**Keywords:** time series classification · ROCKET · dynamic
convolution · dynamic time warping · machine learning

## 1 Introduction

Time series classification is the common theoretical background of countless
recognition tasks in numerous domains, including biology, medicine, healthcare, astronomy, geology, social networks, industry, and finance. These tasks
include biometric person identification, signature verification, speech recognition,
earthquake prediction, the diagnosis of various diseases, and intrusion detection [1,2,6,15,19]. Due to the aforementioned applications and many others,
time series classification has been considered one of the most prominent tasks in
machine learning, for which various approaches have been introduced, including
methods based on neural networks, Bayesian networks, hidden Markov models, genetic algorithms, support vector machines, decision trees, frequent patterns (known as *motifs* or *shapelets*), and hubness-aware classifiers [4,9,10,12–
14,16,23].

N.-T. Nguyen et al. (Eds.): ICCCI 2024, CCIS 2165, pp. 271–282, 2024.
https://doi.org/10.1007/978-3-031-70248-8_21

One of the most surprising early results on time series classification states that the simple $k$-nearest neighbor classifier using dynamic time warping (DTW) as distance measure is competitive to many other classifiers. In particular, Xi et al. compared various time series classifiers and concluded that 1-nearest neighbor with DTW "is an exceptionally competitive classifier" [22], which has been confirmed by many researchers, see e.g. [17,20] and the references therein. The primary reason why DTW is well-suited for time series classification is that DTW is an elastic distance measure in the sense that it allows for shifts and elongations while it matches the two time series to be compared.

Subsequent solutions were based on deep learning techniques [7,11,23,24]. Among the many deep learning-based time series classifiers, fully convolutional networks turned out to be a very strong baseline, outperforming many other models, including residual networks [21].

As discussed in [5], "convolution itself only allows for rigid pattern matching by design." Even if convolutional layers are followed by pooling layers, such as in many popular convolutional neural network architectures, the resulting operation only accounts for translations, but not for elongations of local patterns, and even this ability is limited and somewhat irregular. Therefore, *dynamic convolution* was proposed [5].

Recently, time series classification based on Random Convolutional Kernel Transform (ROCKET) was proposed [8]. While ROCKET outperforms deep learning based techniques, such as the aforementioned fully convolutional neural networks, it uses standard convolution which is an inherent limitation.

In this paper, we aim at alleviating this limitation. In particular, we propose to use dynamic convolution in ROCKET and thus we call the resulting approach DynamicROCKET.

The remainder of the paper is organized as follows. In Sect. 2 we review dynamic convolution and ROCKET. In Sect. 3 we provide details about the incorporation of dynamic convolution into ROCKET. In Sect. 4, we describe the datasets, experimental protocols, and compare our approach, DynamicROCKET with ROCKET. The last section presents the conclusions.

## 2   Background

Next, we provide a short summary of concepts and techniques that are necessary to understand our work.

### 2.1   Problem Formulation

Given a set $\mathcal{C}$ of class labels, and a set $\mathcal{D}$ (called training set) of time series together with their class labels

$$\mathcal{D} = \left\{ \left( x^{(i)}, y^{(i)} \right)_{i=1}^{n} \right\}, \ \ y^{(i)} \in \mathcal{C} \tag{1}$$

where $x^{(i)}$ is a time series, such that $x^{(i)} = (x_1^{(i)}, \ldots, x_l^{(i)})$, each $x_j^{(i)}$ is a real number $(1 \leq j \leq l)$. We aim at finding a model $\mathcal{M}$ that is able to determine the class label $y' \in \mathcal{C}$ of any new (test) time series $x' = (x_1', \ldots, x_l')$.

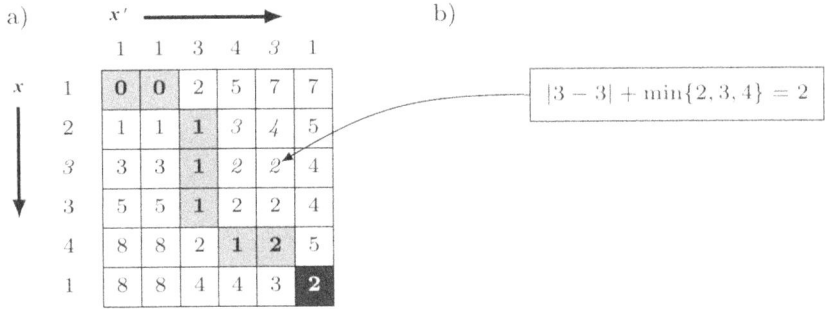

**Fig. 1.** Example of the calculation of the DTW-matrix. a) The DTW-matrix. The time series $x$ and $x'$ are shown on the left and on the top of the matrix. The marked entries of the matrix correspond to the mapping between the two time series. b) The calculation of the value of an entry.

### 2.2 Dynamic Time Warping

Dynamic Time Warping (DTW) is an elastic distance measure for time series based on the paradigm of dynamic programming [18]. DTW allows for shifts and elongations when it compares (matches) two time series.

The calculation of the DTW distance between two time series $x = (x_1, \ldots, x_l)$ and $x' = (x'_1, \ldots, x'_{l'})$ is implemented as filling the entries of an $l \times l'$ matrix. Each entry of the matrix corresponds to the distance between two prefixes, one of these prefixes is from $x$, the another one is from $x'$. In particular, the value in the $i$-th row and $j$-th column, denoted as $d_{i,j}$, corresponds to the distance between $(x_1, \ldots, x_i)$ and $(x'_1, \ldots, x'_j)$, and it is calculated as follows:

$$d_{i,j} = |x_i - x'_j| + \min \left\{ d_{i,j-1}, d_{i-1,j}, d_{i-1,j-1} \right\} \tag{2}$$

where the terms of the minimum correspond to the cases of elongation in $x$, elongation in $x'$ or matching the next elements in both time series.[1]

The entries $d_{i,j}$ of the matrix can be calculated in a column-wise fashion, i.e., in this order: $d_{1,1}, d_{2,1}, \ldots, d_{L_1,1}, d_{1,2}, d_{2,2}, \ldots, d_{L_1,2}, \ldots d_{L_1,L_2}$. The first entry of the matrix, $d_{1,1}$, is initialized as $d_{1,1} = |x_1 - x'_1|$. In the cases, when some of the terms $d_{i,j-1}, d_{i-1,j}$ are $d_{i-1,j-1}$ are not defined, i.e., if $i - 1 = 0$ or $j - 1 = 0$, they are ignored, that is: the minimum in (2) is calculated over the defined terms. Finally, the DTW distance of the two time series is $d_{l,l'}$, the bottom right corner of the matrix in the example in Fig. 1.

### 2.3 Dynamic Convolution

In time series classifiers, such as ROCKET, convolution acts as local pattern detector. As illustrated in Fig. 2, convolution combined with max pooling allows

---

[1] Instead of $|x_i - x'_j|$, one can calculate $(x_i - x'_j)^2$ in Eq. (2). In our study, we used the variant with $|x_i - x'_j|$.

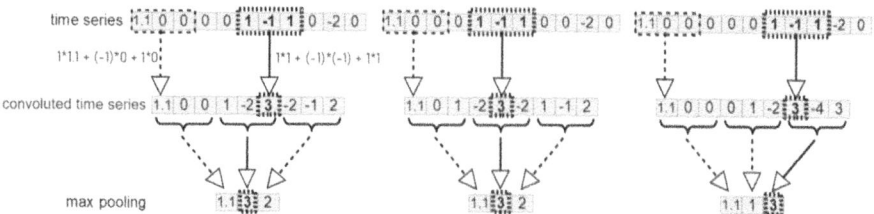

**Fig. 2.** Convolution acts as local pattern detector. The convolutional kernel in the top is expected to detect 'V'-shaped local patterns. In the left, the pattern has been detected at one of the central positions in the time series which is reflected by the high value of max pooling layer at the second position (see the highlighted '3'). In the time series depicted in the center of the figure, the 'V'-shaped pattern has been translated by one position to the left. The output of max pooling remains unchanged indicating the robustness against small translations. On the other hand, in the time series in the right, the same pattern has been translated by one position to the right (compared to its original location in the time series in the left), and the output of max pooling changed.

for some flexibility in pattern matching by hiding the exact location of a pattern within the time series in the sense that the result of max pooling may remain unchanged. However, as shown in Fig. 2, max pooling is only able to establish this robustness in pattern matching if the pattern is shifted within the max pooling window. Furthermore, there may be other types of temporal distortions, such as elongations within local patterns, that can not be taken into account by conventional convolution which motivated the introduction of dynamic convolution.

The main idea behind dynamic convolution is to replace the calculation of dot products (or inner products) in convolution by the calculation of DTW distances between the kernel and time series segments. This is illustrated in Fig. 3.

## 2.4   ROCKET

ROCKET first generates 10,000 random convolutional kernels, i.e., for each of these convolutional kernels, the length of the kernel, its weights, its bias and its dilation are sampled from suitable random distributions. In particular:

- the length $l$ of the convolutional filter is sampled uniformly from $\{7, 9, 11\}$,
- weights are sampled form standard normal distribution (i.e., the mean of the distribution is 0, its standard deviation is 1),
- bias is sampled uniformly from the interval $(-1, 1)$
- dilation is set to $\lfloor 2^d \rfloor$ where $d$ is sampled uniformly from the interval

$$(0, \lfloor log_2((l_{in} - 1)/(l - 1)) - 1 \rfloor)$$

where $l_{in}$ is the length of the time series (i.e. the input of the convolution),

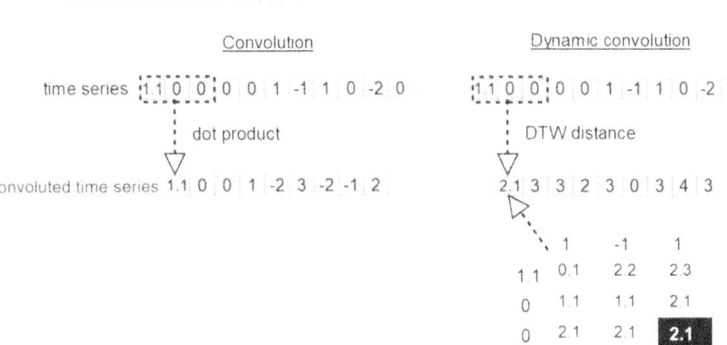

**Fig. 3.** Convolution (left) vs. dynamic convolution (right). In case of dynamic convolution, instead of the dot product (or inner product), DTW distances between the kernel and time series segments are calculated.

- for each of the convolutions, zero padding is performed with a probability of 0.5.

Given a *single* input time series (either training time series or test time series), it is convoluted with all the 10,000 filters, resulting in 10,000 convoluted time series. For each of these 10,000 convoluted time series, ROCKET applies two global pooling operations: global max pooling, and global "PPV pooling". The former refers to the selection of the maximal value, while the later means the calculation of the portion of positive values (PPV) in the convoluted time series. Thus, ROCKET transforms a time series into a feature vector containing 20,000 real values.

This way, Dempster et al. [8] represent each time series of a dataset as a 20,000-dimensional vector. Subsequently, they train ridge regression (if the dataset contains less than 20,000 time series) or logistic regression (otherwise) to classify time series.

Despite its simplicity, ROCKET has been found to outperform various deep learning based techniques, such as the aforementioned fully convolutional neural networks.

## 3 DynamicROCKET

In this section, we describe in detail how the proposed approach, Dynamic-ROCKET, extends ROCKET, see also Fig. 4.

Given ROCKET's 10,000 convolutional filters, we propose to additionally calculate dynamic convolution and to apply global max pooling to those time series that were obtained as the result of dynamic convolution. This results in 10,000 *additional* features, thus the total number of features is $20,000 + 10,000 = 30,000$ in case of DynamicROCKET.

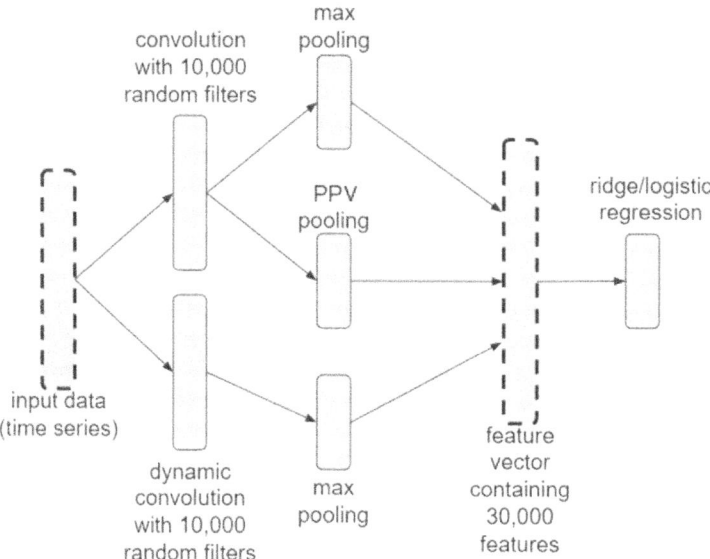

**Fig. 4.** Our approach, DynamicROCKET, for time series classification. It takes a time series as input and calculates convolution and dynamic convolution with 10,000 random filters. In case of conventional convolution, both global max pooling and global PPV pooling are applied, while in case of dynamic convolution, only global max pooling is applied. This results in a vector of 30,000 features in total. This representation is used in ridge regression or logistic regression.

Using this extended representation of time series, according to which each time series is represented as a 30,000-dimensional feature vector, similarly to ROCKET, we train ridge regression or logistic regression to classify time series. Ridge regression is used in case if the number of time series in the dataset is less than the number of features (i.e., 30,000), while logistic regression is used otherwise.

## 4    Experimental Evaluation

The goal of our experiments is to compare the proposed DynamicROCKET with ROCKET in terms of classification accuracy.

### 4.1    Data

We performed experiments on publicly available real-world time series datasets from the UCR time series archive.[2] These datasets are widely used to evaluate time series classification techniques. We selected 10 datasets from the repository

---

[2] https://timeseriesclassification.com/.

that are known to be challenging for ROCKET [8]. The names of datasets used in our experiments are listed in the first column on Table 1, while Fig. 5 shows an example from each of the considered datasets. For more details on each dataset, we refer to the documentation of the repository.

## 4.2 Baselines

We compare the performance of our approach, DynamicROCKET with two variants of ROCKET, denoted as "ROCKET" and "ROCKET-30,000".

ROCKET stands for the original ROCKET classifier as described in Sect. 2.4. In order to make sure that our comparison is fair in the sense that the observed differences can indeed be attributed to the features resulting from dynamic convolution, and not solely to the increased number of features, we also use a modified version of ROCKET which uses 15,000 convolutional filters resulting in 30,000 features (15,000 features using max pooling and 15,000 features using "PPV" pooling). We refer to this classifier as "ROCKET-30,000". The number of features in case of ROCKET-30,000 is equal to the number of features in our approach.

## 4.3 Experimental Protocol

We performed experiments according to the 10-fold cross-validation protocol. That is: we partition the data into 10 disjoint splits, one of these splits is used as test data, while the other 9 splits are used as training data. The experiment (i.e., training and evaluation of the classifier) is repeated 10-times with a different split being used as test data each time.

We used average classification accuracy, i.e., the ratio of correctly classified time series, to assess the performance of the examined classifiers ROCKET, ROCKET-30,000, and DynamicROCKET.

We report classification accuracy averaged over the 10 folds of the cross-validation together with its standard deviation for the examined classifiers, i.e., ROCKET, ROCKET-30,000, and DynamicROCKET. Additionally, we used paired t-test at significance level ($p$-value) of 0.05 in order to assess whether the difference between our approach, DynamicROCKET, and the baselines is statistically significant or not.

## 4.4 Implementation

We implemented the examined classifiers, ROCKET, ROCKET-30,000, and DynamicROCKET in Python using the numpy software library. As none of the datasets contains more than 20,000 time series, we used ridge regression – particularly the "RidgeClassifier" from the scikit-learn machine learning library – throughout the experiments in case of all the three classifier.

To calculate DTW distances quickly, we implemented DTW in Cython [3] and called this function from the Python code. We executed the experiments in

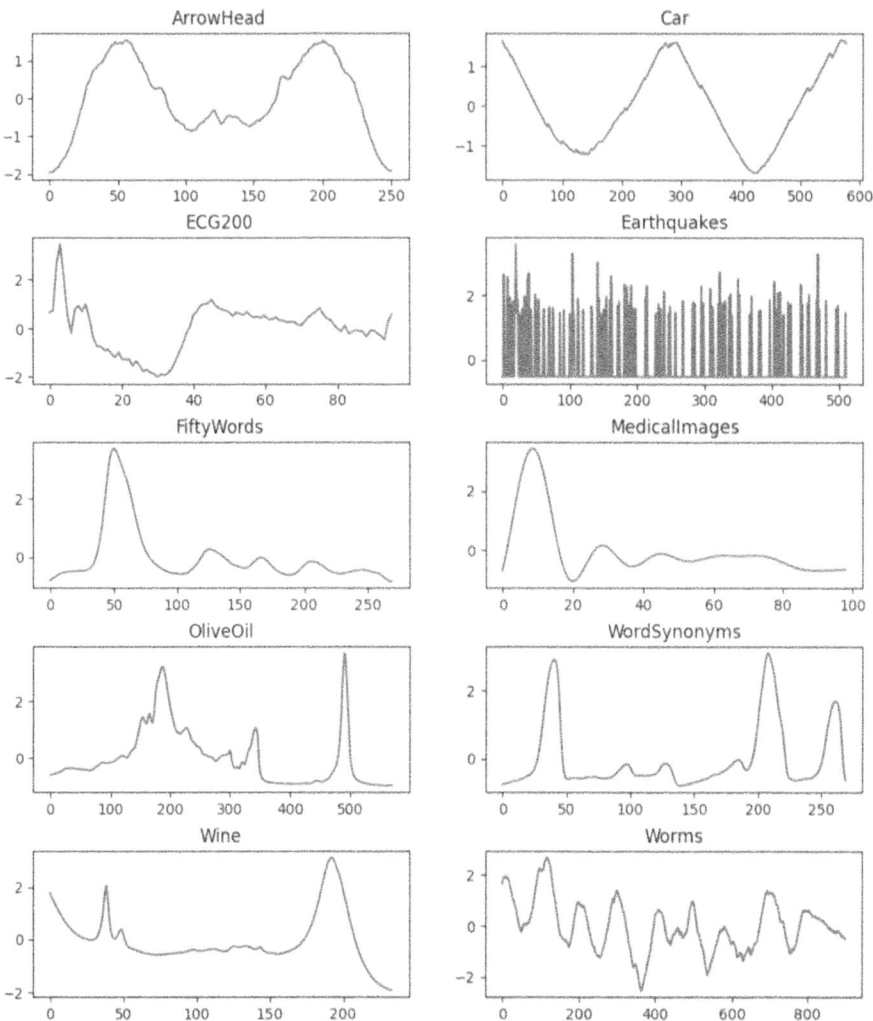

**Fig. 5.** An example time series from each of the considered datasets.

Google Colab.[3] In order to assist reproduction of our work, independent valida-tion of the results and to facilitate follow-up works, we published our code in our GitHub repository

https://github.com/kr7/DynamicROCKET

in form of an IPython notebook that can be directly executed in Google Colab.

---

[3] https://colab.research.google.com.

**Table 1.** Average accuracy ± its standard deviation (calculated over 10 folds) in case of ROCKET, ROCKET-30,000 and our approach, DynamicROCKET. The approach with higher accuracy is <u>underlined</u>. In each case, we also provide two symbols ●/○ denoting if the observed differences between our approach (DynamicROCKET) and the baselines (ROCKET and ROCKET-30,000) are statistically significant (●) or not (○) according to paired $t$-test at significance level ($p$-value) of 0.05.

| Dataset | ROCKET | ROCKET-30,000 | DynamicROCKET (our approach) |
|---|---|---|---|
| ArrowHead | 0.9288 ± 0.0440 ● | 0.9288 ± 0.0440 ● | <u>0.9476 ± 0.0396</u> |
| Car | <u>0.8833 ± 0.0764</u> ○ | <u>0.8833 ± 0.0764</u> ○ | <u>0.8833 ± 0.0850</u> |
| Earthquakes | 0.7418 ± 0.0321 ○ | <u>0.7527 ± 0.0341</u> ○ | 0.7439 ± 0.0415 |
| ECG200 | <u>0.8800 ± 0.0843</u> ○ | 0.8750 ± 0.0783 ○ | <u>0.8800 ± 0.0510</u> |
| FiftyWords | 0.7724 ± 0.0293 ● | 0.7757 ± 0.0249 ● | <u>0.7923 ± 0.0304</u> |
| MedicalImages | 0.7730 ± 0.0369 ○ | 0.7713 ± 0.0327 ● | <u>0.7905 ± 0.0412</u> |
| OliveOil | 0.8833 ± 0.1302 ○ | 0.9000 ± 0.1106 ○ | <u>0.9167 ± 0.1344</u> |
| WordSynonyms | 0.7690 ± 0.0368 ● | 0.7702 ± 0.0245 ● | <u>0.8054 ± 0.0289</u> |
| Wine | 0.9379 ± 0.0686 ○ | 0.9379 ± 0.0686 ○ | <u>0.9561 ± 0.0577</u> |
| Worms | 0.6005 ± 0.1251 ○ | 0.5889 ± 0.1015 ○ | <u>0.6395 ± 0.0787</u> |

### 4.5   Results

Table 1 summarizes our results. It shows, for each dataset, the classification accuracy averaged over 10 folds and its standard deviation after the ± sign both in case of ROCKET, ROCKET-30,000 and our approach, DynamicROCKET. The approach with highest accuracy is <u>underlined</u>. Additionally, we provide two symbols ●/○ denoting if the difference between ROCKET and DynamicROCKET is statistically significant (●) or not (○).

As one can see in Table 1, in the vast majority of the examined cases, our approach, DynamicROCKET, outperforms its competitors or is on par with them. The only exception is the Earthquakes dataset, in case of which ROCKET-30,000 performs best, while DynamicROCKET is only second best. Nevertheless, in case of this dataset, the difference between ROCKET and DynamicROCKET is not significant statistically.

In contrast, considering those cases when ROCKET outperforms its competitors, the difference is statistically significant several times.

### 4.6   Discussion

The observations reported in Table 1 may be attributed to the fact that we have extended the set of features used by ROCKET and these additional features may provide useful information for classification, but in some cases, they do not provide such additional information. The fact that DynamicROCKET is never worse than ROCKET, indicate that in case if the additional features do not provide useful information for classification, they are ignored by the final classifier, in other words: ridge regression assigns low (close to zero) weights to "useless" features.

Nevertheless, we emphasize that, in most cases, the additional features indeed provide additional information based on which classification can be made more accurate.

Furthermore, we point out that in those cases when the additional features do not provide useful information, there is no theoretical guarantee that the classification accuracy will not decrease. In fact, the additional features could introduce noise to the vector representation of time series which could corrupt the final classifier and decrease its accuracy. Therefore, it is important that note that, according to our observations on the Car and ECG200 datasets, the final classifier was indeed able to successfully identify the "useless" features.

## 5    Conclusions and Outlook

In this paper we considered time series classification and extended one of the most prominent recent techniques, ROCKET, by incorporating dynamic convolution into it, thus we called our approach DynamicROCKET. We performed experiments on publicly available real-world time series datasets. The results show that the proposed DynamicROCKET approach may outperform ROCKET, and in several cases, the difference is statistically significant. In order to assist reproduction and independent validation of our work, as well as to facilitate follow-up works, we published our codes.

We point out that DynamicROCKET may be useful in various applications of time series classification, such as handwriting recognition, signature verification, analysis of typing dynamics, or ECG-based diagnosis of various heart diseases.

## References

1. Antal, M., Egyed-Zsigmond, E.: Intrusion detection using mouse dynamics. IET Biometr. **8**(5), 285–294 (2019)
2. Antal, M., Szabó, L.Z., Tordai, T.: Online signature verification on mobisig finger-drawn signature corpus. Mob. Inf. Syst. **2018**, 1–15 (2018)
3. Behnel, S., Bradshaw, R., Citro, C., Dalcin, L., Seljebotn, D.S., Smith, K.: Cython: the best of both worlds. Comput. Sci. Eng. **13**(2), 31–39 (2010)
4. Buza, K.: Asterics: projection-based classification of EEG with asymmetric loss linear regression and genetic algorithm. In: 2020 IEEE 14th International Symposium on Applied Computational Intelligence and Informatics (SACI), pp. 35–40. IEEE (2020)
5. Buza, K., Antal, M.: Convolutional neural networks with dynamic convolution for time series classification. In: Wojtkiewicz, K., Treur, J., Pimenidis, E., Maleszka, M. (eds.) ICCCI 2021. CCIS, vol. 1463, pp. 304–312. Springer, Cham (2021). https://doi.org/10.1007/978-3-030-88113-9_24
6. Buza, K., Nanopoulos, A., Schmidt-Thieme, L., Koller, J.: Fast classification of electrocardiograph signals via instance selection. In: 2011 IEEE First International Conference on Healthcare Informatics, Imaging and Systems Biology, pp. 9–16. IEEE (2011)

7. Chen, W., Shi, K.: A deep learning framework for time series classification using relative position matrix and convolutional neural network. Neurocomputing **359**, 384–394 (2019)

8. Dempster, A., Petitjean, F., Webb, G.I.: Rocket: exceptionally fast and accurate time series classification using random convolutional kernels. Data Min. Knowl. Disc. **34**(5), 1454–1495 (2020)

9. Esmael, B., Arnaout, A., Fruhwirth, R.K., Thonhauser, G.: Improving time series classification using hidden Markov models. In: 2012 12th International Conference on Hybrid Intelligent Systems (HIS), pp. 502–507. IEEE (2012)

10. Fawaz, H.I., Forestier, G., Weber, J., Idoumghar, L., Muller, P.A.: Deep learning for time series classification: a review. Data Min. Knowl. Disc. **33**(4), 917–963 (2019)

11. Guzy, F., Woźniak, M.: Employing dropout regularization to classify recurring drifted data streams. In: 2020 International Joint Conference on Neural Networks (IJCNN), pp. 1–7. IEEE (2020)

12. Hüsken, M., Stagge, P.: Recurrent neural networks for time series classification. Neurocomputing **50**, 223–235 (2003)

13. Jankowski, D., Jackowski, K., Cyganek, B.: Learning decision trees from data streams with concept drift. Procedia Comput. Sci. **80**, 1682–1691 (2016)

14. Lines, J., Davis, L.M., Hills, J., Bagnall, A.: A shapelet transform for time series classification. In: Proceedings of the 18th ACM SIGKDD International Conference on Knowledge Discovery and Data Mining, pp. 289–297 (2012)

15. Okawa, M.: Time-series averaging and local stability-weighted dynamic time warping for online signature verification. Pattern Recogn. **112**, 107699 (2021)

16. Pavlovic, V., Frey, B.J., Huang, T.S.: Time-series classification using mixed-state dynamic bayesian networks. In: Proceedings. 1999 IEEE Computer Society Conference on Computer Vision and Pattern Recognition (Cat. No PR00149), vol. 2, pp. 609–615. IEEE (1999)

17. Radovanović, M., Nanopoulos, A., Ivanović, M.: Time-series classification in many intrinsic dimensions. In: Proceedings of the 2010 SIAM International Conference on Data Mining, pp. 677–688. SIAM (2010)

18. Sakoe, H., Chiba, S.: Dynamic programming algorithm optimization for spoken word recognition. IEEE Trans. Acoust. Speech Signal Process. **26**(1), 43–49 (1978)

19. Szilagyi, S.M., Szilagyi, L., Iclanzan, D., Benyo, Z.: Unified neural network based adaptive ECG signal analysis and compression. Sci. Bull. Politechnica Univ. Timisoara Trans. Autom. Control Comput. Science **51**, 27–36 (2006)

20. Tomašev, N., Buza, K., Marussy, K., Kis, P.B.: Hubness-aware classification, instance selection and feature construction: survey and extensions to time-series. In: Stańczyk, U., Jain, L.C. (eds.) Feature Selection for Data and Pattern Recognition. SCI, vol. 584, pp. 231–262. Springer, Heidelberg (2015). https://doi.org/10.1007/978-3-662-45620-0_11

21. Wang, Z., Yan, W., Oates, T.: Time series classification from scratch with deep neural networks: a strong baseline. In: 2017 International Joint Conference on Neural Networks (IJCNN), pp. 1578–1585. IEEE (2017)

22. Xi, X., Keogh, E., Shelton, C., Wei, L., Ratanamahatana, C.A.: Fast time series classification using numerosity reduction. In: Proceedings of the 23rd International Conference on Machine Learning, pp. 1033–1040 (2006)

23. Zhao, B., Lu, H., Chen, S., Liu, J., Wu, D.: Convolutional neural networks for time series classification. J. Syst. Eng. Electron. **28**(1), 162–169 (2017)
24. Zheng, Y., Liu, Q., Chen, E., Ge, Y., Zhao, J.L.: Time series classification using multi-channels deep convolutional neural networks. In: Li, F., Li, G., Hwang, S., Yao, B., Zhang, Z. (eds.) WAIM 2014. LNCS, vol. 8485, pp. 298–310. Springer, Cham (2014). https://doi.org/10.1007/978-3-319-08010-9_33

# Prediction of the Delay Time of Public Transportation Using Machine Learning

Alicja Piaskowska⬛, Marcin Hernes$^{(\boxtimes)}$ ⬛, Ewa Walaszczyk⬛, Agata Kozina⬛, and Kateryna Czerniachowska⬛

Wroclaw University of Economics and Business, Wrocław, Poland
{alicja.piaskowska,marcin.hernes,ewa.walaszczyk,agata.kozina,
kateryna.czerniachowska}@ue.wroc.pl

**Abstract.** Reliable urban transport is one of the most important elements of an efficiently operating city. However, public transport often disappoints travelers due to a lack of comfort, unpredictability, and frequent delays. For this reason, many people still choose cars, which negatively influence the environment and, as a result, the quality of life in a city. Public transport has many advantages, but undoubtedly, one of its most significant disadvantages is the problem with punctuality. The aim of the research is to develop a machine learning based method for predicting tram delays in Wrocław, Poland. We develop Random Forest Regressor, XGBRegressor, and deep neural network models. Our contribution is related to the integration of the following sets of data: location of public transport vehicles on the route, public transport timetable, weather, and air quality for improving the prediction performance. Such an approach has not yet been developed. The most effective solution was a model based on a random forest regressor. The selected model showed very good results regarding the level of data explanation and low prediction error. Data analysis showed that the average delay time exceeds 1.5 min.

**Keywords:** public transport · delay prediction · machine learning · artificial intelligence

## 1 Introduction

Reliable urban transport is one of the most essential elements of an efficiently operating city. This applies especially to large cities, which are increasingly expanding their borders and gaining new inhabitants [1]. An example of such a city is Wrocław, Poland. Mobility is a critical economic, social, and ecological issue. Residents must travel quickly and comfortably to their workplaces, schools, physicians, or essential service providers. From the point of view of transport efficiency, the optimal solution in such cases is to use public transport. Moreover, public transport is the only possible transportation for certain people (e.g., older people who cannot drive or young people who cannot afford a car yet). Also, many drivers could use public transport more often if they found it to be better for them. Unfortunately, public transport often disappoints travelers due to a

N.-T. Nguyen et al. (Eds.): ICCCI 2024, CCIS 2165, pp. 283–294, 2024.
https://doi.org/10.1007/978-3-031-70248-8_22

lack of comfort, unpredictability, and frequent delays. For this reason, many people still choose cars, which negatively influence the environment and, as a result, the quality of life in a city.

Public transport has many advantages, but undoubtedly, one of its most significant disadvantages is the problem with punctuality. The city does not have an alternative means of transport, such as the underground; therefore, optimizing tram and bus travel is crucial. Improvement in this aspect could have a very positive effect not only on public transport passengers but also on all road users.

The aim of the research is to develop a machine learning-based method for predicting tram delays in Wrocław. The delay time forecast allows you to plan your trip most efficiently by selecting the route with the shortest travel time. By examining the data on tram delays and the results of delay predictions more broadly, you can notice trends and patterns that may explain why delays occur. Our contribution is related to the integration of the following sets of data: location of public transport vehicles on the route, public transport timetable, weather, and air quality for improving the prediction performance. Such an approach has not yet been developed.

The article's structure is as follows: the next section is dedicated to presenting the background of the research literature. Section 3 presents the research methodology, including data processing and methods. The research results are described in Sect. 4. The final section is the conclusion.

## 2  Related Works

Prediction of delays in public transportation is performed using different methods. The paper [2] developed a rule-based system and single-layer neural network for the Deutsche Bahn delay prediction. The spatiotemporal-random-field approach has been used by [3]. Authors based on data from public transport in Warsaw. The developed model incorporates the predictions and generates situation-aware trips, allowing informed travel plan decisions within a smart city. The two-layer artificial neural network (ANN) for delay prediction has been compared with the KNN model by [4]. The authors state that ANN accurately gives an estimate of the arrival time of the bus when compared to KNN. The system based on Automatic Vehicle Location (AVL) has been developed by [5]. The paper [6] proposes a new framework for predicting delays by using General Transit Feed Specification data. The results showed that the neural network-based method outperforms its rivals regarding prediction accuracy. A multivariate spatial-temporal autoregressive (MSTAR) model was developed by [7] to account for transient behavior on the traffic network. The approach provides predictions of speed and volume over 5-min intervals for up to 1 h in advance. The work [8] presents a model for bus travel time prediction. The approach is a multi-output, multi-time-step, deep neural network that combines convolutional and long short-term memory (LSTM) layers. Data from public transport in Greater Copenhagen were used. The study [9] developed a support vector regression model to predict train delays and compare it with the artificial neural networks. The authors stated that the support vector regression outperforms the artificial neural networks. Data was achieved from the Serbian Railways.

Existing approaches, however, did not use weather and air quality data to predict the delay time of public transportation.

# 3   Research Methodology

## 3.1   Data Processing

Data from the Wrocław Open Data collection from December 2, 2022, to January 10, 2023, were used to train the model. The Wrocław Open Data collection is a website through which municipal units, among others, The Wrocław City Hall collect and share public information. One of the institutions that provide data is the Municipal Communication Company. Two data sets were used during the study: the location of public transport vehicles on the route and the public transport timetable.

The first set contains data on the current location of public transport vehicles in Wrocław. Data is available via API or as flat files in CSV format. They are updated by the provider every minute. An example fragment of the file is shown in Fig. 1. Each record has a unique ID corresponding to one vehicle. The name of the line corresponds to the name shown in the timetables. Geographic coordinates are provided: latitude and longitude in standard format. The update date of a given observation is also provided.

| _id | | | | | | | |
|---|---|---|---|---|---|---|---|
| 1 | 0 | None | None | None | 51.1247215270996 | 17.0414524078369 | 2023-09-07 14:32:57.603000 |
| | Side No | Reg No | Brigade | Line Name | stop_lat | stop_lon | update_time |
| 2 | 1900 | None | 1000f | | 51.1685943603516 | 17.0227756500244 | 2023-08-30 14:37:46.417000 |
| 3 | 2206 | None | | | 51.1249618530273 | 17.0399703979492 | 2023-09-09 15:28:15.353000 |
| 4 | 2208 | None | | | 51.1253051757812 | 17.0411186218262 | 2023-09-09 15:28:08.240000 |
| 5 | 2212 | None | | | 51.0794844763184 | 17.0043792724609 | 2023-09-09 15:28:01.563000 |
| 6 | 2218 | None | | | 51.1248207092285 | 17.041877746582 | 2023-09-09 15:28:28.620000 |
| 7 | 2228 | None | | | 51.1241874694824 | 17.0413303375244 | 2023-09-09 15:28:18.437000 |
| 8 | 2237 | None | | | 51.1249656677246 | 17.0398292541504 | 2023-09-09 15:28:22.263000 |
| 9 | 2238 | None | None | None | 51.1238594055176 | 17.0406036376953 | 2023-04-30 02:21:09.067000 |
| 10 | 2239 | None | T202 | T2 | 51.1247520446777 | 17.0383872298584 | 2023-09-09 15:28:25.333000 |
| 11 | 2240 | None | | | 51.1247329711914 | 17.0383892059326 | 2023-09-09 15:28:15.430000 |
| 12 | 2242 | None | | | 51.1252250671387 | 17.0417346954346 | 2023-09-09 15:28:28.377000 |
| 13 | 2252 | None | | | 51.1243896484375 | 17.0413227081299 | 2023-09-09 15:27:54.447000 |
| 14 | 2258 | None | | | 51.1253280639648 | 17.0405902862549 | 2023-09-09 15:27:41.503000 |
| 15 | 2260 | None | | | 51.1248741149902 | 17.0394020080566 | 2023-09-09 15:28:08.390000 |

**Fig. 1.** A fragment of the vehicle location data set. *Source*: Wrocław Open Data.

To maintain the correctness of the results, a column with the date and time of data download was added. The "Brigade" column was transformed by selecting only the last two digits, which enabled us to find the equivalent of the brigade in the timetable data. Additionally, a preliminary selection of data was made by selecting undamaged records. Records that did not have a brigade or tram line number in the "Line Name" column were also rejected. The second set is a collection of files that are compliant with the GTFS standard. It contains 13 files that define the timetable. The data is updated on average once an hour. The most critical files from which data was used for analysis are:

- calendar.txt – table required to select the appropriate schedule depending on the day of the week;
- stops.txt – table containing information about all stops;
- trips.txt – tables with all individual trips;
- stop_times.txt – a table containing information about specific journeys' arrival and departure times at a given stop.

The results were limited only to tram vehicles with a non-empty trip_id. The files have been combined into one file. The fragment of data is shown in Fig. 2.

| | trip_id | arrival_time | stop_id | stop_name | stop_lat | stop_lon | route_id | trip_headsign | direction_id | start_date | end_date |
|---|---|---|---|---|---|---|---|---|---|---|---|
| 2 | 3_11569381 | 07:05:00 | 1684 | BISKUPIN | 51.1012572800 | 17.1091415100 | 1 | POSWIETNE | 1 | 20221201 | 20230108 |
| 3 | 3_11569381 | 07:06:00 | 1626 | Spoldzielcza | 51.1022056800 | 17.1023223300 | 1 | POSWIETNE | 1 | 20221201 | 20230108 |
| 4 | 3_11569381 | 07:07:00 | 1575 | Piramowicza | 51.1031807800 | 17.0963274400 | 1 | POSWIETNE | 1 | 20221201 | 20230108 |
| 5 | 3_11569381 | 07:08:00 | 1372 | Chelmonskiego | 51.1035835600 | 17.0907119500 | 1 | POSWIETNE | 1 | 20221201 | 20230108 |
| 6 | 3_11569381 | 07:09:00 | 1644 | Tramwajowa | 51.1044667800 | 17.0846699700 | 1 | POSWIETNE | 1 | 20221201 | 20230108 |
| 7 | 3_11569381 | 07:11:00 | 1674 | ZOO | 51.1055608100 | 17.0778962000 | 1 | POSWIETNE | 1 | 20221201 | 20230108 |
| 8 | 3_11569381 | 07:12:00 | 1437 | Hala Stulecia | 51.1070882500 | 17.0734645200 | 1 | POSWIETNE | 1 | 20221201 | 20230108 |
| 9 | 3_11569381 | 07:14:00 | 4759 | Kliniki - Politechnika Wroclawska | 51.1097126100 | 17.0651114100 | 1 | POSWIETNE | 1 | 20221201 | 20230108 |
| 10 | 3_11569381 | 07:16:00 | 1751 | PL. GRUNWALDZKI | 51.1111569000 | 17.0614648300 | 1 | POSWIETNE | 1 | 20221201 | 20230108 |
| 11 | 3_11569381 | 07:18:00 | 1571 | Piastowska | 51.1153658200 | 17.0608780600 | 1 | POSWIETNE | 1 | 20221201 | 20230108 |
| 12 | 3_11569381 | 07:20:00 | 1602 | Prusa | 51.1194763700 | 17.0574045500 | 1 | POSWIETNE | 1 | 20221201 | 20230108 |
| 13 | 3_11569381 | 07:21:00 | 1666 | Wyszynskiego | 51.1219185200 | 17.0529416500 | 1 | POSWIETNE | 1 | 20221201 | 20230108 |
| 14 | 3_11569381 | 07:23:00 | 1527 | Nowowiejska | 51.1242391400 | 17.0448911000 | 1 | POSWIETNE | 1 | 20221201 | 20230108 |
| 15 | 3_11569381 | 07:24:00 | 1623 | Slowianska | 51.1235418700 | 17.0402172400 | 1 | POSWIETNE | 1 | 20221201 | 20230108 |

**Fig. 2.** Data on tram departures from specific stops. *Source*: own study.

The most important variables in the set are:

- trip_id – arrival ID,
- arrival_time – arrival time at the stop,
- stop_id – stop identifier,
- stop_name – the name of the stop,
- stop_lat – geographical latitude of the stop,
- stop_lon – longitude of the stop,
- route_id – tram line identifier,
- trip_headsign – direction of travel (name of the last stop),
- direction_id – identifier of the direction of travel,
- start_date – the date from which the schedule is valid,
- end_date – the date until which the schedule is valid.

Weather and air quality data are also used to develop the model. Pollution data comes from the publicly available API of the Chief Inspectorate of Environmental Protection (GIOS). Information is updated hourly. Information about five fundamental indicators, SO2, PM2.5, O3, PM10, NO2, is collected. In addition, the AQI (Air et al.), a standardized air quality indicator suggested by the Chief Inspectorate of Environmental Protection, is calculated. The data used in the research were collected from the GIOS measurement station in Wrocław.

Weather data is downloaded from weather maps for geographical coordinates corresponding to the location of the Wrocław GIOS measurement station. The reading includes temperature, weather categorical classification, air pressure, and humidity. The data was downloaded every ten minutes. At the final stage, all the files mentioned above were combined into one file (a simplified diagram of the connections between data is shown in Fig. 3).

Vehicle location data was grouped by date, line, and brigade to create a sequence of readings for a specific journey. The assumption was made that if the intervals between measurements are longer than 10 min, it is no longer the same trip. The distance between the vehicle's position and the position of the stops was calculated in the timetable data file. In order to exclude measurement errors, it was assumed that if the distance of

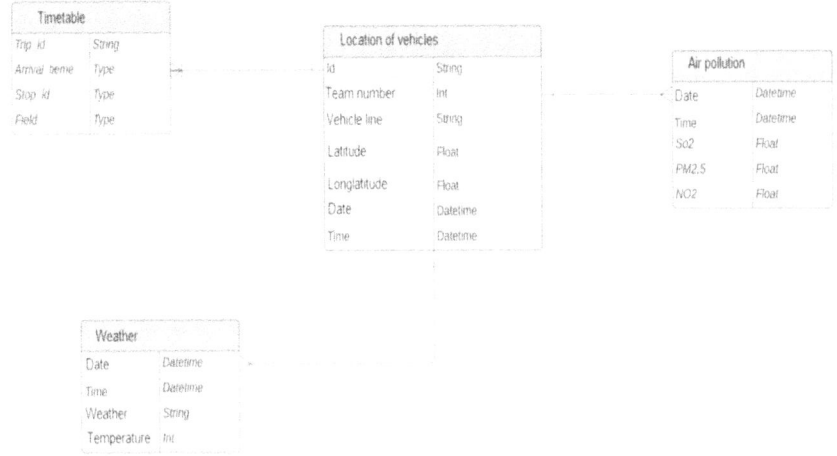

**Fig. 3.** Simplified diagram of relationships between data. *Source*: own study.

vehicles from the stop is less than 300 m, the vehicle is at the stop. The time it takes the tram to reach the stop is calculated. The delay length is the difference between this value and the expected time resulting from the distribution. If this delay is less than a minute, it is interpreted as no delay. However, the record is not considered if it is longer than 30 min. The data is linked by date, time, tram line, and stop. A fragment of the file is shown in Fig. 4.

| | ID | Line Name | Brigade | stop_lon | stop_lat | Trip_id | delay Dat |
|---|---|---|---|---|---|---|---|
| 1 | | | | | | | |
| 2 | 66 | 1 | 1 | 17.0264911651611 | 51.1502990722656 | 21:28:31.250000 02.12.2022 8_11567288 | 0 |
| 3 | 66 | 1 | 1 | 17.0281658172607 | 51.1502723693848 | 21:44:20.253000 02.12.2022 8_11567288 | 0 |
| 4 | 66 | 1 | 1 | 17.030445098877 | 51.1466445922852 | 21:46:37.293000 02.12.2022 8_11567288 | 0 |
| 5 | 66 | 1 | 1 | 17.0323829650879 | 51.1402816772461 | 21:48:31.100000 02.12.2022 8_11567288 | 0 |
| 6 | 66 | 1 | 1 | 17.0362987518311 | 51.1363258361816 | 21:49:30.083000 02.12.2022 8_11567288 | 0 |
| 7 | 66 | 1 | 1 | 17.0366706848145 | 51.1284255981445 | 21:52:25.273000 02.12.2022 8_11567288 | 0 |
| 8 | 66 | 1 | 1 | 17.0361881256104 | 51.1267547607422 | 21:53:34.240000 02.12.2022 8_11567288 | 0 |
| 9 | 66 | 1 | 1 | 17.0357894897461 | 51.1251525878906 | 21:54:36.247000 02.12.2022 8_11567288 | 0 |
| 10 | 66 | 1 | 1 | 17.0419998168945 | 51.1238555908203 | 21:58:21.337000 02.12.2022 8_11567288 | 0 |

**Fig. 4.** A fragment of the final file prepared for analysis. *Source*: own study.

## 3.2  Methods

The delay time prediction model will use the data described in the previous section. This set contained many variables that were duplicated or logically redundant. The selected set contains 38 columns. Categorical variables were recoded into numerical variables. An example of such a variable was the "weather" column, in which values (e.g., "sun" or "clouds") were changed to a numeric representation. This operation was done using the OrdinalEncoder function from the scikit-learn library. A correlation matrix was created to select explanatory variables for the model and learn about the relationships between them. The correlation was calculated using the Pearson method. Two groups of variables that are highly correlated can be observed. The first is data related to the day of the week and trip ID. The second is a group of data related to weather and air pollution. When examining the correlation between variables, the most important thing was to observe the relationship of the explained variable with other variables (Fig. 5). Probably the most essential variables at this stage are "Brigade" (negative correlation) and "Hour" (positive correlation). Additionally, the correlation occurs primarily with the days of the week (positive for Monday, Tuesday, Wednesday, and Thursday, negative for Saturday and Sunday) and air quality indicators.

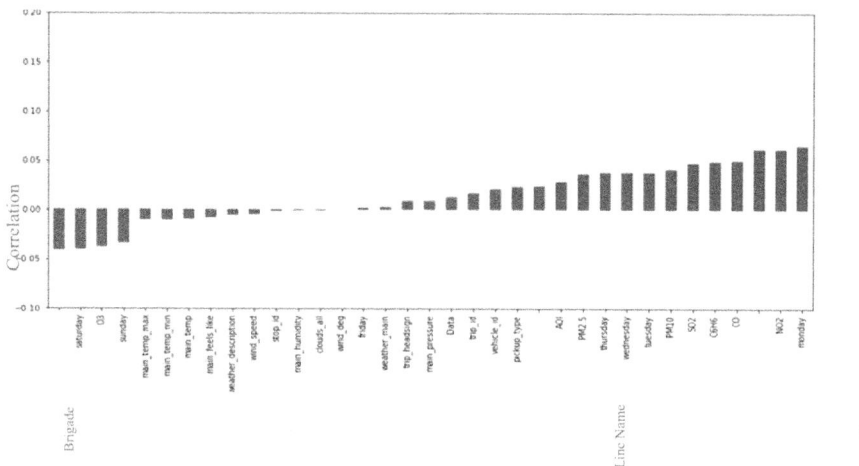

**Fig. 5.** The correlation between explanatory variables and the explained variable. *Source*: own study.

In order to predict delays, three different models were created. Their results were then compared, and the best solution was chosen. The specificity and principle of operation of the models differed significantly. All models used the same data source and division into training and test sets (80% of observations fell into the training set and 20% into the test set). Three regressors were compared, i.e.:

- Random Forest Regressor,
- XGBRegressor,
- deep neural network.

The Random Forest algorithm is a model from the scikit-learn library. It uses a random forest algorithm based on creating many random subsets trained on separate decision trees. Random selection of records into a subset reduces the dependencies between individual trees. The final prediction is the average prediction result from all trees. The use of subsets improves prediction accuracy and helps prevent overfitting, but it is a very slow algorithm, and its training takes much time. The loss function used in the algorithm is the squared error [10]. The algorithm creates 100 decision trees when making predictions. The XGBRegressor algorithm is a solution from the XGBoost library, which implements boosted decision tree algorithms. Its assumption is to combine many weaker models that together provide better results. The model uses an iterative decision tree algorithm with many interdependent decision trees. Each tree learns from all the previous ones. Instead of taking an average score (as with the Random Forest algorithm), the predicted XGBoost score is the sum of all scores. The main advantage of this approach is that it takes a very short time to learn the model. Due to the dependency between successive trees, a significant disadvantage of this approach is low resistance to outlier observations. The algorithm uses a booster based on decision trees using the maximum number of nodes [11].

The third model is a multi-layer perceptron deep neural network [12]. The solution comes from the Keras library. It is a sequential model that contains layers through which the algorithm linearly progresses. The layers contain different activation functions. As in the case of previous models, the network has an imposed objective function to which the model will be optimized. Neural networks perform exceptionally well compared to other algorithms when regressing unstructured data that requires scaling. This solution also has its drawbacks. Training a neural network requires a powerful processor and high computing power. Additionally, the condition for a correct forecast is to provide a large amount of input data to the model. The model is composed of layers with nonlinear and linear activation functions. The Relu function (Rectified Linear Activation Function) is activated only if the signal (the quotient of the input value and the weight) is more significant than zero. In the output layer of the network, the linear function is used, which is linear and provides a linear learning result. The objective function is the mean absolute error. In the training process, the batch size was set to 32 and the number of iterations to 100 (Fig. 6).

```
In [ ]:
        from keras.models import Sequential
        from keras.layers import Dense, Dropout


        model = Sequential()
        model.add(Dense(1000, input_shape=(X_train.shape[1],), activation='relu'))
        model.add(Dense(500, activation='relu'))
        model.add(Dense(250, activation='relu'))
        model.add(Dense(1, activation='linear'))


        model.summary()


        model.compile(optimizer='rmsprop', loss='mse', metrics=['mae'])


        model.fit(X_train, y_train,
                        validation_data = (X_test, y_test),
                        callbacks=[es],
                        epochs=100,
                        batch_size=32,
                        verbose=1)


        NNpredictions = model.predict(X_test)
```

**Fig. 6.** A fragment of the code creating a neural network model. *Source*: own study.

## 4  Results

The process of selecting the best model and assessing its effectiveness is based on comparing the values of efficiency measures. The summary of the measures for all algorithms is shown in Table 1. The determination index $R^2$ explains how well the independent variables explain the variability of the dependent variable. The table shows that The Random Forest and XGBRegressor models have the same $R^2$ value. In both cases, the models explained almost 90% of the data. However, the value of this indicator for the neural network model is surprisingly low. This means that the model had nearly no overall fit to the data.

**Table 1.** Values of efficiency measures for the tested algorithms. *Source*: own study.

| Model | $R^2$ | MAE | MAP |
|---|---|---|---|
| Random Forest | 0.89 | 0.55 | 1.75 |
| XGBRegressor | 0.89 | 1.53 | 7.13 |
| Neural networks | 0.00003 | 2.24 | 15.84 |

In the case of the mean absolute error MAE, the Random Forest Model performed best. A score of 0.55 means that the average error when making predictions was about half a minute. This is excellent information because, according to the assumptions, a delay of less than a minute is not considered an actual delay. The XGBRegressor model achieved the worst acceptable result of 1.53. The neural network model was again the worst, with a score of 2.24. Observing the process of training the network, which optimized its predictions based on this indicator, it was noticeable that it did not manage to go below two at any stage. The MSE measure is more sensitive to misprediction. It can be interpreted as a more restrictive version of the MAE measure. Its values were distributed similarly to the previous measure, although the differences were more significant. For this reason, the indicator is usually used only to compare models with each other and not for independent interpretation. The surprisingly low efficiency of the neural network may have several reasons. First of all, it may result from a type of prediction problem. The training data was provided in the form of a structured table. Typically, neural networks cope much better with data distributed in non-standard formats (e.g., images). On the other hand, the problem of poor results may be due to the design of the network itself. It was simplified and recalculated by 100 epochs. A better adaptation of the network and the use of a device with greater computing power would likely allow for more satisfactory results. Despite acceptable results, the XGBRegressor was a worse solution than Random Forest in every category. One reason may be the sequential transfer of data to the next tree. The new tree is more likely to make incorrect predictions when the prediction from the previous tree is incorrect. The lack of a random element may affect the risk of over-fitting to the training data and, thus, a more significant error when predicting the test data. The Random Forest Model can be considered the best based on the above. This model achieved the best results for the examined problem. It allowed for an efficient model training process. The model performed the best on each measure of all the tested models. Among them, the most significant was the mean absolute error (MEA) value, which was 0.55. It means that the average error of the model was half a minute. Additionally, this regressor had the best prediction performance and very low hardware requirements. More importantly, its results allowed the model to be considered valid and its predictions to be subject to little error. The element of randomness in creating trees allowed for avoiding overfitting while maintaining a satisfactory level of predictions.

The prediction of delays compared to the actual delay values is shown in Fig. 7. The graph shows the data distribution for both groups. It can be seen that the predictor is flatter but nevertheless maintains a similar data distribution.

Apart from the prediction itself, the model also provided very interesting information about the importance of features. The summary of the features' significance is shown in Fig. 8. The analysis showed that the most important features include:

- trip_id (identifier of the journey of a given line on a given day of the week at a given time),
- time,
- brigade,
- line_name,
- trip_heading (vehicle driving direction).

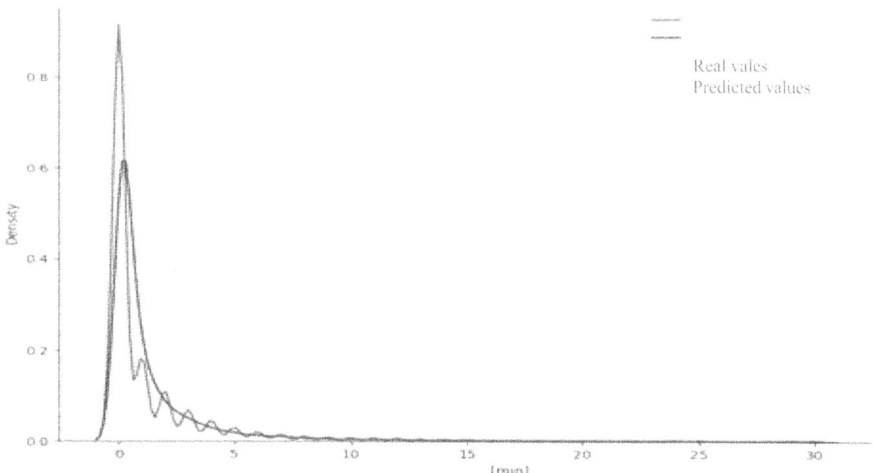

**Fig. 7.** The comparison of predictions with actual values. *Source*: own study.

These dependencies are pretty obvious, logical, and consistent with expectations. This proves that the model correctly captured the essence of the data. The "Hour" and "Brigade" variables were distinguished during correlation research. The remaining variables showed a very low correlation with the explained variable. This may mean that their relationship to the delay time is not linear. Very low significance was shown for days of the week, temperature, and the AQI index. Noteworthy are the $NO_2$ and $CO_2$ indicators, the value of which results directly from air pollution with exhaust gases and

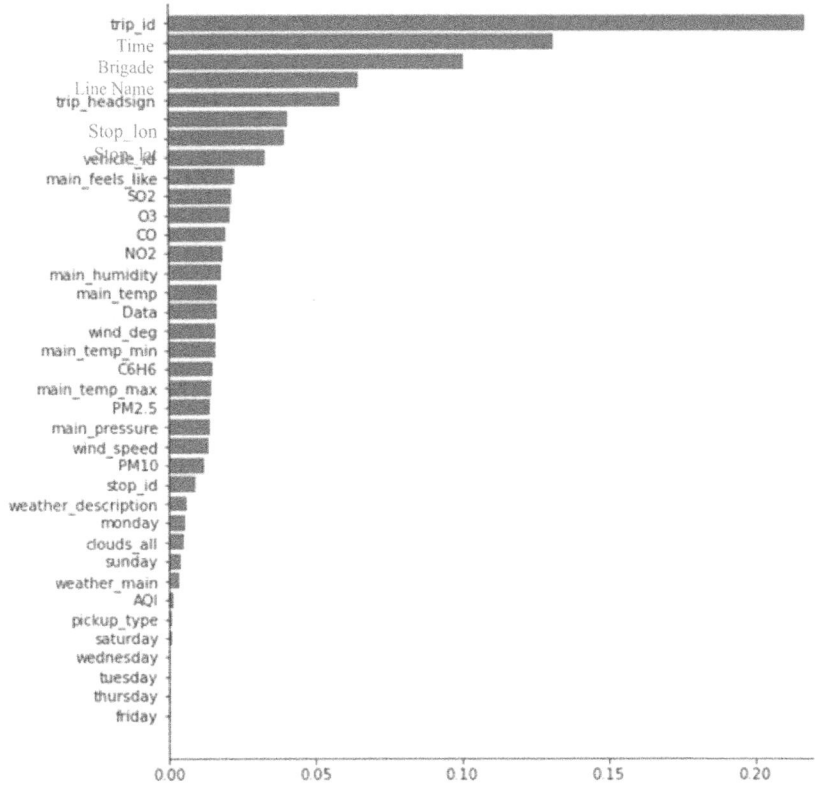

**Fig. 8.** Significance of features according to the predictive model. *Source*: own study.

the generally understood traffic of vehicles powered by combustion engines. This leads to the assumption that the amount of delays is influenced by the number of cars traveling on the tram route at a given moment.

## 5   Conclusion

The research described in this paper included building and testing three models used to predict tram delay times in Wrocław. We integrated the following sets of data: location of public transport vehicles on the route, public transport timetable, weather, and air quality to improve prediction performance.

The prediction was made based on open-source data from the city of Wrocław. Preparing the data allowed us to combine the actual time the vehicle arrived at the stop with the expected arrival time according to the timetable. Some observations were damaged and were removed from the collection. The assumption during the prediction was to select only delays that were more significant than one minute and less than 30 min. Thanks to this, outlier cases were filtered out, which could result from extreme situations, such as accidents or vehicle derailment.

Machine learning methods were used to conduct the analysis. The most effective solution was a model based on random forests. The main assumption of this approach is to create many randomly selected decision trees that will carry out the regression process and then average their results. The selected model showed very good results regarding the level of data explanation and low prediction error. Data analysis showed that the average delay time exceeds 1.5 min. The delay time they were varied depending on the selected tram line and the day of the week. Days off and holidays also influenced punctuality. The most delays occurred in the middle of the week and during holidays. This was probably due to increased traffic and changes in timetables. The lines that had more frequent delays were indicated. One of the reasons for delays may be the tram route, i.e., the condition of tram traction, traffic intensity, or route through congested and collision-prone parts of the city. Vehicle delays may also be influenced by the schedule of stops, the number of passengers, and the number of stops.

The predictive model results provided new information about public transport operations in Wrocław and the potential causes of punctuality problems. This solution can be used in two ways. On the one hand, you can extract direct data from it, allowing you to plan your trip more effectively. On the other hand, it provided premises to investigate bottlenecks and problems on tram routes.

The main limitation of the research is that it does not consider different types of deep neural networks such as convolutional, recurrent, or spiking neural networks. The second limitation is the lack of analysis of the impact of individual variables on the results of machine learning models. Therefore, further research works can be related to developing models based on different types of neural networks and analysis of the interpretability and explainability of developed models.

# References

1. Farahani, R.Z., Miandoabchi, E., Szeto, W.Y., Rashidi, H.: A review of urban transportation network design problems. Eur. J. Oper. Res. **229**(2), 281–302 (2013)
2. Peters, J., Emig, B., Jung, M., Schmidt, S.: Prediction of delays in public transportation using neural networks. In International Conference on Computational Intelligence for Modelling, Control and Automation and International Conference on Intelligent Agents, Web Technologies and Internet Commerce (CIMCA-IAWTIC 2006), vol. 2, pp. 92–97. IEEE (2005)
3. Heppe, L., Liebig, T.: Real-time public transport delay prediction for situation-aware routing. In KI 2017: Advances in Artificial Intelligence: 40th Annual German Conference on AI, Dortmund, Germany, 25–29 September 2017, Proceedings 40, pp. 128–141. Springer, Cham (2017). https://doi.org/10.1007/978-3-319-67190-1_10
4. Jabamony, J., Shanmugavel, G.R.: IoT-based bus arrival time prediction using artificial neural network (ANN) for smart public transport system (SPTS). Int. J. Intell. Eng. Syst. **13**(1), 312–323 (2020)
5. Farooq, M.U., Shakoor, A., Siddique, A.B.: GPS-based public transport arrival time prediction. In: 2017 International Conference on Frontiers of Information Technology (FIT), pp. 76–81. IEEE (2017)
6. Chondrodima, E., Georgiou, H., Pelekis, N., Theodoridis, Y.: Public transport arrival time prediction based on GTFS data. In: International Conference on Machine Learning, Optimization, and Data Science, pp. 481–495. Springer, Cham (2021). https://doi.org/10.1007/978-3-030-95470-3_36

7. Min, W., Wynter, L.: Real-time road traffic prediction with spatio-temporal correlations. Transport. Res. Part C: Emerg. Technol. **19**(4), 606–616 (2011)

8. Petersen, N.C., Rodrigues, F., Pereira, F.C.: Multi-output bus travel time prediction with convolutional LSTM neural network. Expert Syst. Appl. **120**, 426–435 (2019)

9. Marković, N., Milinković, S., Tikhonov, K.S., Schonfeld, P.: Analyzing passenger train arrival delays with support vector regression. Transport. Res. Part C: Emerg. Technol. **56**, 251–262 (2015)

10. Nakashima, H., Arai, I., Fujikawa, K.: The passenger counter is based on a random forest regressor using a drive recorder and bus sensors. In: 2019 IEEE International Conference on Pervasive Computing and Communications Workshops (PerCom Workshops), pp. 561–566. IEEE (2019)

11. Tahseen, S., Danti, A.: Prediction of user's behavior on the social media using XGBRegressor. In: Congress on Intelligent Systems: Proceedings of CIS 2021, vol. 2, pp. 491–502. Springer, Singapore (2022). https://doi.org/10.1007/978-981-16-9113-3_36

12. Kruse, R., Mostaghim, S., Borgelt, C., Braune, C., Steinbrecher, M.:. Multi-layer perceptrons. In: Computational Intelligence: A Methodological Introduction, pp. 53–124. Springer, Cham (2022). https://doi.org/10.1007/978-3-030-42227-1_5

# Processing the 3D Heritage Data Samples Based on Combination of GNN and GAN

Lam Duc Vu Nguyen[1], Sinh Van Nguyen[1(✉)] [ID], Son Thanh Le[1],
Minh Khai Tran[1], and Marcin Maleszka[2]

[1] School of Computer Science and Engineering. International University,
Vietnam National University of Ho Chi Minh City (HCMIU), Linh Trung Ward,
Thu Duc City, Ho Chi Minh City, Vietnam
[2] Wrocław University of Science and Technology, Wrocław, Poland
https://it.hcmiu.edu.vn,https://kis.pwr.edu.pl/en/

**Abstract.** The heritage objects hold historical values because they are associated with culture and have a significant impact on society. 3D reconstruction of the heritage objects is an important research to preserve and promote cultural heritage. A lot of effort has been taken to digitalize the heritages in the researched communities. The state-of-the-art methods that are current on reconstructing the heritage objects based on geometric modeling and computer graphics. Additionally, the methods based on machine learning techniques are widely researched in recent years. However, the obtained results still need to be improved, depending on characteristics of input data. This paper proposes a method for denoising the 3D heritage objects from scanned data based on a combination of Graph Neural Network (GNN) and Generation Adversarial Network (GAN). Our method includes the following steps. We first collect and process data of the real heritage objects. After meshing the object surface, we enrich the data training process by adding a normal vector for each 3D point and face. The information of both positions and normal vector of points and faces can support the denoising process. In the next step, we create a GNN based on the U-Net architecture to extract the features of 3D objects. In order to increase the performance of denoising model, we combine with a GAN framework to generate samples, such that they are as close to ground truth as possible; until the discriminator cannot realize the real or fake objects. Comparing the existing methods, our proposed method obtained better results and is closer to the ground truth data.

**Keywords:** 3D Samples · Objects Denoising · GNN · GAN · Surface Mesh · Heritage Objects · Machine Learning

This research is funded by Vietnam National University Ho Chi Minh City (VNU-HCM) under grant number DS2023-28-01.

# 1    Introduction

Processing the input data is an important step in the scientific research. Data types and their characteristics decide the choice of methods and the solution to the process. Reconstructing or restoring the heritage objects always requires processing of different data. They can be images, video frames, 3D point clouds or a data mesh. It is normal that the obtained data contains problems of missing data, data redundancy or even includes noisy data on the object's surfaces. This leads to interest of researchers in the field of computer graphics, computer vision, image processing and data science recent years. The tools and techniques to process data are classified into two types. The geometric-based method and machine learning-based method. The geometric modeling method processes 3D objects based on computation of geometrical components such as vector, point coordinates, plane, curve, surface and structure of object's shapes [1,2]. The methods for processing 3D data objects follow different steps. First it starts to collect data, simplify data, subdivide object, reconstruct the object surface, fill the holes or refine the reconstructed objects. The obtained results can be adapted to requirements of users if the input data is not too huge and complex. However, the size of input dataset is increasing more and more because of the development of tools and techniques to collect data such as 3D Scanner, Camera with deep sensors, etc. Therefore, processing huge amounts of data leads to increased time consumption, memory space and capacity of computer configuration.

In recent years, the research trends are machine learning-based methods, such as in image processing, object recognition or classification, etc. They are popular and widely used based on the power computation of deep learning models and convolution neural network (CNN) [4]. The accuracy of the obtained results is most important in any machine learning models. A lot of CNN models are proposed and widely applied to train the dataset. Most of them are run on a power server. The hardware devices are more and more developed to adapt processing of big data. For this reason, the configuration of a power computer is now not a problem to the researchers. The method for solving complex cases is now focused on the proposed solution of machine learning. The difficulty is in how to train a model on the input data such that the computer can learn all characteristics of these data; how the computer may understand, estimate, determine or realize the object as it is recognized by humans. The artificial neural networks (ANN) are well known as network models to train data using for machine learning methods. The GAN (Generative Adviseral Network) and GNN (Graph Neural Network) are machine learning methods based on network architecture [9]. GAN can generate new objects such that they looks exactly the same as the real ones. While the GNN are related to data structures that are often used due to their efficiency when processing data structures like mesh.

Processing data of heritages is a challenge in the field of digital heritage that many researchers are facing. The big data can be collected and obtained from big objects. They are architecture buildings, ancient statues with big sizes, the temples or even an old city. In order to simulate or reappear the tangible heritage objects, the data processing step is an important step to restore the objects. Starting from the real 3D objects, they are reconstructed for building a virtual museum, visualizing them on the digital environment or simulating their shape for research and visit goals. This paper proposes a method for generating and denoising data samples of 3D objects from the real statues. The data samples are scanned from the real statues of the history museum on Ho Chi Minh City, Vietnam in our research project [5]. In the previous research [11], we processed data based on geometric modeling method. In this research, we process these data based on combination of GAN and GNN. The GNN is based on the U-Net architecture (it works as a generator of GAN) to extract the features of 3D statues and generate denoised mesh. A discriminator is added to support training process in order to obtain new objects that are close to their ground truth data. This process will be stopped when the discriminator cannot realize the object is real or fake. Our proposed method obtains results that are better than the existing methods and produce new samples close to ground truth data.

The remainder of the paper is structured as follows: Sect. 2 presents several methods for data processing and visualizing the 3D objects with different data types on 3D point clouds, 2D and 3D images, video frames dataset. We detail our method in Sect. 3 including data collection, data processing and data training. Section 4 describes implementation and obtained results. We discuss, evaluate and compare the results with different methods in Sect. 5. The last Section is our conclusion.

## 2   Related Work

This section present the several methods for processing data from different data types such as 3D point clouds, video (or image) files, triangular mesh, etc. Normally, if the input data is large, the proposed methods will be processed to reduce or simplify noise data. In case of missing input data, or the 3D shapes containing problems like holes, the solution is suggested to reconstruct the object's surface. Sinh et al. [3] presented a method for removing the noise data of a 3D point cloud object. The method computed the points density of the whole 3D object based on its geometrical characteristics covering different areas on the object surface. After processing, the real object reduced a large a mount of 3D points, while preserving its 3D shape of original model. The variety of input data is analyzed and presented depending on different context of the each research. The video files are organized in a frame-set structure that was obtained by using a camera in computer vision application. Other data types are collected by using a 3D scanner to hold a set of 3D point clouds [5]. This device can get all data points of the whole object. The variety of data type can be expressed as following. The data is structured based on voxel model represents an object as a 3D grid, where the presence or absence of voxel in the grid encodes the shape of the object. It is the most straightforward representation of 3D data. Voxel works like an extension of the pixel (2D) in a 3D space; Furthermore, Voxel-based data represent complex 3D shapes [6] using a grid of voxels, enabling them to handle tasks such as 3D reconstruction and object classification. Therefore, we can use the technique in image processing to solve the problems with voxels easily. One disadvantage of voxel models is their high memory requirements [7,8], especially for detailed or large-scale objects, which can limit their practical use in certain applications. Another disadvantage is the need for a post-processing step [6] to extract the 3D surface mesh, leading to challenges in direct end-to-end training for 3D surface prediction tasks.

An alternative way to represent 3D geometries is point cloud data. Point cloud representation captures a geometric shape, typically object's surface, as a set of 3D locations [2] in a Euclidean coordinate frame. In 3D, these locations are defined by their $x, y, z$ coordinates, forming a matrix $N \times F$ where $N$ is the number of points, $F$ is the coordinates. They provide a compact representation of surface-based geometry while retaining the ability to express geometric details. The problem is that it is not in a grid structure, therefore we cannot apply convolution method to it [9]. We also need post-processing to get the result as in voxel data structure. Mesh is a collection of vertices, edges and faces that define the shape of their objects. Triangulation is the best 3D representation of 3D objects because of saving memory and computing efficiency [8,10,12]. Normally, 3D data is saved in this type to reduce the memory requirements. Because of its complicated, traditional convolutional method cannot be applied to this data. Moreover, most mesh representations are actually based upon deforming a "template" mesh. In practice, the meshes have a drawback, they cannot encode arbitrary topologies, meaning that they are limited in their ability to accurately represent certain geometries. The researched work in mesh denoising is inspired by image processing. Shachar Fleishman et el. [13] created a mesh denoising algorithm that effectively, simply, and quickly filters vertices in the normal direction using local neighborhoods. It is motivated from the success of bilateral filtering in image denoising, the algorithm adopts this technique for 3D meshes and solves specific issues in the transition from 2D to 3D data. More recent methods focus on normal filtering. Hirokazu [14] introduced frameworks for extending mean and median filtering schemes from image processing to smoothing noisy 3D shapes represented by triangle meshes. The frameworks involve applying mean and median filters to face normals on triangular meshes and editing mesh vertex positions to fit the modified normals. Another research [15] used fuzzy filtering for surface normal smoothing and updating vertex positions based on the least square error criteria. As mentioned above, a mesh is denoted by a number of vertices connecting by the edges and creating a triangular surface which is similar to a graph. Therefore, GNN can be used to deal with a mesh data. One early research using a variant of convolutional neural network (CNN) is called a graph convolutional networks (GCN) [16] that operate directly on graphs for graph-structured data. This approach is motivated by a localized first-order approximation of graph convolutions. The model learns hidden layer representations that encode both local graph structure and node features. To adapt with mesh, the research paper [17] introduces a novel graph-convolution operator for establishing correspondences between filter weights and graph neighborhoods with arbitrary connectivity. Unlike traditional convolutional neural networks (CNNs) that apply to regular grids, this operator dynamically computes correspondences based on features learned by the network, allowing for the analysis of 3D shape meshes and other graph-structured data. The paper demonstrates that the proposed method is effective in learning 3D shape representation without any shape descriptions. The idea of denoising mesh is becoming more and more popular today. Most of them are based on image denoising using GAN and Autoencoders. As mentioned above, the neural network GAN [20, 21] is well-known as machine learning framework to create or produce new instances of the data object that reasonably align with the original dataset. The new data sample is generated based on a Generator and a Discriminator that classifies objects as real or fake. GANs use a smart method to train a generative model. It works as a game with two players: the generator, which makes new examples, and the discriminator, which tells real from fake. These players are trained in a game until the discriminator is fooled about half the time, showing the generator makes realistic examples. In paper [22], authors discussed the use of conditional adversarial networks as a general-purpose solution for problems in image processing, graphics, and vision. The authors propose a common framework using conditional adversarial nets to translate an input image into a corresponding output image through an automatic image-to-image translation task. They explore the effectiveness of conditional adversarial nets on a wide variety of tasks, including synthesizing photos from label maps, reconstructing objects from edge maps, and colorizing images. Aleksandar Bojchevski et al. [23] introduces NetGAN, the first implicit generative model for graphs, aiming to mimic real-world networks. The proposed model is based on learning the distribution of biased random walks over the input graph and is trained using the Wasserstein GAN objective. NetGAN is capable of generating graphs exhibiting well-known network patterns. Inspired from this idea, we can create a mesh-to-mesh GAN and train a denoising generator. Autoencoder is a specific type of feedforward neural networks where the input is the same as the output. They compress the input into a lower-dimensional vector and then reconstruct the output from this representation. The vector is a compact "summary" or "compression" of the input, also called the latent-space representation. An autoencoder consists of 2 components: encoder and decoder. The encoder compresses the input and produces the vector, the decoder then reconstructs the input only using this vector. Learning based methods analyze and learn patterns directly from the data using a neural network, however convolutional operation works well with images may not be directly apply to 3D meshes because 3D meshes are usually irregular. Some researchers solve this problem using intermediate representation [25, 26]. To find a more robust approach, Shen et al. [27], feed mesh data directly to a graph convolutional network (GCN). They discovers that graph convolution operations learn effectively relations between noisy mesh and ground truth. Extending the idea from GCN-Denoiser, GeoBi-GNN [28] use two GCNs to optimize the coordinate of vertex and face normal. The results of experiments demonstrate that the method proposed in the current paper outperforms the existed denoising techniques.

## 3    Proposed Method

**Overview:** In this section, we propose a method for processing data samples of the real statues. Instead of using traditional geometric-based method, we combine GNN and GAN based on supporting of augmented data from scanning 3D point clouds. After scanning the real object to obtain data of 3D point cloud, we triangulate the object surface. In the next step, we create data structure of graphs for the object. They contain coordinates and normal vectors of points and faces. After that, we create a GNN model based on the U-Net architecture to extract the features and regenerate of 3D objects. Within a GAN framework, the GNN is trained to remove noise from the mesh while ensuring the regenerated object meets the criteria. Therefore, its role is considered as a generator in the GAN framework (see Fig. 1).

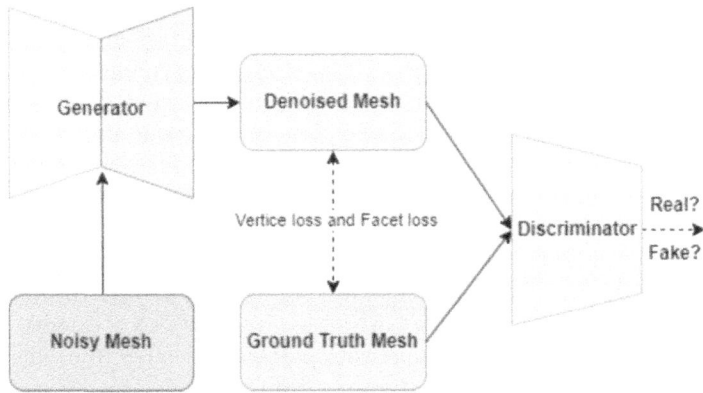

**Fig. 1.** The general architecture of our proposed method.

**Data Collection:** As presented in the previous work [5], in this step we use a 3D scanner to scan and collect data from the real statues. The output is a 3D point cloud for each object. Depending on the status of each object, the scanning process must be repeated multiple times. After that, we triangulate the object's surface from 3D point cloud. To be fair in evaluation and comparison, we use the synthetic datatset in [25] for training and comparison (our dataset is only used for evaluation).

**Data Processing:** To prepare for training dataset, for each object we convert its mesh model to a graph model. We process is as follows: for each vertex, we determine its position and compute its normal vector. Similarly, for each face, we determine the average of its vertices and compute its normal vector. At the end, for each object, we have 2 graphs to serve for training process (one for graph of vertex "Gv" and other one for graph of face "Gf", see Fig. 2).

**Training Dataset:** In the context of dual dataset. We combine two graph neural networks to learn the latent features and reconstruct the mesh. The first module (Module 1) learns the spatial coordinates of vertices, generates predenoised vertices. The second module (Module 2) generates the facet normal from Gf. Simultaneously, they interact with each other to generate the final result. Inspired from the work of [28], our network includes 2 modules with the same architecture as in Fig. 2. The unit layer is Feature-Steered Graph Convolutions [17].

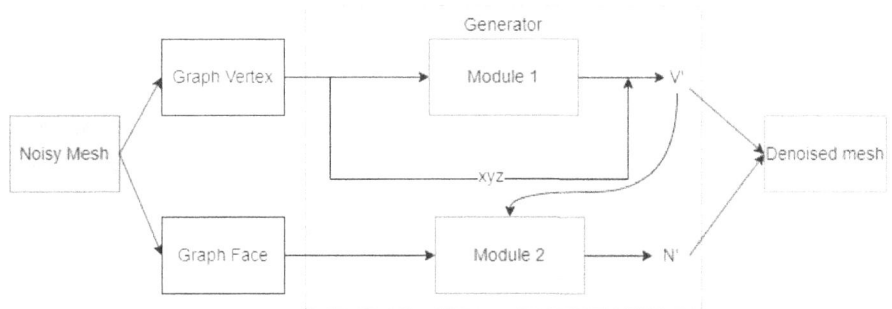

**Fig. 2.** The Generator (GNN) architecture of our proposed method.

The detailed architecture of each module (in the Generator (GNN) architecture) is designed based on the UNet model (see Fig. 3).

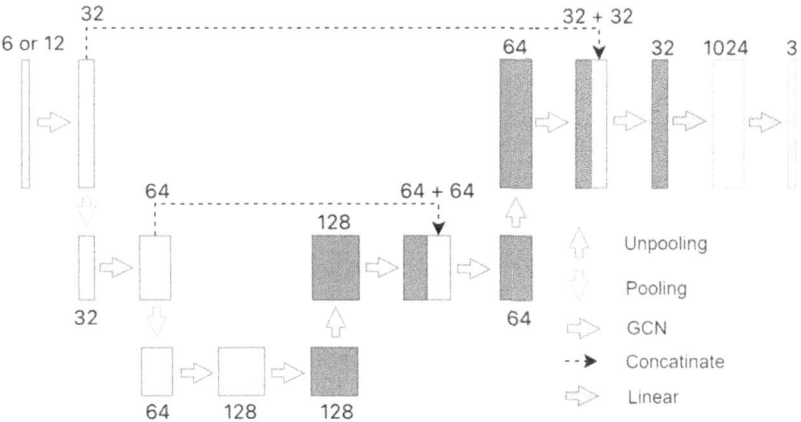

**Fig. 3.** The UNet architecture of the Modules in Generator Architecture.

In the domain of 2D image, GAN shows a robust result in generating a new image. Therefore, we want to improve the mesh denoising performance by combining dual graph architecture with GAN. The dual graph keep the same purpose as before and works as a generator. Then, we add a competitor to the game - a discriminator. Discriminator has the same basic unit as generator. Instead of reconstruct a mesh, it will generate a number close to 0 or 1 to tell us if the mesh is true mesh or fake mesh (generated from generator).

## 4   Implementation and Results

In order to implement our proposed method, we use Pytorch-Geometric [18], an extension of pytorch library and OpenMesh [19] is used for mesh processing task. In the training step, we follow the advisement in [28] to train our network. To ensure stability during training process, we split the object's mesh into smaller parts if its size exceed 20000 faces and train on each part. In order to improve accuracy of our proposed method, we use and adjust the loss function as follows. Since our Generator aims to denoise meshes based on both vertex coordinates and facet normals we use two distinct loss functions ($L_v$ and $L_f$). Both of them are leveraged mean absolute error (L1). $L_v$ focuses on the accuracy of the reconstructed vertices. It calculates the average L1 distance between the predicted vertex coordinates generated by the Generator and the corresponding ground truth coordinates of the clean mesh. $L_f$ evaluates the direction accuracy of the reconstructed facets. It calculates the average L1 distance between the predicted facet normals and the corresponding ground truth normals of the clean mesh. Additionally, we adopt Least Square GAN [30] loss function and modify as below.

$$\min_D V(D) = \mathbb{E}_{v \sim p_{\text{data\_v}}(v)} \left[ (D(v) - b)^2 \right] + \mathbb{E}_{v' \sim p_{v'}(v')} \left[ (D(G(v')) - a)^2 \right]$$

$$+ \mathbb{E}_{f \sim p_{\text{data\_f}}(f)} \left[ (D(f) - b)^2 \right] + \mathbb{E}_{f' \sim p_{f'}(f')} \left[ (D(G(f')) - a)^2 \right]$$

$$\min_G V(G) = \mathbb{E}_{v' \sim p_{v'}(v')} \left[ (D(G(v')) - c)^2 \right] + \mathbb{E}_{f' \sim p_{f'}(f')} \left[ (D(G(f')) - c)^2 \right]$$

where $v$ is the ground truth vertex position, $v'$ is the predicted vertex position, $f$ is the ground truth facet normal and $f'$ is the predicted facet normal. Two numbers: $a$ and $b$ correspond to fake label and real label. Number $c$ works as the target for generator. Since our goal is to make the fake mesh look like the real one, we set $c = b$.

## 5   Discussion, Evaluation and Comparison

We evaluate the training process by benchmarking it against the latest findings in the field, as presented in [28]. Both models are trained and tested on the synthetic dataset. We compare their performance by assessing two types of errors: the distance between vertices and the angle between normal vectors. It means that, we compute the Euclidean distance between the corresponding vertices (the vertices of denoised objects and ground truth objects). Similarity, we also compute the angles between the normal vectors of corresponding faces (the faces of denoised objects and ground truth objects). These metrics help us to evaluate how well each model represents the geometric objects. The distance between vertices measures how accurately the models capture the spatial relationships of points on the surfaces, while the angle between normal vectors measures how well the model predicts face's direction (see Fig. 4).

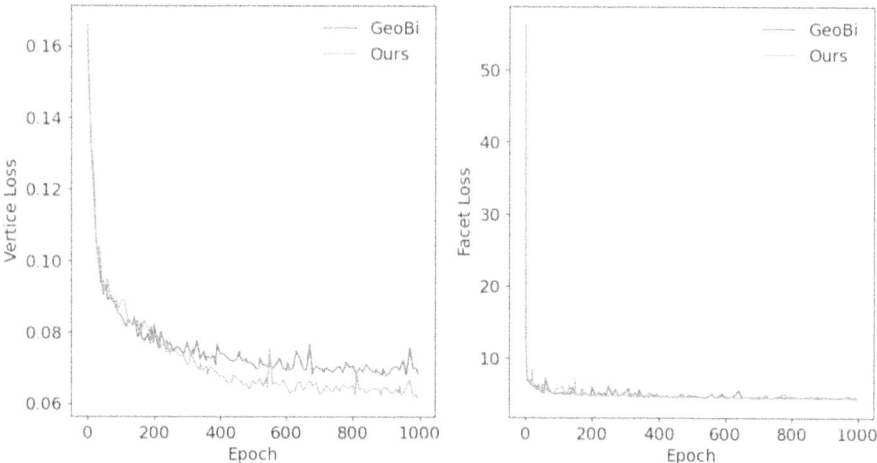

**Fig. 4.** Comparison of the loss functions of the method [28] (blue color) and our method (yellow color) at each epoch. (Color figure online)

During the training process, our model consistently achieves smaller errors in both vertex distance and angle between normal vectors compared to existing approaches. This reliability highlights the effectiveness of our method in accurately representing geometric details and directions within the mesh. Using the same model architecture as in [28] method, our obtained results are better (see Table 1). Subsequently, we select the best model from each training method to evaluate. Additionally, we extend our assessment by comparing the outcomes of these advanced models with those generated using traditional mesh denoising methods [31]. This comprehensive evaluation ensures a understanding of the efficacy and advancements offered by the new techniques. Beyond the training phase, our proposed method stands out by not only excelling in the training process but also yielding a better final model. This indicates that our approach not only learns effectively during training but also generalizes well to produce superior results. We compute the Hausdorff distance (in Meshlab) to measure approximations ($\Delta_{max}$ and $\Delta_{avg}$) of two mesh surfaces. Our results are better the two existing methods (see Table 1). In Fig. 5, we visualize the 3D object to compare the obtained results between the ground truth data, existing methods and our method. For each object, we test with all the methods to obtain results (see Table 1). From the left to the right: (a) Original picture, (b) Ground truth

**Table 1.** Comparison of the accuracy between the methods (each method, we used the same number of vertices of each object).

| 3D Objects | Comparing Units | Method [28] | Method [31] | Our method |
|---|---|---|---|---|
| Ong Dia | Angle | 17.1801 | 22.6473 | 16.7945 |
| | Distance | 0.65346 | 0.76445 | 0.64491 |
| | $\Delta_{max}$ | 2.74683 | 3.14439 | 2.58387 |
| | $\Delta_{avg}$ | 0.41807 | 0.52512 | 0.40841 |
| Chanh Than | Angle | 15.9383 | 22.8448 | 16.0344 |
| | Distance | 0.93172 | 1.06743 | 0.92750 |
| | $\Delta_{max}$ | 3.62061 | 3.65472 | 3.65704 |
| | $\Delta_{avg}$ | 0.60023 | 0.71555 | 0.59496 |
| Armadillo | Angle | 3.93118 | 8.03713 | 3.86480 |
| | Distance | 0.02918 | 0.06991 | 0.02655 |
| | $\Delta_{max}$ | 0.23003 | 0.44799 | 0.25520 |
| | $\Delta_{avg}$ | 0.02388 | 0.06276 | 0.02250 |
| Block | Angle | 1.6964 | 1.82535 | 1.71055 |
| | Distance | 0.02063 | 0.01959 | 0.01929 |
| | $\Delta_{max}$ | 0.22006 | 0.29017 | 0.17214 |
| | $\Delta_{avg}$ | 0.01774 | 0.01624 | 0.01633 |
| Bumpy Torus | Angle | 2.6295 | 6.24732 | 2.60794 |
| | Distance | 0.01615 | 0.03940 | 0.01507 |
| | $\Delta_{max}$ | 0.14210 | 0.23357 | 0.11770 |
| | $\Delta_{avg}$ | 0.01362 | 0.02726 | 0.01277 |
| Bunny | Angle | 3.94195 | 6.41433 | 3.89032 |
| | Distance | 0.00047 | 0.00096 | 0.00044 |
| | $\Delta_{max}$ | 0.00428 | 0.01053 | 0.00457 |
| | $\Delta_{avg}$ | 0.00041 | 0.00082 | 0.00039 |

data, (c) the result of using method [28], (d) the result of using method [31] and (e) the result of our method.

## 6   Conclusion

In this research, we proposed and performed a method for processing heritage data based on combination of GAN and GNN. After collecting data from the real statues of tangible heritage objects, we triangulate the object's surface. These data is used for processing and adding normal vector for each 3D point and face. The information of both positions and normal vector of points and faces are used to support the denoising process. In the next step, we create a GNN based on the U-Net architecture to extract the features of 3D objects. The GNN is

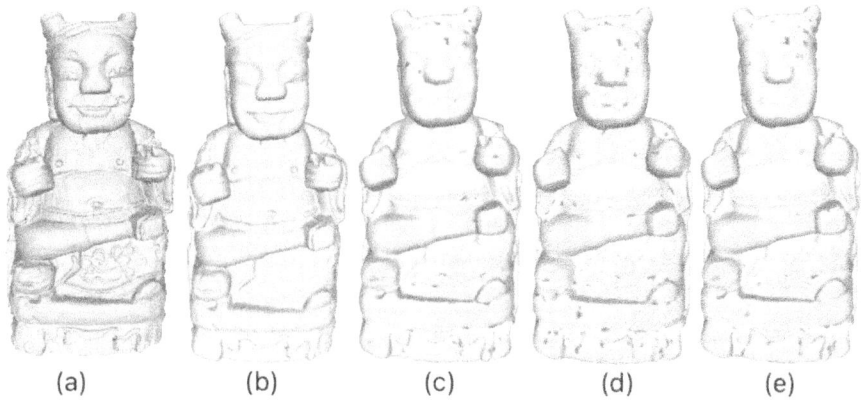

(a)          (b)          (c)          (d)          (e)

**Fig. 5.** Comparison of the existing methods and proposed method. In order to better visualize the vertices error of the output object, we use the colors processing function in the Meshlab (Colorize by quality filter) that maps them into a rather RGB colormap.

used in the GAN framework to generate samples such that they are as close to ground truth as possible; until the discriminator cannot realize the real or fake objects. In comparison with existing methods, our proposed method obtained better results and was closer to ground truth data. We tested our proposed method and obtained the results that are better the other method as presented in Table 1 and Fig. 5 for visualization of the real objects. The future work, we will collect more data from real heritage objects to enrich for the training dataset.

**Acknowledgments.** This research is funded by Vietnam National University HoChiMinh City (VNU-HCM) under grant number DS2023-28-01. We would like to thank for the fund. This research is also supported by The central Interdisciplinary Laboratory in Electronics and Information Technology (AI and Cooperation Robot), International University - VNU-HCM. We would like to thank for supporting machines in experiments.

# References

1. Krawczyk, D., Sitnik, R.: Segmentation of 3D point cloud data representing full human body geometry: a review. J. Pattern Recogn. **139**, 109444 (2023). https://doi.org/10.1016/j.patcog.2023.109444. ISSN 0031-3203
2. Van Nguyen, S., Tran, H.M., Maleszka, M.: Geometric modeling: background for processing the 3d objects. Appl. Intell. **51**(8), 6182–6201 (2021). ISSN: 1573-7497
3. Van-Sinh, N., Alexandra, B., Marc, D.: Simplification of 3D point clouds sampled from elevation surfaces. In: 21st International Conference on Computer Graphics, Visualization and Computer Vision WSCG 2013, Plzen, Czech Republic, pp. 60–69 (2013). ISBN: 978-80-86943-75-6, Rank B
4. Suganthi, S.T., Ayoobkhan, M.U.A., Bacanin, N., Venkatachalam, K., Stepan, H., Pavel, T.: Deep learning model for deep fake face recognition and detection. PeerJ Comput. Sci. **8**, e881 (2022). https://doi.org/10.7717/peerj-cs.881

5. Sinh, N.V., et al.: A solution for building a V-museum based on virtual reality application. In: Advances in Computational Collective Intelligence. ICCCI 2023. Communications in Computer and Information Science, vol. 18647, pp. 597–609. Springer, Cham (2023). https://doi.org/10.1007/978-3-031-41774-0_4

6. Liao, Y., Donne, S., Geiger, A.: Deep marching cubes: learning explicit surface representations. In: 2018 IEEE/CVF Conference on Computer Vision and Pattern Recognition, pp. 2916–2925 (2018). https://doi.org/10.1109/CVPR.2018.00308.

7. Brock, A., et al.: Generative and discriminative voxel modeling with convolutional neural networks. arXiv:1608.04236 (2016). https://doi.org/10.48550/arXiv.1608.04236.

8. Ranjan, A., Bolkart, T., Sanyal, S., Black, M.J.: Generating 3D faces using convolutional mesh autoencoders. In: Ferrari, V., Hebert, M., Sminchisescu, C., Weiss, Y. (eds.) ECCV 2018. LNCS, vol. 11207, pp. 725–741. Springer, Cham (2018). https://doi.org/10.1007/978-3-030-01219-9_43

9. Achlioptas, P., Diamanti, O., Mitliagkas, I., Guibas, L.: Learning representations and generative models for 3D point clouds (2018). https://doi.org/10.48550/arXiv.1707.02392.

10. Kanazawa, A., Tulsiani, S., Efros, A.A., Malik, J.: Learning category-specific mesh reconstruction from image collections (2018). https://doi.org/10.48550/arXiv.1803.07549.

11. Van Nguyen, S., Le, S.T., Tran, M.K., Tran, H.M.: Reconstruction of 3D digital heritage objects for VR and AR applications. J. Inf. Telecommun. **6**(3), 254–269 (2022). https://doi.org/10.1080/24751839.2021.2008133. ISSN: 2475-1839

12. Wang, N., Zhang, Y., Li, Z., Fu, Y., Liu, W., Jiang, Y.-G.: Pixel2Mesh: generating 3D mesh models from single RGB images. In: Ferrari, V., Hebert, M., Sminchisescu, C., Weiss, Y. (eds.) ECCV 2018. LNCS, vol. 11215, pp. 55–71. Springer, Cham (2018). https://doi.org/10.1007/978-3-030-01252-6_4

13. Fleishman, S., Drori, I., Cohen-Or, D.: Bilateral mesh denoising. In ACM SIGGRAPH,: papers, SIGGRAPH 2003, pp. 950–953. Association for Computing Machinery, New York (2003). https://doi.org/10.1145/1201775.882368

14. Yagou, H., Ohtake, Y., Belyaev, A.: Mesh smoothing via mean and median filtering applied to face normals. In: Geometric Modeling and Processing, Theory and Applications. GMP 2002. Proceedings, pp. 124–131 (2002). https://doi.org/10.1109/GMAP.2002.1027503.

15. Shen, Y., Barner, K.E.: Surface denoising with directional fuzzy vector median filtering. In 2003 International Conference on Multimedia and Expo. ICME 2003. Proceedings (Cat. No.03TH8698), p. I-237 (2003). https://doi.org/10.1109/ICME.2003.1220898.

16. Kipf, T.N., Welling, M.: Semi-supervised classification with graph convolutional networks (2017). https://doi.org/10.48550/arXiv.1609.02907.

17. Verma, N., Boyer, E., Verbeek, J.: FeaStNet: feature-steered graph convolutions for 3D shape analysis (2018). https://doi.org/10.48550/arXiv.1706.05206.

18. https://pytorch-geometric.readthedocs.io/en/latest/ . Accessed 05 Mar 2024

19. https://openmesh-python.readthedocs.io/en/latest/ . Accessed 05 Mar 2024

20. Goodfellow, I.J., et al.: Generative Adversarial Networks (2014). https://doi.org/10.48550/arXiv.1406.2661.

21. Radford, A., Metz, L., Chintala, S.: Unsupervised representation learning with deep convolutional generative adversarial networks (2016). https://doi.org/10.48550/arXiv.1511.06434.

22. Isola, P., Zhu, J.-Y., Zhou, T., Efros, A.A.: Image-to-image translation with con-
    ditional adversarial networks. In: 2017 IEEE Conference on Computer Vision
    and Pattern Recognition (CVPR), pp. 5967–5976 (2017). https://doi.org/10.1109/
    CVPR.2017.632.
23. Bojchevski, A., Shchur, O., Zügner, D., Günnemann, S.: NetGAN: generating
    graphs via random walks (2018). https://doi.org/10.48550/arXiv.1803.00816.
24. Kingma, D.P., Welling, M.: Auto-Encoding Variational Bayes (2022). https://doi.
    org/10.48550/arXiv.1312.6114.
25. Wang, P.S., Liu, Y., Tong, X.: Mesh denoising via cascaded normal regression.
    ACM Trans. Graph. **35**(6), 232:1–232:12 (2016). https://doi.org/10.1145/2980179.
    2980232.
26. Zhao, W., Liu, X., Zhao, Y., Fan, X., Zhao, D.: NormalNet: learning-based mesh
    normal denoising via local partition normalization. IEEE Trans. Circuits Syst.
    Video Technol. **31**(12), 4697–4710 (2021). https://doi.org/10.1109/TCSVT.2021.
    3099939
27. Shen, Y., et al.: GCN-denoiser: mesh denoising with graph convolutional networks.
    ACM Trans. Graph. **41**(1), 8:1–8:14 (2022). https://doi.org/10.1145/3480168.
28. Zhang, Y., Shen, G., Wang, Q., Qian, Y., Wei, M., Qin, J.: GeoBi-GNN: geometry-
    aware bi-domain mesh denoising via graph neural networks. Comput. Aided Des.
    **144**, 103154 (2022). https://doi.org/10.1016/j.cad.2021.103154
29. Albawi, S., Mohammed, T.A., Al-Zawi, S.: Understanding of a convolutional neu-
    ral network. In: 2017 International Conference on Engineering and Technology
    (ICET), pp. 1–6 (2017). https://doi.org/10.1109/ICEngTechnol.2017.8308186.
30. Mao, X., Li, Q., Xie, H., Lau, R.Y.K., Wang, Z., Smolley, S.P.: Least squares gen-
    erative adversarial networks. In: 2017 IEEE International Conference on Computer
    Vision (ICCV), pp. 2813–2821 (2017). https://doi.org/10.1109/ICCV.2017.304.
31. Zhang, W., Deng, B., Zhang, J., Bouaziz, S., Liu, L.: Guided mesh normal filtering.
    Comput. Graph. Forum **34**(7), 23–34 (2015). https://doi.org/10.1111/cgf.12742

# Testing the Robustness of Machine Learning Models Through Mutations

Manuel Méndez[(✉)] [iD], Miguel Benito-Parejo [iD], and Mercedes G. Merayo [iD]

Design and Testing of Reliable Systems Research Group, Universidad Complutense de Madrid,
Madrid, Spain
{manumend,mibeni01}@ucm.es, mgmerayo@fdi.ucm.es

**Abstract.** The reliable performance of machine learning algorithms stands as a critical and foundational concern. Usually, scientific attention centres solely on this aspect when selecting among models. However, in real-world scenarios, datasets are vulnerable to human errors during data input. Consequently, algorithms must display consistency and resilience against such errors. We assert that, especially in real-world applications, the resilience is, akin to the performance, a critical characteristic when selecting one algorithm over another. To address this concern, we propose a novel methodology for assessing model robustness by evaluating models both before and after applying mutations to the dataset. To validate the effectiveness of this methodology, we analyse five commonly used machine learning algorithms in a case study concerning traffic flow forecasting in Madrid. In assessing the robustness of the models, we introduce two metrics derived from well-known regression measurements. The results clearly reveal that the random forest model shows the highest robustness, according to our analysis, and that different models can exhibit very different behaviours in terms of this aspect.

**Keywords:** Algorithms Robustness · Data Mutation · Deep Learning · Machine Learning

## 1 Introduction

In recent years, the field of machine learning (ML) has undergone a transformed evolution, fundamentally altering how data is pre-processed and analysed. The continuous advancement and refinement of ML models has revolutionised several fields, such as medicine, autonomous driving technology, weather forecasting, pollutant tracking or facial recognition systems. However, during this technological revolution, the performance and dependability of ML models rely on the quality and reliability of the data. In real-life cases, datasets may contain a variety of errors, ranging from small inconsistencies to omissions, which can affect the accuracy and reliability of predictions. These errors come from many different sources. For example, mistakes when people insert data, optical character recognition problems or accidental damage to files, can make

This work has been supported by the Spanish MINECO/FEDER project AwESOMe (PID2021-122215NB-C31).

data less accurate. Although these errors may seem irrelevant when considered independently, when accumulated they can significantly affect the performance of ML models.

In this work, we analyse the impact of the aforementioned errors on the performance of ML models by inserting mutations in the dataset. We aim at determining the extent to which different ML models are affected by such errors. The ML models that are barely affected by errors will be more robust than the ones that are more sensible to variations in the data. In order to do it, we inject *mutations* in datasets. These mutations simulate potential errors, such as omitting data, skipping decimal points, or repeating previous data points, that may have been introduced in datasets. They will be randomly distributed throughout the predictor variables of the training dataset. Then, we will analyse how different models deal with these kinds of errors. The experiments will be conducted on an existing dataset [19] that was used to predict the hourly traffic flow in a main roadway of Madrid. The *mutated dataset* will be used to train five ML models and the results will be compared by using two different metrics. While our work does not involve the direct application of mutation testing, we are clearly inspired by this technique, particularly in the context of applying mutations to the data. Upon reviewing the scientific literature, we have identified numerous studies that utilise mutation testing in ML, however, they predominantly focus on mutations developed within the code of ML models, as noted by [11,21].

The rest of the paper is organised as follows. In Sect. 2 we review the most relevant and recent contributions related to the topics of the paper. In Sect. 3 we present the dataset in which we will apply our experiments and we briefly describe the ML models that we will compare among them. Moreover, we define the metrics that we will use to assess the quality of the models. In Sect. 4 we describe the mutations and the procedure that we will follow to apply them in our dataset. In Sect. 5 we present the results of our experiments. Finally, in Sect. 6 we present our conclusions and some lines for future work.

## 2    Related Work

Although our work does not directly apply mutation testing in ML, we consider it important to highlight some milestones achieved through the combination of these two fields. Classification algorithms have been employed to detect whether a given test would detect a mutant [23]. In another investigation, they were also used to solve the known Equivalent Mutant Problem directly with classification algorithms [20]. Similarly, in [4], the authors explore the use of ML algorithms to classify mutants into minimal sets or equivalent categories. Conversely, the combination of both fields has also been common, specifically in the application of mutation testing in the code of ML models, as reflected in [21]. Moreover, in another interesting investigation, a tool named DeepMutation++ enables the identification of vulnerable segments of sequential input through runtime analysis in both feed-forward and recurrent neural networks by applying mutation testing [11].

Another testing technique, metamorphic testing [5,18,22], also involves controlled modifications to the dataset. However, these modifications are based on metamorphic relations, which are defined as relationships between the input and output of a system that remain consistent across these modifications. In [25], algorithms like KNN and Naive Bayes are evaluated with metamorphic testing. Metamorphic testing has also been included in the analysis of Support Vector Machine [7], to verify classification models using a set of combined statistical test methods in [27] or even to analyse unsupervised ML models [26].

However, it is noteworthy that none of these investigations evaluate the robustness of a model. To address this, various techniques are employed, such as classical cross-validation or bootstrap methods [14], stratified for imbalanced data [29], bias (highly related to underfitting) and variance (high variance related to overfitting) analysis [6, 30] or analysis of the learning curve. The learning curve [24] allows us to appreciate how the model improves as the dataset increases the number of observations. In [12], error detection is performed by leveraging co-occurrence statistics from large corpora of clean tables.

It is our understanding that the utilisation of mutations within the dataset to assess the robustness of the model represents a novel approach in this work.

Finally, concerning the ML models that we have analysed in this work, they all are widely employed in multitude of fields [8, 10, 13, 16, 19].

## 3   Description of Dataset and Models Employed

In this section, we will outline the main features of the dataset that we have used in our experiments. In addition, we will describe the models used and the metrics against which we will compare them.

The dataset we have used in our experiments is the one utilised in a previous work [19] to forecast hourly traffic flow 12 h in advance on Paseo la Castellana Street, a roadway that runs from north to downtown Madrid, specifically in a north-south direction. This dataset comprises 31380 observations with 9 predictor variables[1] and one target variable associated with each observation. Among the different roadways considered in the previous work, we have chose Paseo la Castellana Street because it experiences the highest traffic volume for most of the day in Madrid. Another reason is that the corresponding dataset contains a very small number of errors, which suits the needs of our experiments. In our case, we want to simulate potential errors in the dataset in a controlled way, not that they are in the original dataset. In Fig. 1, we show the average traffic flow by hour, by day of the week and by month. Note that we omit the first period of the COVID pandemic (from March to June of 2020) because the data during that period were very anomalous.

The experiments have been performed using 5 classical ML models: Linear regression [17], a simple statistical model used to model the relationship between a dependent variable and one or more independent variables by fitting a straight line to the observed data points; Random forest [3], an ensemble learning method that constructs a multitude of decision trees and outputs the mean prediction; Long-short term memory neural networks (LSTM) [28], a type of recurrent neural network architecture designed to capture long-term dependencies in sequential data; Bidirectional LSTM (BiLSTM) [9], a variation of LSTM that processes input sequences in both forward and backward directions

---

[1] The characteristics of predictor variables can be found at [19].

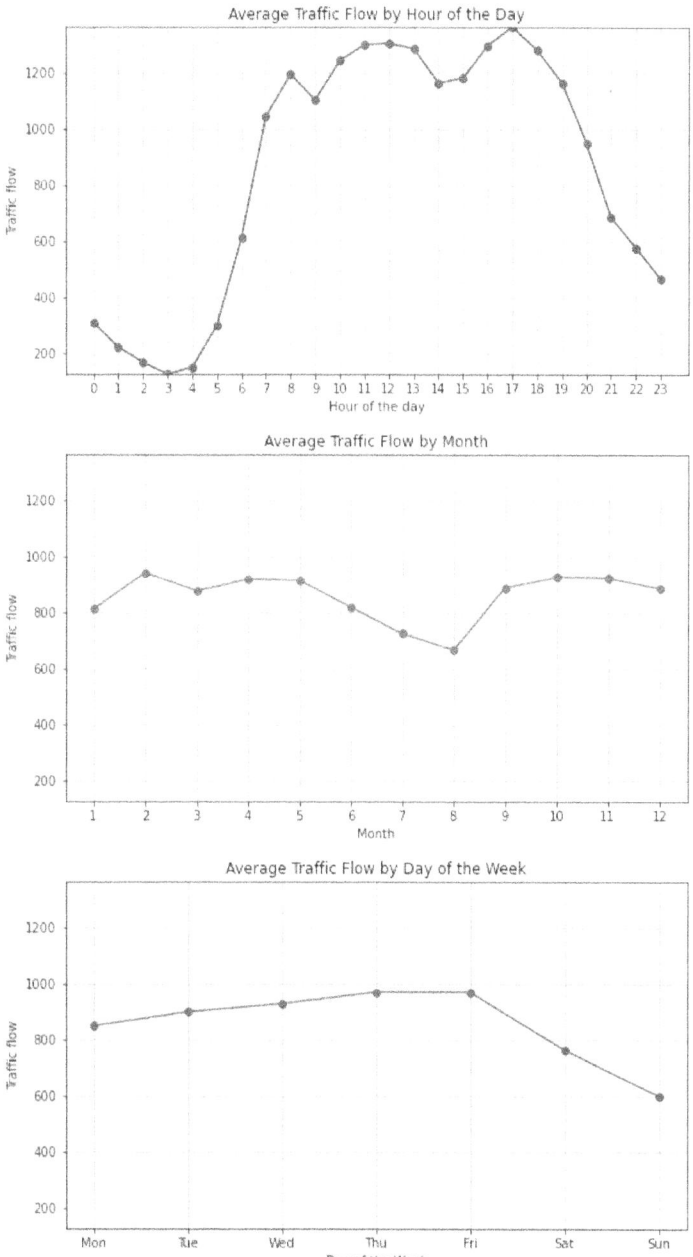

**Fig. 1.** P/Castellana (N-S) average traffic flow by hour (blue), by month (red) and by day of the week (green), excluding the first quarantine period. (Color figure online)

and the hybrid model developed in [19], Convolutional neural network combined with a Bidirectional LSTM (CNN-BiLSTM), that captures spatial features using convolutional layers and temporal dependencies using bidirectional recurrent layers. In order to assess the robustness of the models, we will utilise two metrics: the relative variance of mean absolute error (MAE) and the relative variance of mean squared error (MSE). In other words, we will examine the percentage variability of each of these metrics before and after applying mutations.

## 4  Mutations

In this section, we describe the different mutations that we have selected to apply to the dataset, as well as the procedure carried out to inject them. It is worth noting that the mutations have been applied to all the predictor variables in the training set.

In order to define formally the application of the different mutations we have considered to the dataset, we will used the following notation.

Let $(x_1, \ldots, x_p)$ and $(y)$ be the predictor variables and target variable, respectively, of an observation of the training set. Let us consider that the training set contains $n$ observations. We denote by $x_{i,j}$, where $1 \leq i \leq n$ and $1 \leq j \leq p$, the value of the $j$-th predictor variable associated with the $i$-th observation and by $x'_{i,j}$ the mutation of that value.

Next, we introduce the 4 kinds of mutations that we have used in our experiments and define how we applied them to the dataset.

- *DEC*: The first mutation simulates mistakes corresponding to either, the omission or the incorrect position of a decimal point. The application of this mutation consists in replacing the original value, $x_{i,j}$, by this value multiplied by a power of 10. This mutation will be only applied on decimal numbers. We have to take into account that if $|x_{i,j}|$ is higher than 1, then $|x'_{i,j}|$ must also be higher than 1. In other case, the error would not be about misplacing or omitting the decimal point, but rather about inserting an additional digit, specifically 0. For example, if you have $x_{i,j} = 18.27$ and the position of decimal point is omitted or changed, then $x'_{i,j}$ could take the values: $1.827, 182.7$ (changing the position of the decimal point) or $1827$ (omitting the decimal point). However, if $|x_{i,j}|$ is less than 1, this cannot happen, because if the decimal point is moved to the left, no action (removing the 0) is necessary for the number to have value. For example, if $x_{i,j} = 0.23$ and the decimal point is omitted, then $x'_{i,j}$ will be $023$, which represents the same value as $23$. In the case that the position of the decimal point is changed and $x'_{i,j} = 02.3$, this value also represents the value $2.3$ and it is not necessary to change it. This is the reason that the exponent of the power of 10 that is used to inject the mutations must be less than or equal to the length of the fractional part of $x_{i,j}$. Next we define formally the application of the mutation to the dataset.

  Let $x_{i,j}$ be the value corresponding to a predictor variable of an observation. Let $\alpha$ and $\beta$ be the length of the whole part and the fractional part of $x_{i,j}$, respectively. Given $e \in \mathbb{Z}$, randomly chosen, such that $-(\alpha - 1) \leq e \leq \beta$ and $e \neq 0$, we define $x'_{i,j} = x_{i,j} \cdot 10^e$.

– *IMP*: The second mutation is related to the values that can be assigned to a variable when the corresponding data is missing. Specifically, there are three options regarding the assigned value: 0, the previous one or the average of the previous data in the same column. Formally, given $x_{i,j}$ the mutated value $x'_{i,j}$ is randomly chosen among the following options:

$$0, \quad x_{i-1,j}, \quad \frac{1}{i-1} \cdot \sum_{k=1}^{i-1} x_{k,j}$$

– *SGN*: This mutation simulates a common mistake: the omission or incorrect placement of the sign. Formally, given $x_{i,j}$ the mutated value $x'_{i,j} = -x_{i,j}$

– *DIG*: The last mutation corresponds to placing a wrong digit in a number. This mutation is only applied to the whole part of the number. This is due to the fact that we consider that if this error affects the fractional part it is irrelevant. However, it would be trivial to extend the application of this mutation to the fractional part if it is necessary, for example, when dealing with normalised data that take values close to 0. Formally, given $x_{i,j}$, Let $\alpha$ be the length of the whole part of $x_{i,j}$ and $d_s$, where $1 \leq s \leq \alpha$, be the digit corresponding to the $s$th position of the whole part of $x_{i,j}$, respectively. In this case, we consider $d_1$ the least significant digit of the whole part and $d_\alpha$, the most significant digit of the whole part. Given $1 \leq s \leq \alpha$ and $m \in \mathbb{N}$ such that $0 \leq m \leq 9$ and $m \neq d_s$ we have that

$$x'_{i,j} = \begin{cases} x_{i,j} - d_s \cdot 10^{s-1} + m \cdot 10^{s-1} & \text{if } x_{i,j} \geq 0 \\ -x_{i,j} + d_s \cdot 10^{s-1} - m \cdot 10^{s-1} & \text{if } x_{i,j} < 0 \end{cases}$$

Next, we will describe the procedure that we will follow to apply these mutations. In the training set, we will randomly apply mutations to 500, 5000, and 20000[2] (respectively, to the 0.059%, 0.59% and 2.36% of the total) numerical cells. We will apply the same number of mutations of each type. Specifically, in the case of the *IMP* mutation, we will apply the possible values randomly, therefore, we will not have the same number of mutations for each of them. The code has been written in Python and is available in https://github.com/MMH1997/MutationOnDataset.

## 5   Experiments

In this section, we will describe our experiments. First of all, it is important to remark that in some rare cases, the mutations may not be applied as expected. For instance, if the *IMP* mutation is applied to a value equal to 0, no change will occur. Another example, if the *IMP* mutation uses the value corresponding to the previous observation and this value is equal to the value to which the mutation is being applied, no change will be observed. Similarly, if the value to apply in the mutation is the average of the values of the previous observations, the original value will not exhibit any change. Moreover,

---

[2] In accordance with the explanations provided in Sect. 5, when we refer to "20000 mutations", we are describing the application of 2000 mutations to one-tenth of the entire dataset, with the outcomes scaling equivalently to applying 20000 mutations across the entire dataset.

when a value is single-digit, the *DEC* mutation is not possible. These situations only occur about 10% of cases according to our experiments.

The selected hyperparameters for the models are similar but not exactly the same ones as those employed in [19]. In the Random forest model, we will use 25 trees, a maximum depth of 10, and a minimum number of 10 samples in a node before splitting it. In the neural network models, we use the same hyperparameters and structure as in the aforementioned work, except for the number of units in the LSTM or BiLSTM layers. In the previous work, the number of units was 500, but in this analysis, we use 50. In both models we use a seed in order to ensure that the results of the experiments are consistent and comparable across different executions. Finally, the linear regression model does not have any hyperparameter to consider.

We conducted the experiments after preprocessing the data, following nearly all the steps outlined in [19]. These steps involve transforming all data into numerical format. Subsequently, the data is converted from a sequence to pairs of input (predictor variables) and output (target variable), effectively transforming a time series problem into a supervised ML problem. Following this modification, the total number of cells in which mutations can be applied rises to 847152. After conducting preliminary experiments, we observed that the normalisation of data did not significantly impact the results. Therefore, we decided to avoid the normalisation step performed in the original work. This ensures that mutations directly affect the raw data.

In all the experiments, we utilised the same original and mutated training sets, as well as the same testing set, with a $75 - 25$ split. The results will be presented by model according to the predefined metrics. For the initial set of 500 mutations, we conducted 10 experiments for each model to ensure the consistency of our assessment. Regarding the second set of 5000 mutations, we performed 4 experiments. This distribution is based on the need to analyse whether the mutations directly affect the model, in the first case, while in the subsequent cases, the impact is already established, and we focused on understanding how it affects the model. Finally, for the last set, we also conducted 4 experiments. However, after confirming that the results before and after the mutations are scalable, we decided to proceed with this approach. That is, we randomly selected one-tenth of the dataset and apply 2000 mutations instead of 20000. This allowed us to reduce computational time while obtaining optimal results.

It is worth emphasising that our evaluation focuses on the robustness of the models rather than their performance. While one model may exhibit a lower (better) mean absolute error than another model initially, the percentage difference in mean absolute error of the first model after applying mutations may be greater than that of the second. In such instances, we would consider the second model to be more robust.

### 5.1 Linear Regression

In the case of the linear regression model, it is noteworthy to mention that due to the large number of predictor variables, the model does not perform well with this data. As can be seen in Fig. 2, the initial set of mutations has minimal impact on the metrics with an increment just due to randomness. However, with 5000 mutations, the model, on average, performs 4.257% worse in MAE and 4.994% in MSE than the original model. This can be considered a significant change. Finally, when the number of mutations is

| Linear Regression | | | |
|---|---|---|---|
| MAE | | MSE | |
| Av. | Std. | Av. | Std. |
| 500 | 0.516 | 0.217 | 0.205 | 0.159 |
| 5000 | 4.257 | 2.538 | 4.994 | 3.104 |
| 20000 | 9.568 | 4.972 | 14.530 | 7.863 |

**Fig. 2.** Percentage difference in the MAE and MSE (average and standard deviation) obtained by Linear Regression model concerning the number of mutations applied to the dataset.

20000, these values are almost tripled. In conclusion, even with less than around $1\%$ of the data mutated, this algorithm exhibits significant changes.

## 5.2  Random Forest

| Random Forest | | | |
|---|---|---|---|
| MAE | | MSE | |
| Av. | Std. | Av. | Std. |
| 500 | 0.116 | 0.212 | -0.837 | 0.442 |
| 5000 | -0.704 | 0.683 | -2.278 | 1.541 |
| 20000 | 2.016 | 1.034 | 3.811 | 2.715 |

**Fig. 3.** Percentage difference in the MAE and MSE (average and standard deviation) obtained by Random Forest model concerning the number of mutations applied to the dataset.

Unlike the previous case, the random forest model performs well with this data. Figure 3 shows that the application of the two initial sets of mutations has no impact on the performance of the model. In fact, the impact is so small that in some cases, the average value after these applications is better than the value in the original dataset. After the application of 20000 mutations (mutations in around $2.35\%$ of the total data), the change starts to be significant, with a MAE $2.016\%$ and a MSE $3.811\%$ worse than in the original case. In summary, the significant resistance to overfitting, the tree diversity, and the effective handling of noisy or outlier-laden data due to the ability to average the results of this model, make it highly robust.

## 5.3  LSTM Neural Network

The LSTM model performs well with this data. As shown in Fig. 4, a pattern can be detected: the results obtained with the original dataset are better than the ones produced in the case of the mutated dataset regardless of the number of mutations. This implies that even when mutating only $0.059\%$ of the data, the performance of the model undergo significant modifications. The changes observed in MSE are consistently higher than those observed in MAE. We conclude that the robustness of the LSTM model is extremely poor. The high sensitivity to hyperparameters, combined with its propensity for overfitting, may be the reasons behind the poor robustness exhibited by LSTMs.

| | LSTM | | | |
| | MAE | | MSE | |
| | Av. | Std. | Av. | Std. |
|---|---|---|---|---|
| **500** | 8,816 | 7,906 | 17,001 | 14,053 |
| **5000** | 15,097 | 11,783 | 29,040 | 23,335 |
| **20000** | 9,544 | 10,352 | 17,133 | 14,380 |

**Fig. 4.** Percentage difference in the MAE and MSE (average and standard deviation) obtained by LSTM model concerning the number of mutations applied to the dataset.

## 5.4    BiLSTM Neural Network

| | BiLSTM | | | |
| | MAE | | MSE | |
| | Av. | Std. | Av. | Std. |
|---|---|---|---|---|
| **500** | 2,177 | 5,342 | 4,971 | 9,473 |
| **5000** | 1,986 | 1,468 | 1,914 | 1,551 |
| **20000** | 5,298 | 3,438 | 6,086 | 8,114 |

**Fig. 5.** Percentage difference in the MAE and MSE (average and standard deviation) obtained by BiLSTM model concerning the number of mutations applied to the dataset.

The BiLSTM also performs well with this data. Initially, in the case of the initial set of mutations, Fig. 5 exhibits certain inconsistencies because the results when 5000 mutations are applied are better than the ones obtained when 500 mutations are applied. However, this can be explained by overfitting, which also accounts for the high standard deviations, and randomness. These results indicate that although BiLSTM model shows worse results when 500 mutations are applied than in the original case, the variation between 500 and 20000 mutations is not significantly high, indicating a moderate variation. This variation is much lower than in the case of LSTM, likely due to the ability of the bidirectional layer to capture complex contextual information, thereby enhancing robustness compared to LSTM.

## 5.5    CNN-BiLSTM Neural Network

This model presents the best performance among those analysed [19]. Similarly to the previous case, Fig. 6 shows that the increase in percentage difference in MAE and MSE is not directly proportional to the increase in the number of mutations applied to the dataset. However, in this case, the values are smaller. The high robustness of CNN-BiLSTM compared to BiLSTM, or even more so compared to LSTM, can be attributed to the improvement obtained by the CNN layer in capturing local characteristics in the data.

| | CNN+BiLSTM | | | |
| | MAE | | MSE | |
| | Av. | Std. | Av. | Std. |
|---|---|---|---|---|
| 500 | 1.993 | 3.460 | 3.970 | 2.469 |
| 5000 | 2.712 | 2.621 | 3.691 | 3.757 |
| 20000 | 4.477 | 3.823 | 3.459 | 4.031 |

**Fig. 6.** Percentage difference in the MAE and MSE (average and standard deviation) obtained by the hybrid CNN-BiLSTM model concerning the number of mutations applied to the dataset.

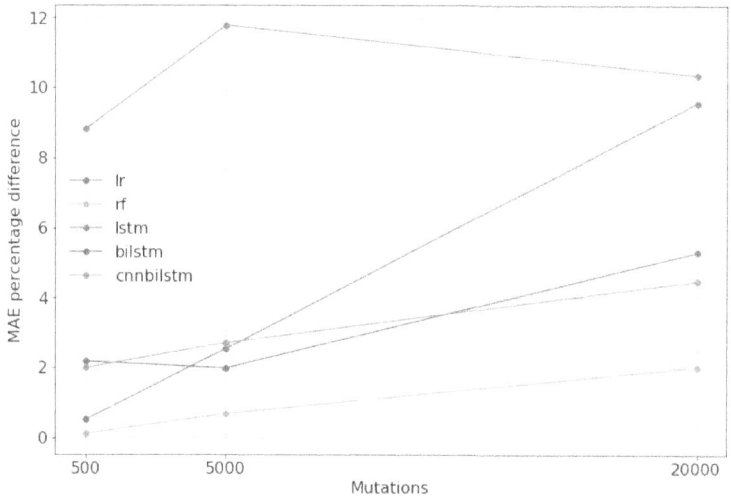

**Fig. 7.** MAE percentage difference by number of mutations applied and models.

## 5.6 Discussion

In this section, we compare the results obtained from our experiments for the different models. As depicted in Fig. 7 and Fig. 8, that summarise the percentage difference of MAE and MSE results, respectively, we observe a relative correlation between both metrics. However, there are variations in the results of different models.

In the case of linear regression, the percentage difference increases linearly as the number of mutants increases. Random forest remains almost constant, and the increment of the percentage difference as the number of mutants increases is very slight, not exceeding 2 in the MAE case and 4 in the MSE. LSTM obtains the highest percentage difference in both MAE and MSE. These values are very high, making it the least robust model among those analysed. BiLSTM and CNN-BiLSTM show similar behaviours, with slightly smaller percentage differences in the hybrid case.

In conclusion, LSTM is clearly the least robust model, regardless of the number of mutations applied to the data. Linear regression shows a high dependence on the number of mutations, being very robust when the number of mutations applied is small and much less robust when it is high. Finally, random forest, BiLSTM, and CNN-BiLSTM

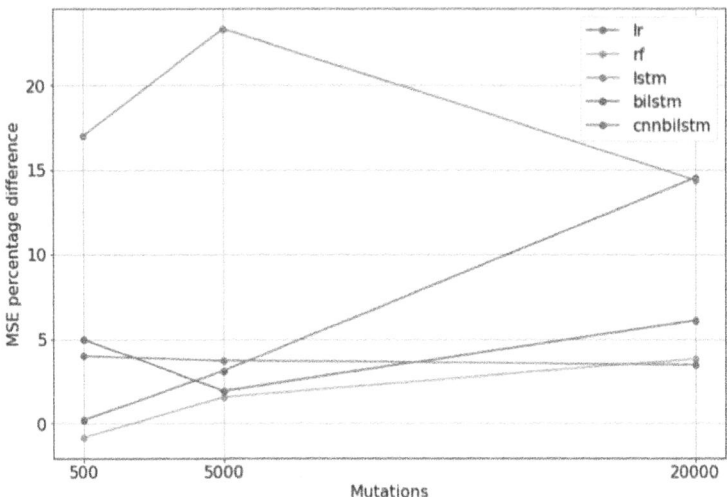

**Fig. 8.** MSE percentage difference by number of mutations applied and models.

do not seem to have much dependence on the number of mutations applied, at least in the terms we are considering. Random Forest is the most robust model, followed by the hybrid CNN-BiLSTM and the BiLSTM model.

## 6     Conclusions and Future Work

In this paper, we have proposed a method to evaluate the robustness of machine learning models by applying mutations to the datasets on which the models are trained. Our experiments show that there are significant differences in the robustness among different models. Across the models analysed, we conclude that Random Forest is the most robust model, while LSTM is the least robust.

We consider several avenues for future work. First, we aim to extend this study to other models by analysing datasets of different types and conducting a larger number of experiments. Additionally, we seek to enhance the characteristics of the mutations performed, adapting them to real-world scenarios more effectively. For instance, we would prioritise errors where a '3' is mistaken for an '8' over errors where a '3' is mistaken for a '7'. Similarly, we would prioritise errors where the decimal point is omitted or shifted slightly from its original position over errors where the decimal point is shifted by several places from the original. Moreover, it would be interesting to differentiate between mutations in terms of their detrimental effects in order to delve deeper into the behaviour regarding errors of the models.

Finally, we would like to translate this research into real-world applications [15], as well as formal validation ones [1,2]. Specifically, we aim to develop an interface where users can choose between different models, weighing the well-performance and robustness of the analysed models according to their preferences.

# References

1. Benito-Parejo, M., Merayo, M.G.: An evolutionary algorithm for selection of test cases. In: 2020 IEEE Congress on Evolutionary Computation (CEC), pp. 1–8 (2020)
2. Benito-Parejo, M., Merayo, M.G.: Using genetic algorithms to select test cases for finite state machines with timeouts. In: 2021 IEEE Congress on Evolutionary Computation (CEC), pp. 2403–2410 (2021)
3. Breiman, L.: Random forests. Mach. Learn. **45**, 5–32 (2001)
4. Brito, C., Durelli, V.H.S., Durelli, R.S., de Souza, S.R.S., Vincenzi, A.M.R., Delamaro, M.E.: A preliminary investigation into using machine learning algorithms to identify minimal and equivalent mutants. In: 2020 IEEE International Conference on Software Testing, Verification and Validation Workshops (ICSTW), pp. 304–313 (2020)
5. Chen, T.Y., et al.: Metamorphic testing: a review of challenges and opportunities. ACM Comput. Surv. **51**(1) (2018)
6. Dietterich, T.G.,, Kong, E.B.: Machine learning bias, statistical bias, and statistical variance of decision tree algorithms. Technical report, Department of Computer Science, Oregon State University, Corvallis (1995)
7. Dwarakanath, A., et al.: Identifying implementation bugs in machine learning based image classifiers using metamorphic testing. In: 27th ACM SIGSOFT International Symposium on Software Testing and Analysis, ISSTA 2018, pp. 118–128. ACM (2018)
8. Galkina, A., Grafeeva, N.: Machine learning methods for earthquake prediction: a survey. In: Proceedings of the Fourth Conference on Software Engineering and Information Management SEIM 2019 (2019)
9. Graves, A., Fernández, S., Schmidhuber, J.: Bidirectional LSTM networks for improved phoneme classification and recognition. In: Duch, W., Kacprzyk, J., Oja, E., Zadrożny, S. (eds.) ICANN 2005, pp. 799–804. Springer, Heidelberg (2005)
10. Himeur, Y., et al.: AI-big data analytics for building automation and management systems: a survey, actual challenges and future perspectives. Artif. Intell. Rev. **56**, 1–93 (2022)
11. Hu, Q., Ma, L., Xie, X., Yu, B., Liu, Y., Zhao, J.: Deepmutation++: a mutation testing framework for deep learning systems. In: 2019 34th IEEE/ACM International Conference on Automated Software Engineering (ASE), pp. 1158–1161 (2019)
12. Huang, Z., He, Y.: Auto-detect: data-driven error detection in tables. In: Proceedings of the 2018 International Conference on Management of Data, SIGMOD 2018, pp. 1377–1392. Association for Computing Machinery, New York (2018)
13. Kaur, K., Kaur, P.: The application of AI techniques in requirements classification: a systematic mapping. Artif. Intell. Rev. **57**, 02 (2024)
14. Lasfar, R., Tóth, G.: The difference of model robustness assessment using cross-validation and bootstrap methods. J. Chemomet. **2014**, e3530 (2024)
15. Méndez, M., et al.: Combining metamorphic testing and machine learning to enhance openstreetmap. IEEE Trans. Reliab. 1–15 (2024)
16. Méndez, M., Merayo, M.G., Núñez, M.: Machine learning algorithms to forecast air quality: a survey. Artif. Intell. Rev. **56**(9), 10031–10066 (2023)
17. Montgomery, D.C., Peck, E.A., Vining, G.G.: Introduction to Linear Regression Analysis. Wiley (2021)
18. Méndez, M., Benito-Parejo, M., Ibias, A., Núñez, M.: Metamorphic testing of chess engines. Inf. Softw. Technol. **162**, 107263 (2023)
19. Méndez, M., Merayo, M.G., Núñez, M.: Long-term traffic flow forecasting using a hybrid CNN-BiLSTM model. Eng. Appl. Artif. Intell. **121**, 106041 (2023)
20. Naeem, M.R., Lin, T., Naeem, H., Liu, H.: A machine learning approach for classification of equivalent mutants. J. Softw. Evolut. Process **32**(5), e2238 (2020)

21. Panichella, A., Liem, C.C.S.: What are we really testing in mutation testing for machine learning? a critical reflection. In: 2021 IEEE/ACM 43rd International Conference on Software Engineering: New Ideas and Emerging Results (ICSE-NIER), pp. 66–70 (2021)

22. Segura, S., Fraser, G., Sanchez, A.B., Ruiz-Cortés, A.: A survey on metamorphic testing. IEEE Trans. Softw. Eng. **42**(9), 805–824 (2016)

23. Strug, J., Strug, B.: Machine learning approach in mutation testing. In: Testing Software and Systems, vol. 7641, pp. 200–214. Springer, Heidelberg (2012)

24. Viering, T., Loog, M.: The shape of learning curves: a review. IEEE Trans. Pattern Anal. Mach. Intell. **45**(6), 7799–7819 (2023)

25. Xie, X., Ho, J.W.K., Murphy, C., Kaiser, G., Xu, B., Chen, T.Y.: Testing and validating machine learning classifiers by metamorphic testing. J. Syst. Softw. **84**(4), 544–558 (2011). The Ninth International Conference on Quality Software

26. Xie, X., Zhang, Z., Chen, T.Y., Liu, Y., Poon, P.-L., Xu, B.: Mettle: a metamorphic testing approach to assessing and validating unsupervised machine learning systems. IEEE Trans. Reliab. **69**(4), 1293–1322 (2020)

27. Xu, L., Towey, D., French, A.P., Benford, S., Zhou, Z.Q., Chen, T.Y.: Using metamorphic relations to verify and enhance artcode classification. J. Syst. Softw. **182**, 111060 (2021)

28. Yu, Y., Si, X., Hu, C., Zhang, J.: A review of recurrent neural networks: LSTM cells and network architectures. Neural Comput. **31**(7), 1235–1270 (2019)

29. Zeng, X., Martinez, T.R.: Distribution-balanced stratified cross-validation for accuracy estimation. J. Exp. Theor. Artif. Intell. **12**(1), 1–12 (2000)

30. Zhou, Y., Wu, J., Wang, H., He, J.: Adversarial robustness through bias variance decomposition: a new perspective for federated learning. In: Proceedings of the 31st ACM International Conference on Information and Knowledge Management (CIKM 2022), pp. 2753–2762. Association for Computing Machinery, New York (2022)

# Outlier Detection in Human Activity Recognition Systems

Agnieszka Duraj$^{(\boxtimes)}$ and Daniel Duczymiński

Institute of Information Technology, Lodz University of Technology,
al. Politechniki 8, 93-590 Łódź, Poland
`agnieszka.duraj@p.lodz.pl`

**Abstract.** The paper focuses on the detection of outliers in human motion phases. The aim was to find the most effective machine learning method to detect anomalous segments within physical activities. The article investigates the effectiveness of machine learning algorithms in detecting outlier activities within datasets. A novel approach employing nested binary classifier models is proposed to enhance outlier detection. The models' nested binary classifiers were evaluated for accuracy and precision in identifying outlier activities and compared with classifiers k-nearest neighbor, support vector machine, CART decision trees, and naive Bayes classifier. The nested models, iteratively refined through multiple nesting levels, demonstrate improved accuracy compared to standalone classifiers, particularly in identifying outlier activities. Results indicate varying performance across datasets and nesting levels, highlighting the potential of nested models in enhancing anomaly detection in diverse applications.

**Keywords:** outliers · anomalies · human activity recognition · machine learning · classification · human activity recognition

## 1 Introduction

Detecting outliers is essential, as an outlier can significantly influence subsequent stages of intelligent data analysis. Such outliers may precipitate classification or grouping errors or unveil novel phenomena, genes, diseases, or entities with previously undisclosed characteristics. Within scholarly literature, many outlier definitions and methodological approaches are delineated. The definition of an outlier varies markedly across applications, spanning from the characterization of satellite image attributes to the identification of network congestion or intrusions within computer networks. Outliers may manifest as indicators of malfunctioning machinery, measurement inaccuracies, or distinctiveness, exhibiting unique features or attribute values hitherto unobserved.

The rapid development of new technologies used in everyday life, especially sensors in wristbands, smartwatches, and mobile phones, has many applications in everyday life and routine medical activities. Fast, real-time intelligence analysis of data from various sensors monitoring everyday human activities enables productive planning of learning, training, diet, and monitoring of human activity and health.

A commonly used approach to The detection of phases of human movements is to employ a support vector machine. For example, [11] showed that a multiclass SVM machine produces accurate results classifying human activity. The authors of [15] proposed the SVM algorithm for recognizing twelve human movements. The support vector machine in human motion classification tasks was also used in [3,14]. In turn, the works [5,12,13] propose methods based on the k-nearest neighbors algorithm (k-NN). On the other hand, you can also find works presenting the use of other classification algorithms, such as the naive Bayes algorithm, decision trees, random forests, or hybrid methods combining several classification algorithms at various stages. See, for example, the combination of decision trees and the Bayes classifier in [7], or convolutional neural networks and the Bayesian classifier in [1]. In the work [9], the authors used the k-nearest neighbors algorithm (k-NN) and a random forest. In [8], the authors presented a novel technique for detecting anomalous segments in raw time sequences. the authors focus on finding anomalous data samples that were probably generated by another mechanism. They used the LSTM network. When it comes to identifying different phases of human movement, we're essentially tasked with categorizing time series data. This data, used for classification, is gathered from dedicated sensors as well as common devices like smartphones and smartwatches. The process involves preprocessing the input data and generating feature vectors by analyzing overlapping time series windows of a set size. Ultimately, classifier models are trained using these feature vectors.

The authors conducted comprehensive experiments to assess the performance of specific machine learning algorithms in identifying outlier segments associated with human activities. The research involved three datasets, which were meticulously preprocessed to remove irregular patterns such as missing, incorrect, or empty records. The article emphasizes detecting uncommon activities and unforeseen events that occur sporadically. This approach to defining and detecting outlier patterns represents a novel aspect of the research topic undertaken by the authors.

The paper presents the Nested Binary Classifier (NestBC) method, a straightforward and efficient approach based on a series of binary classifiers that iteratively recognize individual activities. Each classifier is relatively simple, reducing computational requirements and allowing efficient implementation even on resource-constrained devices. NestBC's nested structure can be easily extended with new classes without training the entire model from scratch. Each new classifier is added independently, increasing the method's flexibility. The binary nature of each classifier makes NestBC's decisions more straightforward to understand and track. At each classification stage, we can see what features were crucial for recognizing a given activity. NestBC can be scaled to handle large numbers of classes by adding additional levels of classifiers. This approach allows for more precise management of classification complexity.

Advanced techniques such as convolutional neural networks (CNN) require extensive computational resources and training data. Using heavy models may be impractical for many applications, especially those running on resource constrained devices such as smartphones or IoT devices. NestBC's nested classifier was compared with deep learning in [4] so this article compared simple machine learning algorithms with NestBC to discover the method's limitations, advantages, and disadvantages.

The article has the following structure: Sect. 1 describes human activities in HAR systems and outlier detection. Next, Sect. 2 defines the problem discussed in the paper, and outliers in the context of human movement phases are explained. The approaches proposed for the outlier activity detection method are in Sect. 3. The proposed nested algorithm is described in 3.2. The experiments and results are then listed under Sect. 4. The paper ends in a discussion and a conclusion in Sect. 6.

## 2    Outliers in the Context of Human Movement Phases

Activity recognition is an unconventional classification task due to the existence of temporal sequences of data points. Articles on this topic discuss two types of datasets used to train activity recognition models. The first type includes numerical data, such as measurements from inertial sensors attached to the object under test. These sensors are dominated by accelerometers and gyroscopes, often integrated with modern mobile devices. They enable the determination of linear and angular accelerations as well as the orientation of the device in three-dimensional space. The second datasets consists of image data, encompassing videos and photographs. This article did not consider it.

In human movement phases, an outlier can be construed as an observation markedly deviating from the norm. In numerical datasets, this could manifest as an exceptionally high or low acceleration value recorded within a specific millisecond, possibly attributed to equipment over sensitivity or noise. Conversely, in image data, an outlier might manifest as an isolated pixel displaying a distinct color, potentially resulting from subpar video transmission quality.

This paper delves into anomalies in a broader sense, encapsulating activities nested within other activities and data features that diverge from the dataset's general characteristics. Such anomalies pose a reasonable likelihood of stemming from entirely different activities.

For instance, consider the scenario of an individual jogging on a rugged terrain. While a well-trained model typically identifies the action of jogging, an instance where the individual momentarily leaps over an obstacle might mislead the tracking system into registering this transient deviation as a distinct activity-jumping, in this case.

# 3    Approaches Proposed for Outlier Activity Detection

## 3.1    Supervised Learning Algorithms

This article delves into the crucial task of detecting unique activities in HAR systems. To tackle this, we have selected supervised learning algorithms that are highly relevant to this field. Supervised learning is a type of machine learning where an algorithm learns from labeled data, making it particularly useful in HAR systems where the activities are known and labeled. The selected algorithms include the k-nearest neighbors (kNN), the CART decision tree algorithm (CART), the naive Bayes classifier (NB), and the support vector machine (SVM).

The k-nearest neighbors (kNN) method is based on classification based on the values of the set elements in the immediate vicinity. Based on the $k$ number of nearby samples, the algorithm determines what class to assign a specific case to. The new observation is classified into the class to which most belong. The Euclidean distance measure, a common method for calculating the distance between two points in a multi-dimensional space, was used to determine the nearest neighbors. This distance measure is particularly useful in HAR systems as it allows for the comparison of different activity patterns based on their feature values.

Given that we are dealing with streaming data, the CART classification tree, a decision tree algorithm, is a handy tool. In this approach, data splitting is based on one decision variable and continues until the answer to a given criterion generates further questions, making it ideal for handling continuous data streams. The CART algorithm is particularly useful in HAR systems as it can handle sequential data and make decisions based on the order of questions, which aligns with the nature of many activities in HAR.

In the case of Bayes classification, the probability of occurrence of all classes is calculated, and then the class that achieves the highest result is assigned to a given observation. To simplify the calculations, it is assumed that the events are independent. This independence assumption, which means that the occurrence of one event does not affect the occurrence of another, is a key feature of the naive Bayes classifier. The classifier is called naive because it is most often not satisfied, but it can still provide accurate results in many cases.

The support vector machine is based on mapping training examples as points in space. Each point is marked as belonging to one of two categories. On their basis, a linear classifier is created, which then assigns new examples to a specific category. The support vector machine (SVM) determines a hyperplane separating samples from two groups with the most significant possible margin. The multiclass problem is solved by reduction to a binary classification problem. This approach, however, makes it necessary to reduce the dimensionality, a process of reducing the number of variables in a dataset, for example, by the PCA algorithm. This dimensionality reduction is crucial in HAR systems as it allows for a more efficient and accurate classification of activities.

The selection of machine learning algorithms such as kNN, CART, Naive Bayes, and SVM was driven by their practicality and suitability to the problem. The kNN algorithm, for instance, is advantageous for complex and heterogeneous HAR data, as it doesn't require any assumptions about the data distribution. It also performs well in situations where the number of examples is not huge, which is often the case in activity detection in HAR systems. CART can handle a data stream. It is suitable for problems where decisions are made based on sequential questions, consistent with the nature of many activities in HAR. These sequential questions, based on the values of specific features, allow the algorithm to make more accurate and context-aware decisions, improving the overall performance of the activity recognition system.

Naive Bayes, on the other hand, uses probability, which is helpful in situations with uncertainty about the activity classification. It scales well and can be easily extended to large data sets. The algorithm calculates the probability of an activity given a set of observed features and then assigns the activity with the highest likelihood to the observation. This probabilistic approach allows for a more nuanced and flexible classification process. SVM is known for its high performance in classification problems, especially for high-dimensional data. It seeks to maximize the margin, the distance between the decision boundary, and the nearest data point from either class. This can lead to better generalization in activity classification, allowing for a wider separation between different activity classes.

### 3.2 Nested Binary Classifier

In the study, a method called a Nested Binary Classifier (NestBC) was introduced to identify human movement patterns. The level of nesting within the classifier depends on the variety of activities it can discern.

The procedural steps of this algorithm are outlined as follows:

Let $A = \{a_1, a_2, ..., a_k\}$ represent the collection of $k$ activities (e.g., *running*, *walking*, *lyingdown*, etc.). Construct a training set $Z_{training}$ and a test set $Z_{test}$, each comprising labeled classes $C = \{C_1, C_2, ..., C_k\}$ corresponding to the activities contained in the dataset being analyzed.

1. Duplicate the sets $Z_{training}$ and $Z_{test}$, denoted as $Z_{training-ki}$ and $Z_{test-ki}$ respectively, where $ki$ denotes the copy number (nesting).
2. Randomly select one activity $a_i$ from $A$.
3. In the new sets $Z_{training-ki}$ and $Z_{test-ki}$, identify class labels associated with activity $a_i$, labeling $C_1 = 1$ for activity $a_i$ and $C_0 = 0$ for the remaining activities $a_{i-1}$.
4. Conduct binary classification to recognize activity $a_i$.
5. Remove records from sets $Z_{training}$ and $Z_{test}$ labeled with the class corresponding to activity $a_i$.
6. Iterate steps 1–5 until the final activity $a_k$ is reached.
7. The remaining records, unrecognized in subsequent activity nestings, are identified as outliers.

The explanation of how the algorithm works is explained with an example.

*Example No 1*

Consider a set $A$ containing $k = 3$ activities defined by an expert $\{a_1, a_2, a_3\}$, where $a_1 = running$, $a_2 = walking$, $a_3 = lying$.

We create a training set $Z_{training0}$ and a test set $Z_{test0}$. These sets are labeled according to the selected activities. We therefore have classes $C = \{C_1, C_2, C_3\}$ corresponding to the activities $\{a_1, a_2, a_3\}$ contained in the analyzed data set. Then we create a copy of the test and training sets, obtaining the sets $Z_{training1}$ and $Z_{test1}$. For the selected activity, e.g. $a_1 = running$ in the sets $Z_{training1}$, $Z_{test1}$, we perform new labeling. Class $C_1$ will receive the label *one* $C_1 = 1$ for activity $a_1$ and class *zero* $C_0 = 0$ for the remaining activities $a_2, a_3$. Then, we perform binary classification to recognize the activity of $a_1$. After performing the classification, we remove the recognized activity $a_1$ from the initial sets $Z_{training0}$, $Z_{test0}$ and randomly select another activity. Let the next activity drawn be $a_2 = walking$. We make a copy of the sets $Z_{training0}$, $Z_{test0}$, obtaining the sets $Z_{training2}$ and $Z_{test2}$. This time, the class has a label of 1 if we are dealing with activity $a_2$ and 0 for $a_3$. After classification, we remove all recognized $a_2$ activities and repeat all steps. If all activities have been removed from the $Z_{training0}$, $Z_{test0}$ sets and these sets are not empty, there are unrecognized activities in the set, called outliers.

## 4    Experimental Studies

### 4.1    Characteristics of Data Sets

Three distinct data sets presenting physical activities as streaming time series have been selected for research.

- Inertial Sensors for Human Activity Recognition (INERTIA) [2] comprises data pertaining to eleven different physical activities. Contains 23,346 samples.
- WISDM [6] encompasses over 1 million samples. It includes raw inertial data collected from 36 mobile phone users.
- UCI HAR [10] consists of two sets of samples, available in raw and processed forms, each containing just over 10,000 pieces.

Each data stream is characterized by varying numbers of samples and descriptive features. Predominantly, the observations comprise raw data, indicating direct readings from inertial sensors. Due to the data's streaming nature, the sets were compressed into sets of overlapping time windows of fixed sizes. This maintained continuity, and segments of potential anomalous activity were obtained. For the INERTIA set, each window overlapped the next one by 50%. This set is small and characterized by a low frequency of collected samples. Each data frame contained 40 data. For the WISDM set, 80% overlapping windows were created with a size of 144 samples (approximately 7 s). For the UCI HAR set, windows overlapping by 50% were used. Each window contains 128 samples, so its length is approximately 2.56 s. The distribution of classes in individual sets (INERTIA, WISDM, UCI HAR) and the number of windows created for a given class are given in Table 1.

Moreover, the signals within each data set have been decomposed into smaller components; for instance, acceleration has been separated into body and gravitational acceleration. Furthermore, additional statistical measures, such as mean or standard deviation, have been computed for each variable. Consequently, the feature vector encompasses over 560 attributes.

**Table 1.** Distribution of classes in the sets INERTIA, WISDM, UCI HAR.

| Class name | Number of windows (percentage of total) | | |
|---|---|---|---|
| | INERTIA | WISDM | UCI HAR |
| going | 70 (~15,18%) | 20 369 (~39,50%) | 1 722 (~16,72%) |
| seat | 256 (~55,53%) | 2 864 (~5,55%) | 1 777 (~17,25%) |
| standing | | 2 360 (~4,58%) | 1 906 (~18,51%) |
| running | 13 (~2,82%) | 15 152 (~29,39%) | 1 944 (~18,88%) |
| walking up the stairs | 72 (~15,62%) | 5 940 (~11,50%) | 1 544 (~14,99%) |
| walking down the stairs | 50 (~10,85%) | 4 878 (~9,44%) | 1 406 (~13,65%) |
| Total | 461 | 51 563 | 10 299 |

The quality of the classifiers was assessed based on known measures such as accuracy (ACC) and precision (PP). These measures were defined by the Formulas (1, 2), where TP is true-positive, i.e., correctly assigned to a given class, FP is false-positive, i.e., incorrectly classified to a given class, TN is true-negative, i.e., correctly rejected by the algorithm, FN is false-negative, i.e. incorrectly rejected by the algorithm.

$$ACC = \frac{(TP + TN)}{(TP + TN + FP + FN)} \tag{1}$$

$$PP = \frac{TP}{TP + FP} \tag{2}$$

In the case of anomaly detection, an additional criterion was taken into account when assessing the effectiveness of the algorithms used. An important factor was the number of correctly diagnosed anomalous segments.

### 4.2   Experimental Research and Tests

The research was divided into two stages. In the first stage, the accuracy and precision of the kNN, SVM, CART, and NB classifiers were checked for the detected outlier activities. In the second stage, four models of a nested binary classifier were checked: Nested k-nearest neighbor binary classifier (NestBC_kNN), nested CART binary classifier (NestBC_CART), nested support vector machine (NestBC_SVM), nested Bayes classifier (NestBC_NB).

In the first stage, three different experimental tests were performed on each of the three prepared datasets. Each of them tested the ability of individual classifier models to recognize anomalous activity segments. All available activities from each datasets were used for learning. The same activities were marked as anomalous and retried using ML algorithms. Algorithms are characterized by low stability. For this reason, majority voting was additionally used to increase the prediction efficiency. Majority voting involves giving each test sample the label predicted by the more significant number of algorithms participating in the voting. Every vote is equal, so no algorithm is favored or ignored. In many cases, this allowed us to select the correct unique segments. Finally, the obtained results were compared again with the correct answers.

In the second step, in the case of the new nested binary classifier proposed in this work, five simple activities were selected for each set, of which one (or two) was considered anomalous. Four classifier models were prepared for all machine learning algorithms in the next step. Each of them was taught to recognize one of the different activities. They used the same training sets but marked differently than the others when learning. Windows representing valid activity are labeled with their name, while those not are labeled as unknown. Then, the first model attempted to predict labels on a previously prepared test set. Correct samples were rejected and incorrect (anomalous) samples were transferred to the next level. At each subsequent level of nesting, the model checked whether the observations considered anomalous by its predecessor did not coincidentally concern its activity and rejected them. The last segments obtained this way (i.e., by elimination), and still listed as outliers potentially caused by other activity were compared with the actual data. The test was repeated with all algorithms.

## 5  Results

The first experiment investigated the effectiveness of machine learning algorithms such as kNN, SVM, CART, and NB in detecting outlier activities for the INERTIA, WISDM, and UCI HAR datasets. For the INERTIA dataset, the number of actual anomalies, i.e., outlier activities, was five. In the WISDM dataset, there were 5219 anomalies, while in UCI HAR, there were 1069.

For the INERTIA dataset, the best results for the k-nearest neighbors classifier were obtained for k=7. The kNN classifier correctly assigned activity segments in 92.09% of cases. Unfortunately, the classifier's precision for outlier activities was very low. Only one out of five anomalous segments were detected. The CART classifier achieved an accuracy of 97.12%. It correctly detected three out of five actual anomalies. The missing two segments were incorrectly assigned to the "walking" class. A decrease in sensitivity measures for the outliers class was observed. The Support Vector Machine classifier achieved an accuracy of 84.17%. Unfortunately, not a single anomalous segment was detected. It was noted that the SVM model had difficulty distinguishing dynamic activities. The most significant difficulties were observed in the class of stair movement. For the Naive Bayes classifier, the accuracy was 95.68%. Despite class imbalances and a small number of samples, the NB classifier detected four out of five outlier segments.

For the WISDM dataset, using the kNN classifier resulted in an accuracy of 97.55%. The precision for any class did not drop below 91%, sensitivity below 94%, and specificity below 73%. The precision for the outlier segment class was 98.33%. The CART algorithm correctly assigned 94.54% of the samples. The accuracy measure significantly decreased for the remaining two classifiers. The accuracy of the SVM method is equal, ACC = 85.58%, and for NB, only ACC = 81.37%. For the WISDM dataset, the kNN and CART algorithms demonstrated the best accuracy and precision in recognizing outlier segments. SVM and NB classifiers yielded the worst results.

For the UCI HAR dataset, the best results were achieved using Support Vector Machines, with an accuracy of 95.04%. On the other hand, kNN achieved an accuracy of 90.60%, while NB achieved only 77.02%. Most of the misclassifications revolved around the dynamics of the studied activities. The accuracy and precision results of the k-NN, CART, SVM, and NB classification algorithms for each dataset are presented in Table 2.

**Table 2.** The accuracy and precision measures for classifiers kNN, CART, SVM, NB for three datasets INERTIA, WISDM, UCI HAR.

|  | INERTIA | | WISDM | | UCI HAR | |
|---|---|---|---|---|---|---|
| Classifier | Accuracy | Precision | Accuracy | Precision | Accuracy | Precision) |
| kNN | 92.09 | 33.33 | 97.55 | 99.53 | 90.6 | 84.09 |
| CART | 97.12 | 100 | 94.54 | 98.33 | 87 | 80.21 |
| SVN | 84.17 | 0 | 85.58 | 96.31 | 95.04 | 90.79 |
| NB | 95.68 | 80 | 81.37 | 87.18 | 77,02 | 80.38 |

The second experiment checked the correct operation of four nested binary classifier models. Table 3 shows the obtained accuracy values of the nested binary classifier models for each dataset at each nesting level.

Based on Table 3, it is evident that for the INERTIA dataset, the nested binary classifier kNN (Nest_kNN) incorrectly classified activity segments at the III and IV levels. This is likely because the detected activities at these levels had the highest correlation coefficient. The decrease in accuracy at the last two levels was caused by the high similarity between the detected activities, ascending and descending stairs. NestBC_kNN detected 11 activities as outliers, of which 3 were actual anomalies. The missing two outliers were classified into the "ascending stairs" class. For NestBC_CART, three out of five anomalies were detected in this dataset. For NestBC_SVM, there was also a decrease in accuracy at the third and fourth levels, indicating model overfitting. All five actual anomalies were found among the 23 activities diagnosed as outlier, while the remaining 17 belonged to other activities. Remarkably, 11 of them originated from the main class of the last nested model. The final nested model, the Naive Bayes classifier, showed signs of underfitting, achieving the lowest accuracy and precision.

**Table 3.** The accuracy values of the nested binary classifier models for each datasets at each nesting level.

| Nested classifier | Level of nesting | INERTIA | WISDM | UCI HAR |
|---|---|---|---|---|
| NestBC_kNN | I | 98.56 | 99.68 | 95.45 |
| | II | 100 | 99.09 | 96.62 |
| | III | 89.74 | 98.11 | 96.57 |
| | IV | 80 | 97.67 | 96.25 |
| NestBC_CART | I | 99.28 | 99.75 | 93.7 |
| | II | 100 | 96.87 | 94.4 |
| | III | 97.22 | 94.38 | 93.66 |
| | IV | 88.46 | 94.67 | 90.5 |
| NestBC_SVM | I | 98.56 | 99.45 | 97.5 |
| | II | 100 | 89.06 | 96.3 |
| | III | 79.49 | sie.37 | 97.67 |
| | IV | 65.63 | 85.35 | 96.65 |
| NestBC_NB | I | 97.12 | 95.92 | 65.8 |
| | II | 84.48 | 73.65 | 51.74 |
| | III | 46.87 | 98.94 | 89.68 |
| | IV | 100 | 94.95 | 92.86 |

For the WISDM dataset, with each subsequent nesting, the accuracy of the NestBC_kNN model slightly decreased but remained at a high level. The classifier detected 5,287 activities as outlier, correctly identifying 5,158 as actual outlier activities. An additional 129 activities were misclassified. For NestBC_CART, it was observed that the number of false-positive and false-negative observations increased in dynamic activities. Accuracy did not drop below 94% in any of the nestings. A total of 5,105 actual outlier observations were detected, while 663 were excessively matched. NestBC_SVM correctly selected all activities at the I level, but the classifier's effectiveness decreased at subsequent levels. Ultimately, the anomalous class dominated at the III and IV levels. In the end, 5,173 anomalies were correctly detected, but 2,829 samples were misclassified. For NestBC_BN, 4,221 correct outlier segments were detected, while 998 outliers were not detected.

For the third UCI HAR set, the NestBC_kNN classifier correctly detected 1,032 real outliers, and 130 activities were overfitted as outlier. In turn, NestBC_CART detected 998 of 1069 anomalous segments. Three hundred sixty-seven activities were over-matched. This classifier achieved the lowest accuracy value at the last nesting level. The NestBC_SVM model correctly diagnosed 1,041 segments as unique. This model incorrectly classified 131 segments as outliers. In the case of the NestBC_BN model, the accuracy began to increase with subsequent nestings, contrary to the other models. However, it detected only 78 unique segments.

To compare the machine learning algorithms kNN, CART, SVM, and NB with nested models detecting outlier activities, Table 4 and Fig. 1. The table 4 includes the average accuracy achieved at four levels for nested models. It should be noted that the nested model shows better accuracy in each case and more precisely detects outlier activities.

**Table 4.** Accuracy of classifiers obtained for individual sets. The average accuracy of four levels is reported for nested models.

|            | INERTIA | WISDM | UCI HAR |
|------------|---------|-------|---------|
| kNN        | 92.09   | 97.55 | 90.6    |
| CART       | **97.12** | 94.54 | 87    |
| SVN        | 84.17   | 85.58 | 95.04   |
| NB         | 95.68   | 81.37 | 77.02   |
| NestBC_kNN | 92.08   | 98.64 | **96.22** |
| NestBC_CART| 92.26   | **98.66** | 95.79 |
| NestBC_SVM | 92.26   | 98.10 | 95.23   |
| NestBC_NB  | 94.13   | 97.17 | 94.50   |

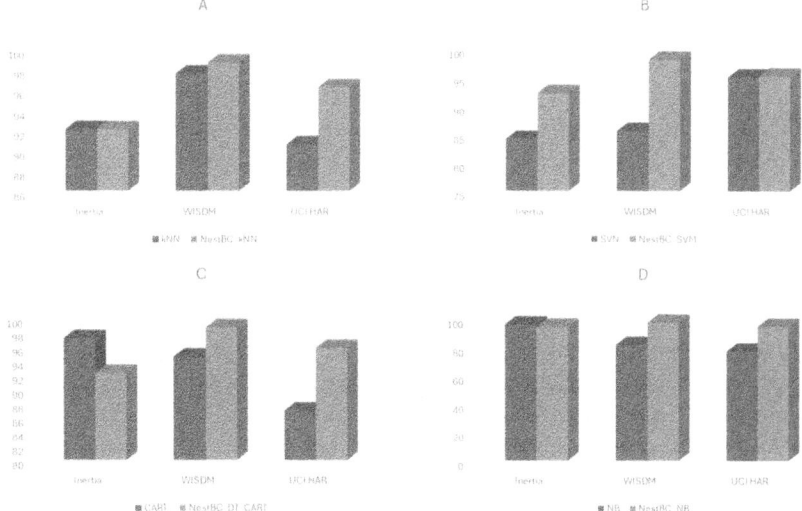

**Fig. 1.** Comparison of the accuracy of individual classifiers with their nested versions.

# 6   Conclusion

Based on the experimental research and tests conducted, both stages of the study provided valuable insights into the effectiveness of various machine learning algorithms in detecting outlier activities for three datasets.

Several important observations were made in the first stage, which focused on evaluating the accuracy and precision of individual classifiers such as kNN, SVM, CART, and NB to identify outlier activities. While each algorithm demonstrated strengths in certain aspects, such as kNN's high accuracy in some datasets and NB's resilience to class imbalances, limitations were observed, particularly in detecting outlier activities. For instance, SVM struggled with distinguishing dynamic activities, leading to low detection rates for anomalies in specific datasets.

In response to these challenges, a second stage was devised involving nested binary classifier models to enhance the accuracy of outlier detection. These nested models, namely NestBC_kNN, NestBC_CART, NestBC_SVM, and NestBC_NB, was designed to iteratively refine the detection process through multiple levels of nesting.

The results from the nested models demonstrated notable improvements in accuracy compared to standalone classifiers, especially in identifying outlier activities. For example, NestBC_kNN consistently achieved high accuracy levels across different datasets and nesting levels, effectively detecting actual anomalies while minimizing false-positives. Similarly, NestBC_CART showed promising performance, correctly identifying a significant portion of anomalies across datasets.

However, it is essential to note that each nested model exhibited unique strengths and weaknesses, as evidenced by variations in accuracy levels and detection rates across nesting levels and datasets. For instance, NestBC_SVM initially showed strong performance but experienced diminishing accuracy at deeper nesting levels, indicating potential overfitting issues.

In summary, the experimental research and tests highlighted the efficacy of nested binary classifier models in enhancing the detection of outlier activities within datasets. By iteratively refining the classification process through multiple nesting levels, these models offer a promising approach for improving the accuracy and precision of anomaly detection in various applications. Further investigation and optimization may be necessary to address specific challenges and ensure robust performance across different datasets and scenarios.

The Nested Binary Classifier (NestBC) method was tested, especially in the context of unique and outlier activities where computational resources are limited, and the interpretability of the results is crucial. Thanks to its simplicity, flexibility, and modularity, NestBC provides a practical and effective solution that can be used in a wide range of HAR applications.

The authors plan to extend our research with additional, newer datasets to test how the proposed methods perform in different contexts and with different datasets. In future research, we will compare the NestBC method with modern deep learning and ensemble learning methods.

# References

1. Ali, A., Samara, W., Alhaddad, D., Ware, A., Saraereh, O.A.: Human activity and motion pattern recognition within indoor environment using convolutional neural networks clustering and naive bayes classification algorithms. Sensors **22**(3), 1016 (2022)
2. Anguita, D., Ghio, A., Oneto, L., Parra, X., Reyes-Ortiz, J.L., et al.: A public domain dataset for human activity recognition using smartphones. In: ESANN, vol. 3, p. 3 (2013)
3. Batool, M., Jalal, A., Kim, K.: Sensors technologies for human activity analysis based on SVM optimized by PSO algorithm. In: 2019 International Conference on Applied and Engineering Mathematics (ICAEM), pp. 145–150. IEEE (2019)
4. Duraj, A., Duczymiński, D.: Nested binary classifier as an outlier detection method in human activity recognition systems. Entropy **25**(8), 1121 (2023)
5. Ferreira, P.J., Cardoso, J.M., Mendes-Moreira, J.: k nn prototyping schemes for embedded human activity recognition with online learning. Computers **9**(4), 96 (2020)
6. Lockhart, J.W., Weiss, G.M., Xue, J.C., Gallagher, S.T., Grosner, A.B., Pulickal, T.T.: Design considerations for the wisdom smart phone-based sensor mining architecture. In: Proceedings of the Fifth International Workshop on Knowledge Discovery from Sensor Data, pp. 25–33 (2011)
7. Maswadi, K., Ghani, N.A., Hamid, S., Rasheed, M.B.: Human activity classification using decision tree and Naive Bayes classifiers. Multim. Tools Appl. **80**, 21709–21726 (2021)
8. Munoz-Organero, M.: Outlier detection in wearable sensor data for human activity recognition (HAR) based on DRNNS. IEEE Access **7**, 74422–74436 (2019)
9. Neira-Rodado, D., Nugent, C., Cleland, I., Velasquez, J., Viloria, A.: Evaluating the impact of a two-stage multivariate data cleansing approach to improve to the performance of machine learning classifiers: a case study in human activity recognition. Sensors **20**(7), 1858 (2020)
10. Newman, D., Hettich, S., Blake, C., Merz, C.: UCI repository of machine learning databases. Disponível on-line em (1998). http://www.ics.uci.edu/~mlearn/MLRepository.html. University of California, Irvine, Dept. of Information and Computer Sciences
11. Palaniappan, A., Bhargavi, R., Vaidehi, V.: Abnormal human activity recognition using SVM based approach. In: 2012 International Conference on Recent Trends in Information Technology, pp. 97–102. IEEE (2012)
12. Patro, S.G.K., et al.: A hybrid action-related k-nearest neighbour (har-knn) approach for recommendation systems. IEEE Access **8**, 90978–90991 (2020)
13. Ravindran, S., Aghila, G.: A data-independent reusable projection (DIRP) technique for dimension reduction in big data classification using k-nearest neighbor (k-nn). Natl. Acad. Sci. Lett. **43**(1), 13–21 (2020)
14. Wang, S., Yang, J., Chen, N., Chen, X., Zhang, Q.: Human activity recognition with user-free accelerometers in the sensor networks. In: 2005 International Conference on Neural Networks and Brain, vol. 2, pp. 1212–1217. IEEE (2005)
15. Wu, H., Pan, W., Xiong, X., Xu, S.: Human activity recognition based on the combined SVM&HMM. In: 2014 IEEE International Conference on Information and Automation (ICIA), pp. 219–224. IEEE (2014)

# Using Brain-Computer Interface and Artificial Intelligence Algorithms for Language Learning

Marta Kuchciak⬤, Macin Sieradzki⬤, Wojciech Cebula, Katarzyna Bialas⬤, and Michal Kedziora(✉)⬤

Wroclaw University of Science and Technology, Wroclaw, Poland
Michal.kedziora@pwr.edu.pl

**Abstract.** The research paper aims to introduce a methodology and platform that integrates a brain computer interface (BCI) and artificial intelligence (AI) to improve the process of learning a foreign language. Through the use of a BCI, our solution can accurately gauge the bioelectrical activity of the user's brain to determine whether the user has learned a piece of knowledge, allowing for a more personalized and efficient learning experience. The application's AI model, trained via backpropagation through time (BPTT) to handle temporal dependencies in the data, was optimized using the Huber loss function and the "Adam" optimizer, though it faced challenges of overtraining due to limited data. In this paper, we present a novel approach that leverages the synergy between BCIs and AI to offer a more effective and personalized language learning experience. We detail the process, from the initial design to the deployment of the platform, including the creation of a custom AI model trained on brainwave data. Our proposed data collection and cleanup process was also presented.

**Keywords:** BCI · AI · Neural Network · EEG

## 1 Introduction

In an era where the rapid acquisition of new languages is both a personal advantage and a professional necessity, the exploration of innovative learning methodologies is paramount. Traditional language learning tools, while beneficial, often fall short in providing personalized, efficient, and engaging methods for learners. The integration of technology in education, specifically through brain-computer interfaces (BCIs) and artificial intelligence (AI), opens new horizons in understanding and enhancing cognitive processes involved in language learning. The application of BCIs, coupled with AI algorithms, offers a groundbreaking approach to tailor language learning experiences to the individual's cognitive states, potentially revolutionizing how we acquire new languages.

This paper focuses on the practical application of BCIs and AI algorithms within a proposed platform designed for language learning. Traditional flashcard methods, while effective for memorization, lack dynamic adjustment to the

© The Author(s), under exclusive license to Springer Nature Switzerland AG 2024
N.-T. Nguyen et al. (Eds.): ICCCI 2024, CCIS 2165, pp. 334–343, 2024.
https://doi.org/10.1007/978-3-031-70248-8_26

learner's engagement and retention level. By employing a BCI to monitor the user's bioelectrical brain activity, our proof-of-concept application intelligently adjusts the learning pace and content based on real-time cognitive feedback. This approach aims to optimize learning efficiency by ensuring that the learner is always engaged at an optimal level of challenge.

In this paper, we present a novel approach that leverages the synergy between BCIs and AI to offer a more effective and personalized language learning experience. We detail the development process, from the initial design to the deployment of the platform, including the creation of a custom AI model trained on brainwave data to predict user familiarity with specific flashcards. Our proposed platform not only facilitates the management of flashcard sets but also dynamically adapts learning sessions based on the user's cognitive state, thereby enhancing the learning process.

While the use of BCIs and AI in educational tools is not entirely new, our application differentiates itself through its specific focus on language learning and the direct feedback mechanism from brain activity to content delivery. Previous works have explored various aspects of BCIs and AI in education but seldom in direct application to language acquisition tools. Our approach builds upon these foundations by integrating real-time cognitive state analysis into a practical language learning application, setting a new benchmark for personalized education technologies.

The remainder of this paper is structured as follows: Sect. 2 provides a detailed review of the literature, highlighting previous efforts in BCIs, AI in education, and their application to language learning. Section 3 describes the methodology, including system design, the development of the AI model, and the integration with the BCI. Finally, the last section discusses the implications of our findings, and potential areas for future research and concludes the paper.

## 2 Related Works

In [13] the author emphasizes the expanding role of brain-computer interface (BCI) technology beyond medical applications, particularly its potential to transform educational platforms and assistive technologies. It suggests that BCIs can enhance learning strategies and cognitive capabilities, necessitating a review of current BCI technologies and their relevance to education [12]. By highlighting the need for understanding educational and social complexities, the paper calls for strategic implementation of BCIs that could lead to improved academic outcomes and reduced achievement gaps, thereby contributing valuable insights into the integration of BCI in educational settings for language learning and beyond The paper [9] highlights the pivotal role of artificial intelligence (AI) in enhancing English language teaching through electronic means, emphasizing AI's potential to mimic human cognitive processes in decision making and learning environments. It underscores the significance and prospects of AI applications in revolutionizing electronic English language education, suggesting that these technological advancements can significantly contribute to the development and efficiency of language learning methodologies. The synergy between advancements in artificial intelligence (AI), including machine learning and deep learning, and brain computer interface (BCI) technology, particularly for enhancing EEG based applications in visual, linguistic, and motion fields, thereby highlighting AI and algorithms crucial role in evolving interdisciplinary BCI research was described in [4].

On the subject of combining flashcards with a brain-computer interface, we are forerunners, for this reason, we have not been able to compare our way of connecting the BCI with other solutions. We derive our knowledge of the interface from the documentation of the device [10], the work of Katarzyna Bialas, M.Sc., titled: "Study of the possibilities of controlling a mobile application using the NeuroSky MindWave brain-computer interface" [2] and direct contact with the author of the aforementioned work in a related field [5,6] especially [3]. These sources allowed us to decide on the method of collecting learning data and connecting the device to the application.

## 3   Methodology

The NeuroSky MindWave Mobile 2 [10] is an EEG headset that collects brain waves. The main part of the device is the frontal sensor, whose task is to collect the user's brain activity, and a built-in chip that, based on built-in algorithms, calculates values such as focus or meditation. The set connects to a computer using Bluetooth technology.

The parameters representing brain waves and signal quality are transmitted as soon as the device is inserted every one second. A few seconds later, attention and meditation parameters additionally appear, which are calculated by built-in algorithms on the device's processor. Parameters such as mental effort and task familiarity are the most complex, for which the device takes about a minute to tune into the user and start providing these parameters. In addition, mental effort and task familiarity are provided once every few seconds. The blink strength parameter is sent when the device detects a blink.

The modeling component is an excluded component, since it was used only to train the model. It combined the features of Science and Learning with BCI in the following way. There were two buttons to indicate whether the user knew a particular flashcard, or not, while, as in Science with BCI, data from the BCI device were sent to the API, i.e. what waves were recorded while looking at a given flashcard (date parameter). In addition, there was sent information about what the user marked (known/unknown - parameter correct), which side of the flashcard he looked at when registering a given flashcard (turned over parameter added to the data parameter when reading data from the device), the registration timestamps for a given flashcard (start_timestamp and finish_timestamp parameters) and the identifier of a given flashcard (flashcard parameter).

Training data for our project: We did not find any work in any source that, on the basis of electroencephalography, determines the user's familiarity with the displayed word. For this reason, we decided to collect the learning data set ourselves. We hope that with the help of the collected data we will be able to train a neural network to determine the user's familiarity with the vocabulary word. A phenomenon called a P300 episode may help with this. Our brain responds to a stimulus at a rate of 200 milliseconds, causing a noticeable excitation in the alpha band. It occurs with all types of stimuli and is a natural reaction of the body. However, it turns out that when we see something important, the agitation is more noticeable. It is called P300 because it occurs 300 milliseconds after the appearance of the stimulus.

The characteristics of the project involve collecting brain waves that will be classified into one of two categories: brain waves while reviewing a known flashcard and brainwaves while reviewing an unfamiliar flashcard. The data-acquisition app slightly modifies the records received by the devices: it adds to the record the date and time of its received, and changes the Poor Signal Level parameter to Signal Strength, as the quality of the connection expressed as a percentage (Fig. 1).

| | | | | | | | | | | | | | | | | |
|---|---|---|---|---|---|---|---|---|---|---|---|---|---|---|---|---|
| 196 | 07:05:08 | 100 | 10805,00 | 15268,00 | 13173,00 | 6783,00 | 9327,00 | 5542,00 | 7577,00 | 6963,00 | 70,00 | 64,00 | -1,00 | 0,00 | 0,00 |
| 197 | 07:05:09 | 100 | 15062,00 | 10662,00 | 1707,00 | 2863,00 | 3367,00 | 1239,00 | 938,00 | 1090,00 | 57,00 | 75,00 | -1,00 | 0,00 | 0,00 |
| 198 | 07:05:10 | 100 | 82160,00 | 3748,00 | 1001,00 | 348,00 | 2488,00 | 905,00 | 272,00 | 209,00 | 63,00 | 61,00 | 46,00 | 0,00 | 0,00 |
| 199 | 07:05:11 | 100 | 114067,00 | 26173,00 | 4202,00 | 4943,00 | 2211,00 | 2809,00 | 2945,00 | 785,00 | 47,00 | 53,00 | -1,00 | 0,00 | 0,00 |
| 200 | 07:05:12 | 100 | 40662,00 | 21946,00 | 9717,00 | 9522,00 | 3962,00 | 4884,00 | 6762,00 | 3252,00 | 43,00 | 53,00 | -1,00 | 0,00 | 0,00 |
| 201 | 07:05:13 | 100 | 7165,00 | 9121,00 | 247,00 | 3238,00 | 3027,00 | 1624,00 | 1505,00 | 841,00 | 48,00 | 51,00 | -1,00 | 0,00 | 0,00 |
| 202 | 07:05:14 | 100 | 4122,00 | 10289,00 | 1247,00 | 2123,00 | 1061,00 | 2130,00 | 1906,00 | 1198,00 | 47,00 | 54,00 | -1,00 | 2,27 | 9,20 |
| 203 | 07:05:15 | 100 | 1729114,0 | 121714,00 | 9682,00 | 10557,00 | 9790,00 | 5276,00 | 3851,00 | 4090,00 | 40,00 | 48,00 | -1,00 | 0,00 | 0,00 |
| 204 | 07:05:16 | 100 | 17229,00 | 19139,00 | 7717,00 | 11546,00 | 23980,00 | 4653,00 | 7763,00 | 2834,00 | 41,00 | 44,00 | -1,00 | 0,00 | 0,00 |
| 205 | 07:05:17 | 100 | 221367,00 | 73359,00 | 4825,00 | 16251,00 | 12374,00 | 5929,00 | 2587,00 | 1657,00 | 35,00 | 43,00 | 50,00 | 0,00 | 0,00 |
| 206 | 07:05:18 | 100 | 1226185,0 | 282490,00 | 17762,00 | 22218,00 | 18793,00 | 15283,00 | 17637,00 | 3775,00 | 24,00 | 38,00 | -1,00 | 0,00 | 0,00 |
| 207 | 07:05:19 | 100 | 27285,00 | 28867,00 | 14764,00 | 10628,00 | 7578,00 | 7017,00 | 6292,00 | 5189,00 | 35,00 | 50,00 | -1,00 | 0,00 | 0,00 |
| 208 | 07:05:20 | 100 | 12645,00 | 11001,00 | 10433,00 | 7754,00 | 2771,00 | 2379,00 | 1407,00 | 2473,00 | 30,00 | 57,00 | -1,00 | 0,00 | 0,00 |
| 209 | 07:05:21 | 100 | 12918,00 | 22697,00 | 3032,00 | 11860,00 | 8303,00 | 10020,00 | 2847,00 | 2282,00 | 43,00 | 63,00 | -1,00 | 0,00 | 0,00 |
| 210 | 07:05:22 | 100 | 35799,00 | 31146,00 | 5533,00 | 9288,00 | 9140,00 | 4830,00 | 4907,00 | 2273,00 | 50,00 | 70,00 | 49,00 | 0,00 | 0,00 |
| 211 | 07:05:23 | 100 | 275107,00 | 30369,00 | 1803,00 | 1912,00 | 2353,00 | 1406,00 | 1687,00 | 702,00 | 40,00 | 56,00 | -1,00 | 0,00 | 0,00 |
| 212 | 07:05:24 | 100 | 3783,00 | 8392,00 | 819,00 | 2798,00 | 5596,00 | 6951,00 | 4584,00 | 6849,00 | 54,00 | 40,00 | -1,00 | 2,26 | 9,00 |
| 213 | 07:05:25 | 100 | 211018,00 | 116124,00 | 25888,00 | 7176,00 | 8083,00 | 4139,00 | 1915,00 | 585,00 | 38,00 | 38,00 | 58,00 | 0,00 | 0,00 |
| 214 | 07:05:26 | 100 | 114814,00 | 19513,00 | 5842,00 | 1148,00 | 1719,00 | 2519,00 | 787,00 | 241,00 | 38,00 | 38,00 | 64,00 | 0,00 | 0,00 |

**Fig. 1.** A section of raw EEG data results from the device.

A value of -1 (indicator 1) for the blink strength parameter means no blink in a given second. The frame marked with indicator 2 indicates records received immediately after turning on the device - these readings are not accurate and should be discarded. Indicator 3 indicates the values of mental effort and task familiarity, which are calculated by the device once every few seconds (Fig. 2).

Preliminary data collection. In order to collect data, we prepared 100 English vocabulary words, which were equally divided into 2 sets: an easy set and a difficult set. The main criterion for selecting the vocabulary was that the vocabulary belonging to the easy set was known by everyone, while the vocabulary words in the difficult set were unknown to everyone (we later verified this) (Fig. 3).

The course of the study was as follows:

1. Introducing the subject to the general sense of the study.
2. Placing the BCI device on the subject's head.
3. Turning on the BCI device and connecting to the NeuroSky MindWave Reader application.
4. Tuning of the device to the user.

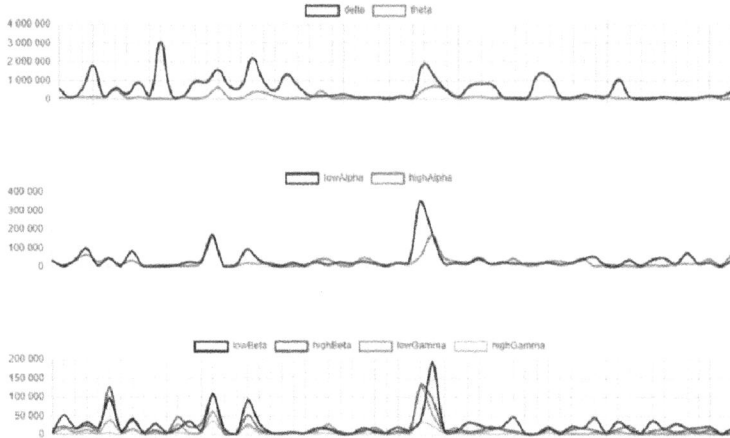

**Fig. 2.** The received data is presented in the application in the form of charts and various indicators, thanks to which the user can watch and check their brain activity in real time.

**Fig. 3.** A diagram for the proposed system.

5. Beginning of the subject's review of the flashcard from the easy set (noting the time of start).
6. Review of 50 flashcards by the test subject in order: word in English, word in Polish.
7. Completion of the review of the flashcards from the easy set (noting the time of completion).
8. Exporting the collected data to a wave file, while reviewing the known flashcards.
9. Beginning of the review of flashcards from the difficult set by the subject (noting the time of start).
10. Review of 50 flashcards by the test subject in order: word in English, word in Polish.

11. Completion of the review of flashcards from the difficult set (noting the time of completion).
12. Exporting the collected data to a wave file, while reviewing the unknown flashcards.
13. Completion of the survey by the subject.

Next step was performing data cleanup:

1. Remove records from the file that were received outside of the flashcard review time by the person examined (based on the recorded time by the examiner).
2. Fill in the mental effort and task familiarity parameters with the last value, in case this one is 0.
3. Categorizing each record into one of two categories (1 - known, 0 - unknown) based on the reviewed set of vocabulary words.

In this way, we surveyed 13 people, which translated into 26 csv files (13 each for the easy and difficult set). After cleaning the data, we obtained 4784 records. The biggest problem with a survey done this way is undoubtedly the assumption that the user knew/didn't know all the vocabulary words from the easy/difficult set. To verify our assumption, we created a survey that, among other things, asked users about this.

The results of the survey showed that preparing vocabulary that no one knows is a very difficult challenge, which we did not manage. Our observations also showed that those surveyed quickly noticed that each of the following vocabulary words would be easy, which resulted in less focus when reviewing the flashcard.

Having learned from our mistakes during the first data collection, we decided to create our own proper software for collecting learning data. This is how the application was created "Neurofiszka Bridge" (which was originally an EEG Bridge script) and a modeling mode on the side of the web application. EEG Bridge processed the records in such a way that it was no longer necessary to clean the data (e.g. completing mental effort and task familiarity parameters). The modeling mode collected the waveforms transmitted by EEG Bridge, which were only transmitted when the user looked at the flashcard. In addition, it added to the record such information as the side of the of the card the user was looking at (English or Polish), the card's id and the user's response (based on the based on the user's pressing the "known"/"unknown" button) (Fig. 4).

This way of collecting data eliminates all the problems that occurred when collecting data in the first approach. This time we prepared 200 vocabulary words of varying levels of difficulty, but the goal of the subject tested was not to review all 200 flashcards, but to do it for a set time (10–12 min).

In total, the subjects reviewed 2856 flashcards (of which 49.16% of the flashcards they knew), which translated into 10140 records.

Brain waves collected via an electroencephalograph are indirectly correlated with the current mental state of the subject. Consequently, we may be able to extract interesting information directly from the received waves, and this, in turn, gives us the opportunity to assess the knowledge of the of the presented flashcard. This is a classic classification task, which involves assigning a class

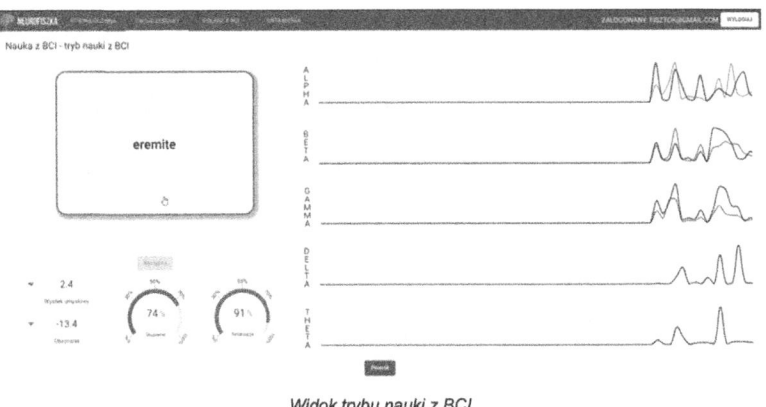

*Widok trybu nauki z BCI*

**Fig. 4.** BCI learning mode view from the system.

(label) for the presented input. The set of labels is predetermined - in the case of our problem, the set consists of only two classes (can't - 0, can - 1). A classification consisting of only two classes we call a binary classification. In the following section, the classification trainers are explained.

### 3.1 Multilayer Perceptron

The multilayer perceptron (MLP) is a type of feedforward neural network that consists of multiple layers of interconnected neurons. The architecture includes an input layer, hidden layers, and an output layer. Each neuron of MLP is given input from the neurons in the previous layer. Then it applies an activation function to the sum of the inputs and passes the result to the next layer. Passing input through the network layer by layer is called forward propagation.

$$z^{i+1} = a^i w^{i+1} + b \qquad (1)$$

$$a^i = f(z^i) \qquad (2)$$

Each layer transforms input using its weights and biases, which are learned during training using algorithms such as backpropagation.

### 3.2 Recurrent Neural Network

A recurrent neural network (RNN) [11] is a type of neural network that can handle sequential data, where the order of the inputs matters. RNN has a "memory" that allows it to process sequences of variable length and retain information about the previous inputs. The "memory" is represented by the hidden state vector that contains information about the previous inputs in the sequence. At each time step, the RNN takes an input vector and a hidden state vector. It produces an output vector and a new hidden state vector.

Forward propagation formulas:

$$h^t = f(Ux^t + Wh^{t-1} + b) \tag{3}$$

$$y^t = g(Vh^t + c) \tag{4}$$

where:

$h$ - hidden state vector

$y$ - output vector

$U, W, V$ - weight matrices

$b, c$ - bias vectors

$f, g$ - activation functions

For training RNN it is needed to take into account the dependencies between the hidden state vectors across time steps. To accomplish this a backpropagation through time (BPTT) algorithm is being used [8]. BPTT computes the gradients of the loss function concerning the weights and biases at each time step and updates the weights and biases using a gradient descent algorithm.

Tests were conducted that allowed us to determine the optimal hyperparameters and structure of the network recursive network. The model was trained using the Huber loss function [7] and the "Adam" optimizer [1] (Fig. 5).

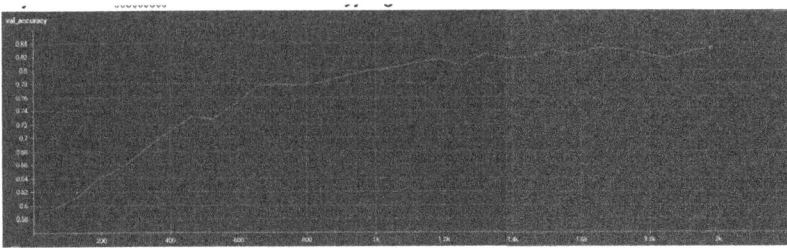

**Fig. 5.** F1 score chart for the validation set

The model starts to overtrain rather quickly, despite its small size, which is probably caused by the small number of data collected. Presented below are graphs of the loss function and F1 Score, which is responsible for the quality of our model [14].

## 4    Conclusion and Future Work

The project successfully achieved its aim of developing a platform that integrates a brain computer interface (BCI) and artificial intelligence (AI) to enhance the process of learning foreign language. Through the use of a BCI, the application can accurately gauge the bioelectrical activity of the user's brain to determine their familiarity with specific flashcards, allowing for a more personalized and efficient learning experience. The application's AI model, trained via backpropagation through time (BPTT) to handle temporal dependencies in the data, was optimized using the Huber loss function and the "Adam" optimizer, though it faced challenges of overtraining due to limited data. Despite this, the initial tests confirmed the model's potential, with satisfactory performance metrics in both loss function and F1 Score. The POC solution delivers on its fundamental promises, offering CRUD operations on flashcard sets and varied learning modes, while indicating room for future enhancements to expand its capabilities and effectiveness.

**Acknowledgments.** The work was supported by the project Minigrants for doctoral students of the Wroclaw University of Science and Technology.

# References

1. Barron, J.T.: A general and adaptive robust loss function. In: Proceedings of the IEEE/CVF Conference on Computer Vision and Pattern Recognition, pp. 4331–4339 (2019)
2. Bialas, K.: Study of the possibilities of controlling a mobile application using the NeuroSky MindWave brain-computer interface. Master's thesis, Wroclaw University of Science and Technology (2020)
3. Białas, K., Kedziora, M., Chałupnik, R., Song, H.H.: Multifactor authentication system using simplified EEG brain-computer interface. IEEE Trans. Hum.-Mach. Syst. **52**(5), 867–876 (2022). https://doi.org/10.1109/THMS.2022.3196142
4. Cao, Z.: A review of artificial intelligence for EEG based brain computer interfaces and applications. Brain Sci. Adv. **6**(3), 162–170 (2020). https://doi.org/10.26599/BSA.2020.9050017
5. Chalupnik, R., Bialas, K., Jozwiak, I., Kedziora, M.: Acquiring and processing data using simplified EEG-based brain-computer interface for the purpose of detecting emotions. In: The Fourteenth International Conference on Advances in Computer-Human Interactions (ACHI 2021), Nice (2021)
6. Chalupnik, R., Bialas, K., Majewska, Z., Kedziora, M.: Using simplified EEG-based brain computer interface and decision tree classifier for emotions detection (paper in review). In: AINA: The 36th International Conference on Advanced Information Networking and Applications (AINA-2022). University of Technology Sydney (UTS), Sydney (2022)
7. Dwaram, J.R., Madapuri, R.K.: Crop yield forecasting by long short-term memory network with Adam optimizer and Buber loss function in Andhra Pradesh, India. Concurr. Comput.: Pract. Exp. **34**(27) (2022). https://doi.org/10.1002/cpe.7310
8. Lillicrap, T.P., Santoro, A.: Backpropagation through time and the brain. Curr. Opin. Neurobiol. **55**, 82–89 (2019). https://doi.org/10.1016/j.conb.2019.01.011
9. Mijwil, M.M., Abdulrhman, S.H., Abttan, R.A., Faieq, A.K., Alkhazraji, A.: Artificial intelligence applications in English language teaching: a short survey. Asian J. Appl. Sci. **10**(6) (2023). https://doi.org/10.24203/ajas.v10i6.7111
10. Mindwave: Technical Specs (2015). https://store.neurosky.com/pages/mindwave
11. Sherstinsky, A.: Fundamentals of recurrent neural network (RNN) and long short-term memory (LSTM) network. Physica D **404**, 132306 (2020). https://doi.org/10.1016/j.physd.2019.132306

12. Rebolledo Font de la Vall, R., González Araya, F.: Exploring the benefits and challenges of AI-language learning tools. Int. J. Soc. Sci. Human. Invent. **10**(01), 7569–7576 (2023). https://doi.org/10.18535/ijsshi/v10i01.02
13. Wegemer, C.: Brain-computer interfaces and education: the state of technology and imperatives for the future. Int. J. Learn. Technol. **14**(2), 141 (2019). https://doi.org/10.1504/ijlt.2019.101848
14. Yacouby, R., Axman, D.: Probabilistic extension of precision, recall, and f1 score for more thorough evaluation of classification models. In: Eger, S., Gao, Y., Peyrard, M., Zhao, W., Hovy, E. (eds.) Proceedings of the First Workshop on Evaluation and Comparison of NLP Systems, pp. 79–91. Association for Computational Linguistics (2020). https://doi.org/10.18653/v1/2020.eval4nlp-1.9

# Social Networks and Intelligent Systems

# From Detection Through Display to Understanding: Bridging AI and UI in Disinformation and Fake News Analysis

Rafal Kozik[1], Aleksandra Pawlicka[2], Marek Pawlicki[1], and Michał Choraś[1(✉)]

[1] Bydgoszcz University of Science and Technology, Bydgoszcz, Poland
chorasm@pbs.edu.pl
[2] University of Warsaw, Warsaw, Poland

**Abstract.** Fake news detection is a hot topic in the current times, and there has been a huge rise in publications presenting fake news detection methods. However, the vast majority of papers present AI-based technical and mathematical approaches and tabular results on benchmark datasets, while the systems, working prototype and user interfaces are not touched upon. In this work, we discuss and more importantly present proposed fake news detection systems focusing on user interfaces and working prototypes in order to fill the above-mentioned gap.

**Keywords:** Fake news detection · Disinformation · UI/UX design

## 1 Introduction and Rationale

Whenever any computer system or AI-based system is to be widely used, it needs acceptance from the users and relevant communities. Often, what decides about such acceptance is the user experience (UX) and user interface. The same concerns fake news (disinformation) detection systems or any other system based on AI technologies, like NLP (Natural Language Processing).

Fake news detection is a hot topic in the current times, and there has been a huge rise in publications presenting fake news detection methods. Unfortunately, most papers present just theoretical foundations, models, algorithms and results, without any discussion or even a thought on end-users, applications and system interfaces.

Currently, there are no standardised solutions for presenting results of fake news detection, gathering datasets or even what should be the input for analysis in such a system. It is clear that such systems cannot reach societies and high TRLs (Technology Readiness Level), without proper design of user interfaces and a proper discussion on user experience.

Therefore, in this work, we decided to close this crucial gap. The major contribution of this paper is the critical discussion on our design and development of various interfaces and approaches for fake news detection systems with several visual examples.

© The Author(s), under exclusive license to Springer Nature Switzerland AG 2024
N.-T. Nguyen et al. (Eds.): ICCCI 2024, CCIS 2165, pp. 347–357, 2024.
https://doi.org/10.1007/978-3-031-70248-8_27

The paper is structured as follows: in Sect. 2, the state of the art presenting relevant systems and interfaces is given. In Sect. 3 we present the SocialTruth project result, namely Digital Companion Light, to show our design of a user interface for a fake news detection system. Another design and UI coming from a current SWAROG project is presented and discussed in Sect. 4. Discussion (e.g. on xAI) and conclusions are provided thereafter.

## 2    Relevant Systems and Prototypes

In the field of fake news detection, the importance of user interfaces is often overshadowed by a focus on algorithms. The design and implementation of UIs in these systems are not only frequently overlooked but also lack standardized methodologies. The scarcity of detailed UI design principles in fake news detection highlights a critical gap in the field. In the following section, a selection of scientific works have been presented the authors of which did take the aspect of UI into consideration when designing their fake news detection systems.

In their work, Gupta et al. [4], describe the user interface they developed for their Fake News Detector application. The interface, which the authors call intuitive, allows users to insert a link to a news article for verification, customize the collection of reliable sources and receive visual feedback for the queried article. In the verification process, users select news sources to match against, and upon clicking the verification button, the backend conducts an analysis, displaying results as a list of potentially matching articles based on semantic similarity. If no matches are found, the UI indicates that the article might potentially be fake [4].

Purificato et al. [14] have developed an explainable user interface for a fake news detection system. They justify the need for making the systems' decisions interpretable and transparent due to the human supervisors who must ensure the systems navigate the nuanced differences between the news items of different levels of fabrication. The authors even suggest that in many cases the creation of an explainable user interface might be more important than implementing the system as such. This is why the interface they designed is meant to present to the user the parts of a news item that are classified as possibly fake along with the explanation why. The interface, inspired by the Harry Potter franchise, resembles a newspaper in which a wizard reveals (By means of the "Revelio" spell) the truth in the provided content, whether it be a URL or a manually typed text. The results have been colour-coded, i.e., the potentially fake content is marked red, while the passages classified as true are green. Additionally, the colour gradient is added, meant to reflect the similarity score between a fragment of the text and the attributed source. The feature contributions to the prediction are also made available to the users, along with the indication of the strength of the contribution [14].

Another group of researchers who took the user interface into particular consideration whilst designing their fake news detection solution were Desai et al. [2] The authors call their interface "friendly". It allows the user to enter the text and receive a prediction of whether a piece of news is fake or real, as well as provide feedback to indicate if the prediction was correct or not. The developers also added a dark mode feature to the interface [2].

The work by Stissi also focused on equipping an automatic fake news detection system with an interface which is "functional, accessible and human-centred", as well as "robust, accurate and versatile" [15]. This would be achieved by the possibility of adding the best methods of detection to the back-end framework; the interface would also be developed with accessibility in mind, e.g., catering to sight, audio and other cognitive impairments [15].

Nevertheless, although in the aforementioned works the user interface of fake news detection systems has been given high priority, none of the papers mentions a methodology, taxonomy or other scientific background on what makes an interface friendly or intuitive.

In the work by Bunde et al. [1], the initial cycle of development of a fake news detection service focusing on the user interface is described. The authors wished to uncover which features should be implemented in the design so as to help users decide whether a piece of news and its source are genuine or fake. They say they had found it particularly challenging, pointing to the fact that the studies of UIs for services supporting users in assessing news content were non-existent at the time when the studies were performed. Thus, the authors developed their own methodology - design knowledge, by first exploring the challenges and contributions within the fake news-related research. Then, they developed design requirements and mapped the challenges to them. This allowed establishing design principles. Lastly, this process led to the establishment of design features, which pertain to the interface providing:

- the source at the top of the screen
- visualized source credibility
- visualized likelihood of the classification
- the processed news article
- a feature to discover other topics
- as well as the sources of disproof [1].

Based on this research, the authors designed prototypical user interfaces for the services which help in assessing the credibility of news items. The qualitative assessment of the interface was conducted with human participants; the interface was mostly perceived as useful. The respondents also expressed their preference for a more lightweight design, as well as being provided with explanations/justifications for the systems' classification results. Lastly, the participants wished to be given the source of the news [1].

As clearly presented in this Section, there is not much work and significant results yet reported on designing suitable and friendly interfaces for fake news detection systems.

In our opinion, without friendly and trustworthy user interfaces such methods and systems will not be used in practice, and those will only be limited to laboratory usage by scientists. Such a situation will promote disinformation and favour those who create fake news.

Therefore, in the next sections, we will take you on the journey and insights of our experience in developing user dashboards and interfaces for fake news detection.

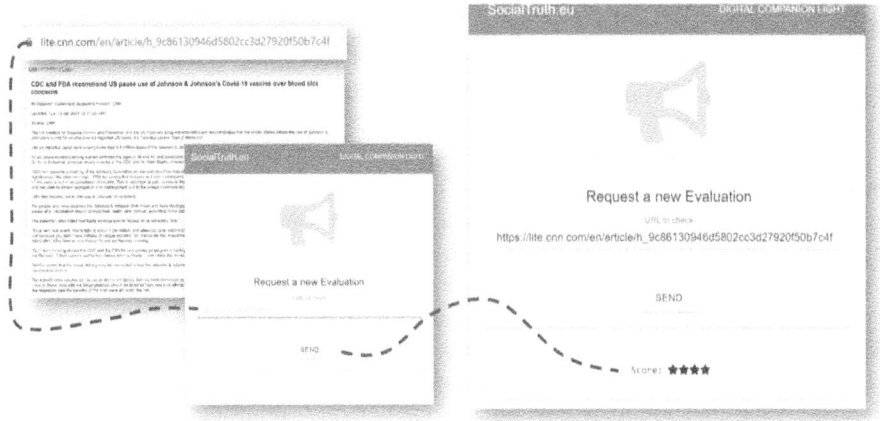

**Fig. 1.** Information flow and UI of SocialTruth Lightweight interface

## 3 SocialTruth Approach to User Interface in Fake News Detection System

Our first important results and gained experience on designing large-scale distributed fake news detection systems come from the H2020 project SocialTruth (Open Distributed Digital Content Verification for Hyper-connected Sociality).

### 3.1 SocialTruth UI/UX Design Choices and Examples

In this project, our team created a lightweight web-based interface (we called it Digital Companion Light) dedicated to the end-users of the project, namely journalists, educationists, and citizens (e.g. foundations promoting health and diet). It is worth mentioning that the consortium also created a Firefox plug-in version of the Digital Companion that also served as the gateway to our NLP-based analyzers. However, in this paper, we focus on our own work, and we present the Digital Companion Light version, which is a web-service not connected to any browser, etc.

Therefore, our goal was to design a simple interface for Digital Companion Light, as presented in Fig. 1.

When a user wanted to check textual content, the URL had to be provided/copied as the input data for our system. Then, the content was analyzed

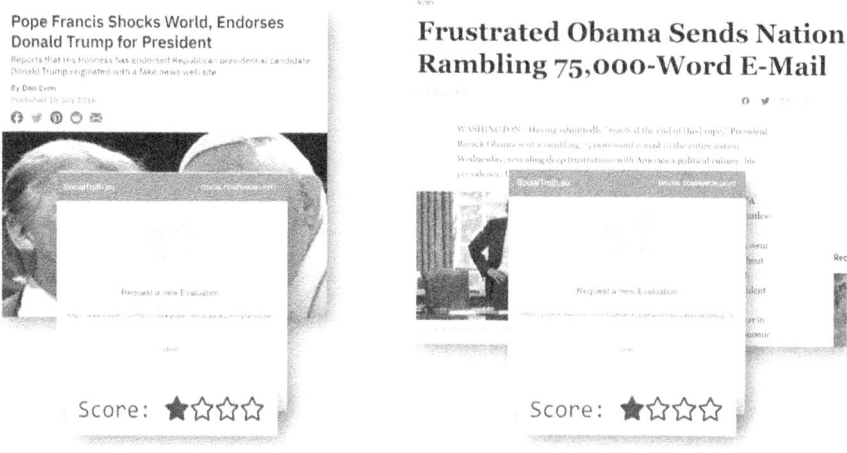

**Fig. 2.** Example of well-detected fake news analyzed in our system

by AI-based NLP solutions (see our publications on textual content analyzers, e.g. [8,10,11,11]) and the result is provided to the user.

In SocialTruth, we decided to present results in a non-binary and non-numerical fashion and we decided to use the '4 stars' approach chosen by the project consortium. In this approach, 4 stars meant reliable and true content, while 1-star meant a high probability of fake news/disinformation.

In other words: 4 stars is better than 1-star evaluation result. An example of well-known fake news flagged by our system with 1-star evaluation results is shown in Fig. 2.

In SocialTruth, we decided to use URL as input data and 4-stars notification as output. After evaluation and testing, we learnt that 4-stars notification was not clear to all users, while the URL input approach also had several drawbacks. It is obvious that various websites and content types contain not only text to be checked by the system, but also images, commercials, dynamic content (scripts) etc. which made analyses longer and less reliable.

Therefore, as presented in the next Section, we decided to change the approach and user interface, and we proposed another UI/UX in the SWAROG project products presented in Sect. 4.

### 3.2   Clarification on What is Evaluated and How?

Hereby, we would like to clarify that our NLP-based system (as described e.g. in [10,12]) is evaluating the content from the style, semantics, sentiment etc. point of view.

In SocialTruth, we did not consider what the truth is or not (there is no single truth, and there are many contradicting (even scientific) opinions and interpretations possible on a given topic).

As an example and clarification, we would like to bring the news on the energy crisis (blackout) in the USA (Texas). In Fig. 3, one can see the evaluation of two contradicting opinions by experts, both from reliable sources (newspapers).

Some energy experts claim that the blackout was due to renewable energy, while others claim that the blackout was caused by a lack of renewable energy sources and traditional energy solutions.

Since our system analyses the text characteristics, while it is not an energy expert (those have different opinions anyway), both of those news entries are classified as 'true' ( 3).

It is important to emphasize that in SocialTruth we analyzed the way the text is written and its characteristics, not the content/knowledge itself (while we will pursue slightly another approach in the SWAROG project where we provide more details to the user).

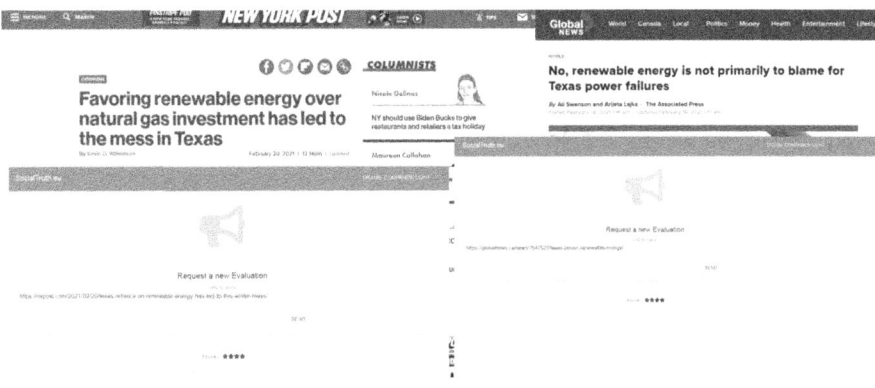

**Fig. 3.** Contradicting opinions on energy blackout, both evaluated as 'true' based on the characteristics of the text.

## 4    SWAROG Approach to the User Interface in Fake News Detection System

SWAROG is a currently active national project funded in the INFOS-TRATEG programme by the Polish National Centre of Research and Development (NCBiR) where we also work on textual content analysis for disinformation detection [3,5]

In SWAROG, we have further improved AI-based text analyzers for fake news detection, e.g. in [7,9], but in this paper, we focus only on the presentation of the results, not AI methods for text analysis.

Hereby, we decided to significantly change the user interface (we do not use URL as input and 4-stars notation anymore), and our major choices for a lightweight SWAROG user interface are:

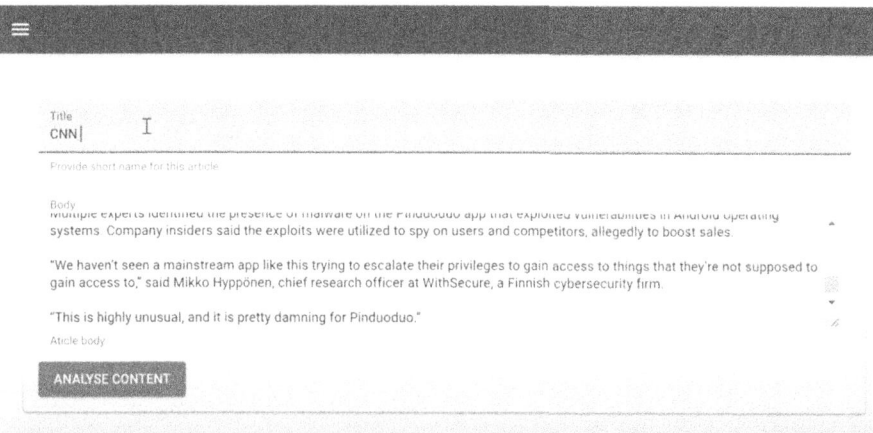

**Fig. 4.** Landing page for prompting or pasting the text, which is later sent to ML service.

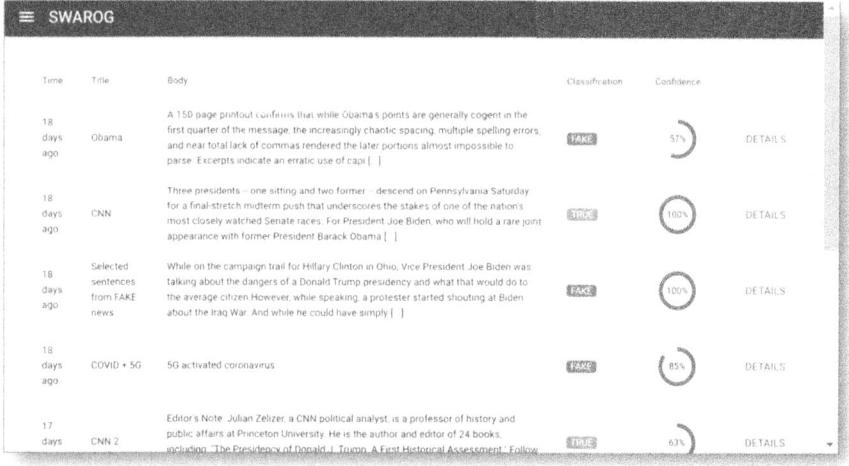

**Fig. 5.** Graphical interface and main dashboard presenting results of text analysis to fake or true with given confidence score.

- text is used as input (users are invited to copy&paste textual content to be checked)
- as for the results, we use fake/true classification with related percentages and colour (red/blue)
- we provide additional features to explain and better justify our results in a visual manner

**Fig. 6.** Example of the screen presenting more details and explanations for the evaluated textual content.

As can be seen in Fig. 4, the user can directly copy&paste the text to be evaluated into the system, which clearly indicates that we analyze the textual content.

In our new interface, we provide richer evaluation results, including e.g.:

– the information if the text is analyzed as 'fake' or 'true' with colours (red for fake, green for 'true') - see Fig. 5
– the information on confidence of the system expressed in percentages
– we present the results from previously analyzed content
– we offer users to check even more details such as text categories, information on similar content (similar news), and even information on important words from the evaluated text contributing to the analysis (see Fig. 6 and Fig. 7).

We believe that the approach taken in the SWAROG project is better, clearer, more reliable and friendly for end-users (e.g. journalists).

## 5   Discussion and Future Work

In this paper, we presented UI design and visualizations that should help end-users understand the results of the system. Another aspect (not covered here) is the explanation (so-called xAI) of the used AI methods such as deep neural networks and NLP solutions.

Indeed, in our interfaces, we offer the capability for data scientists or end-users to check explanations (xAI) for the methods used in the system [16]. Those figures or tables present the inner workings of advanced AI solutions and are not

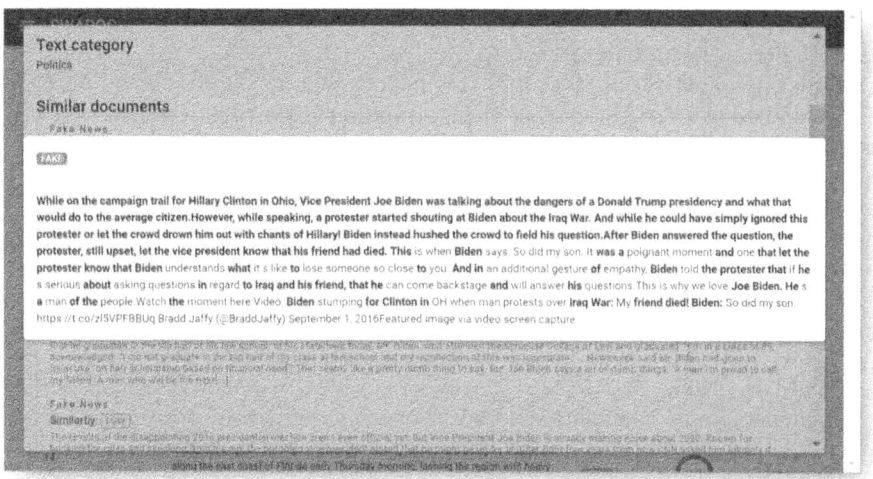

**Fig. 7.** Visualising important words to provide more details and explanations.

**Fig. 8.** SHAP values visualisation for sample sentences.

useful or natural for users who are not familiar with e.g. SHAP, LIME and other xAI approaches [13] (Fig. 8).

Moreover, we believe that such capability should not be offered openly without strict control, since as we proved recently in our work, xAI can be also used to fool fake detection systems [6].

## 6    Conclusions

In this paper, we proudly presented the evolution and several approaches to present fake news analysis and detection results for end-users in intuitive user interfaces and dashboards. Hereby, we presented the SocialTruth approach and the improved approach we designed in the SWAROG project. We believe that the evolution of the interface is beneficial a for wide set of users (citizens, journalists, homeland security agencies etc.).

More importantly, our work and this paper open up the discussion and give real examples to close the gap in the yet undiscovered topic of advanced user interfaces for AI-based fake news and disinformation detection systems.

**Acknowledgement.** This publication is funded by the National Center for Research and Development within INFOSTRATEG program, number of application for funding: INFOSTRATEG-I/0019/2021-00.

# References

1. Bunde, E., Kühl, N., Meske, C.: Fake or credible? Towards designing services to support users' credibility assessment of news content. In: Proceedings of the Annual Hawaii International Conference on System Science, vol. 2022, pp. 1883–1892 (2022). https://doi.org/10.24251/hicss.2022.237
2. Desai, A.K., Gunderson, Z., Prabhu, Y., Yan, E.: Fake News Detector Stop fake news in its tracks with our AI-powered detector. Devpost. https://devpost.com/software/fake-news-detector-p2b86i
3. Gackowska, M., Katek, G., Srutek, M., Kozik, R., Choras, M.: Document annotation tool for news content analysis. In: Burduk, R., Choras, M., Kozik, R., Ksieniewicz, P., Marciniak, T., Trajdos, P. (eds.) Progress on Pattern Classification, Image Processing and Communications - Proceedings of the CORES and IP&C Conferences 2023, Wrocław. LNNS, vol. 766, pp. 211–217. Springer, Cham (2023). https://doi.org/10.1007/978-3-031-41630-9_21
4. Gupta, V., Beckh, K., Giesselbach, S., Wegener, D., Wirtz, T.: Supporting verification of news articles with automated search for semantically similar articles (2021). http://arxiv.org/abs/2103.15581
5. Kozik, R., Komorniczak, J., Ksieniewicz, P., Pawlicka, A., Pawlicki, M., Choraś, M.: Swarog project approach to fake news detection problem. In: García Bringas, P., et al. (eds.) International Joint Conference 16th International Conference on Computational Intelligence in Security for Information Systems (CISIS 2023) 14th International Conference on EUropean Transnational Education (ICEUTE 2023), pp. 79–88. Springer, Cham (2023)
6. Kozik, R., Ficco, M., Pawlicka, A., Pawlicki, M., Palmieri, F., Choraś, M.: When explainability turns into a threat - using XAI to fool a fake news detection method. Comput. Secur. **137**, 103599 (2024). https://doi.org/10.1016/j.cose.2023.103599
7. Kozik, R., Mazurczyk, W., Cabaj, K., Pawlicka, A., Pawlicki, M., Choraś, M.: Combating disinformation with holistic architecture, neuro-symbolic AI and NLU models. In: 2023 IEEE 10th International Conference on Data Science and Advanced Analytics (DSAA), pp. 1–9 (2023). https://doi.org/10.1109/DSAA60987.2023.10302543
8. Ksieniewicz, P., Choraś, M., Kozik, R., Woźniak, M.: Machine learning methods for fake news classification. In: Yin, H., Camacho, D., Tino, P., Tallón-Ballesteros, A.J., Menezes, R., Allmendinger, R. (eds.) Intelligent Data Engineering and Automated Learning - IDEAL 2019, pp. 332–339. Springer, Cham (2019)
9. Ksieniewicz, P., Zyblewski, P., Borek-Marciniec, W., Kozik, R., Choraś, M., Woźniak, M.: Alphabet flatting as a variant of n-gram feature extraction method in ensemble classification of fake news. Eng. Appl. Artif. Intell. **120**, 105882 (2023). https://doi.org/10.1016/j.engappai.2023.105882

10. Ksieniewicz, P., Zyblewski, P., Choraś, M., Kozik, R., Giełczyk, A., Woźniak, M.: Fake news detection from data streams. In: 2020 International Joint Conference on Neural Networks (IJCNN), pp. 1–8 (2020). https://doi.org/10.1109/IJCNN48605.2020.9207498

11. Kula, S., Choraś, M., Kozik, R., Ksieniewicz, P., Woźniak, M.: Sentiment analysis for fake news detection by means of neural networks. In: Krzhizhanovskaya, V.V., et al. (eds.) Computational Science - ICCS 2020, pp. 653–666. Springer, Cham (2020)

12. Kula, S., Kozik, R., Choraś, M.: Implementation of the Bert-derived architectures to tackle disinformation challenges. Neural Comput. Appl. (2021). https://doi.org/10.1007/s00521-021-06276-0

13. Pawlicka, A., Pawlicki, M., Kozik, R., Kurek, W., Choraś, M.: How explainable is explainability? Towards better metrics for explainable AI. In: Visvizi, A., Troisi, O., Corvello, V. (eds.) Research and Innovation Forum 2023, pp. 685–695. Springer, Cham (2024)

14. Purificato, E., Shahania, S., Luca, E.W.D.: Tell me why it's fake: developing an explainable user interface for a fake news detection system. In: Proceedings of the 3rd Italian Workshop on Explainable Artificial Intelligence co-located with 21th International Conference of the Italian Association for Artificial Intelligence (AIxIA 2022), pp. 51–63 (2022)

15. Stissi, S.: A Functional and Scale-able User Platform for Automatic Fake News Detection. Semantic Scholar (2020)

16. Szczepański, M., Pawlicki, M., Kozik, R., Choraś, M.: New explainability method for bert-based model in fake news detection. Sci. Rep. **11**(1), 23705 (2021). https://doi.org/10.1038/s41598-021-03100-6

# Social Attraction Mutation: A Novel Method for Mutation Based on Attraction

Márk Domonkos[(✉)] ⓘ, Zhang Huanpeng, Natabara Máté Gyöngyössy ⓘ, and János Botzheim ⓘ

Department of Artificial Intelligence, Faculty of Informatics, ELTE Eötvös Loránd University, Pázmány P. Sétány 1/A, Budapest 1117, Hungary
{domonkos,czdmgt,natabara,botzheim}@inf.elte.hu

**Abstract.** In the field of Computational Intelligence, Evolutionary Algorithms are considered a widely known and used group of methods. In this paper, we present a novel type of mutation operator, called Social Attraction Mutation, based on Random Deviation Mutation and inspired by the Social Attraction of belonging to a leader group. We carry out tests on our operator with a wide range of parameter sets and execution configurations, on four continuous benchmark functions, and two classical control problems from Gymnasium (successor of OpenAI Gym). The results indicate that this kind of mutation operation can lead to better convergence during the search. However, it is too greedy to use it alone, without combining it with another, probabilistic mutation operator.

**Keywords:** Genetic Algorithms · Mutation Operator · Evolutionary algorithms · Social Attraction Mutation

## 1 Introduction and Related Works

Evolutionary Algorithms (EAs) are a group of biologically inspired algorithms in Computational Intelligence. Such algorithms are widely used for problems that conventional search algorithms appear to be difficult to solve. The group of problems mentioned involves various real-life problems from engineering, finance, science, and design. The power of EAs lies in the trade-off between the quality of the solution found and the resources allocated to find that solution.

From the historical point of view, we must mention John Holland and his students for forming one of the main directions that started the research on EAs, which are the Genetic Algorithms (GAs) [7,8].

The research in GAs currently has a wide range of applications and to tackle the problems raised by these applications, researchers constantly trying to extend the field's capabilities. Research also investigated the opportunities in multimedia and wireless network applications of GAs in the context of its past, present, and potential future possibilities [10]. In [1] the usage of genetic algorithms was demonstrated, where the GA was responsible for the feature selection in the classification of electromyographic data with the help of a support vector

N.-T. Nguyen et al. (Eds.): ICCCI 2024, CCIS 2165, pp. 358–370, 2024.
https://doi.org/10.1007/978-3-031-70248-8_28

machine in the 250 ms signal segment. In another article [6], authors examined the potential of GAs in the field of photovoltaic energy production systems. The method called hybrid maximum power point tracking enabled a nearly 3% increase in efficiency alongside other benefits. To make safer transactions in the cryptocurrency market, in [2], a fraudulent transaction detection method was developed, where the Deep Learning model was optimized with a Genetic Algorithm - Cuckoo Search hybrid. Three experimental cases were assessed involving laser sintering and melting processes in [5], with the use of a virus evolutionary genetic algorithm, which outperformed other widely used ones.

Use of group making in the evolutionary algorithms (and thus counting in social aspects) is a relatively low emphasized direction in the field. Works in this direction include leveraging the k-means algorithm for initialization of GA [11], or using for combinatorial problems (Traveling Salesman Problem) splitting to smaller bits [12], the integration of a social interaction mechanism to the GA flow was utilized for generation for software test data in [15]. The use of game theory and social aspects were also used in [16, 18] to optimize the algorithm's efficiency.

Recently, there has been a higher emphasis on the further development of the operators used in evolutionary methods. In [9], a special mutation operator was evaluated through extensive computational experiments, and compared for non-linear numerical functions, outperforming the standard GA for solving the categorization of fixed points as solutions. Researchers developed a two-parameter mutation operator for inductive modeling using combinatorial-genetic algorithms in [14]. An automatic GA mutation was developed in [4] for molecules, by the usage of masked language model.

The structure of the paper is as follows: In Sect. 2 we will introduce the main inspiration and the functioning of our proposed method. In Sect. 3 we will present our experimental methodology with the configurations, and benchmark problems used. In Sect. 4 we will show the results obtained during the experiment, and finally in Sect. 5 we conclude our presentation of the experiment with the summary of the results.

## 2   Proposed Algorithm

Mutation operation is one of the operators used during the execution of a Genetic Algorithm. It is widely known that the mutation operator can add new information to the population.

The algorithmic inspiration of our proposed method is a mutation operation called Random Deviation Mutation (RDM), or Gaussian Mutation, also used in standard GAs [8]. When using RDM we simply add a random vector to the solution's (individual's) position (genome) in the search space. Usually (but not necessarily), a zero-center normal distribution is used to modify the genes. In our experiments, we use a standard deviation of 0.5. This modification is thus probabilistic.

The social inspiration for the modification was taken from the widely known basic human need, as a social being, of belonging to a group [3]. This kind of

sociological attraction is the key idea of our new method, and we wanted to
assess this kind of idea as part of an evolutionary operator.

In our method, the mutation is divided into two parallel processing branches
on the elite and non-elite populations. In the first parallel branch, we calculate
the centroid of the elite population which is a subgroup of the most superior
individuals. This centroid is called the Social Attraction Point (SAP), which is
the mean of the geneset of the elite individuals in the case of a single cluster. In
the second parallel branch, the non-elite individuals undergo crossover.

After the two parallel operations conclude each individual is traversed
(mutated) sequentially. Before each individual's move, it is assigned a step size,
calculated as a parametrizable fraction (which we set to 0.1) of the distance
from the center point, then the selected individuals are moving towards the SAP
according to Eq. (1). The whole process is depicted in Fig. 1.

$$\mathbf{x}(t+1) = \mathbf{x}(t) + (\mathbf{SAP}(t) - \mathbf{x}(t)) \cdot step\_ratio, \tag{1}$$

where $\mathbf{x}(t+1)$ is the mutated genome of an individual, $\mathbf{x}(t)$ is the original genome,
$\mathbf{SAP}(t)$ is the social attraction point in the current generation, and $step\_ratio$
controls the step size of the mutation will move the individual.

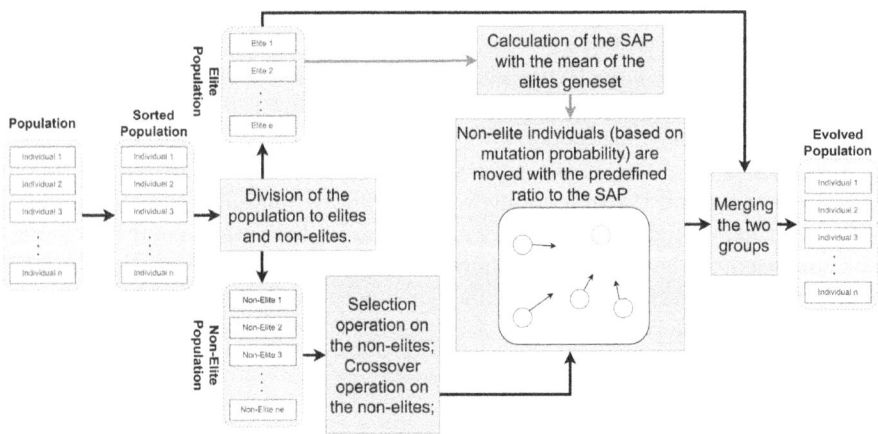

**Fig. 1.** Visualization of the Social Attraction Mutation method. Calculation of the
SAP and application of operators on the non-elite population are independent. Their
execution could be parallel or serial depending on the exact implementation.

## 2.1   Mutation Variants Used

– **Social Attraction Mutation 0.05/0.1/0.2/0.4 (SAM_0.05/SAM_0.1/
  SAM_0.2/SAM_0.4):** Before the selection process, the population's individuals are arranged in order of their fitness, and a top scoring subset of elite

individuals is chosen based on a predetermined ratio (5%/10%/20% / 40% of the population). In the mutation stage, the mean (centroid for a single cluster) of the elite individuals' geneset is determined, identifying the SAP of the elite subset. Subsequently, all non-elite individuals will move toward the center point with the predefined step ratio according to Eq. (1).

– **Random Deviation Mutation (RDM):** It entails the stochastic alteration of the selected genes within an individual by applying small random deviations to their current values.
– **Combined    SAM    0.05/0.1/0.2/0.4    (CSAM_0.05/CSAM_0.1/ CSAM_0.2/CSAM_0.4):** Before a predefined changing point in the generations, the SAM mutation is used, after that the RDM.

For all the methods above we used a mutation probability of 0.2 in our experiments. In the case of SAM we consider all elites to belong to a single cluster, thus centroid calculation can be simplified to taking their gene-wise mean.

## 3    Methodology

To measure the efficiency of the Social Attraction Mutation, we tested it (under different elite ratios) and compared it with the Random Deviation Mutation on four continuous benchmark problems and two classical control problems from Gymnasium [17] (formerly known as OpenAI Gym). Because of the stochastic nature of the algorithms 30 independent runs are executed for the continuous benchmark problems and 10 for the control problems.

We made the tests for each problem type in the configurations described in Table 1.

**Table 1.** Configurations used. Parameters (left to right): population size, number of generations, runs per experiment, ratio of the population to be included in each round of the tournament selection, crossover probability, linear combination parameter for the crossover, mutation probability, mutation standard deviation.

|  | Population | Generation | Runs | Tourn. size | P_Cross | $\alpha$ | P_Mut | Mut_std |
|---|---|---|---|---|---|---|---|---|
| **Continuous Benchmark Problems** | | | | | | | | |
| Config1 | 100 | 100 | 30 | 15% | 0.8 | 0.2 | 0.2 | 0.5 |
| Config2 | 200 | 100 | 30 | 15% | 0.8 | 0.2 | 0.2 | 0.5 |
| **Control Problems** | | | | | | | | |
| Config | 100 | 50 | 10 | 15% | 0.8 | 0.2 | 0.2 | 0.5 |

### 3.1 Selection and Crossover Operations Used Throughout the Experiment

For selection, we use traditional tournament selection on both benchmark functions and classical control problems, with a tournament size of 15% of the individuals. It is a widely used method in evolutionary algorithms [13]. This approach involves randomly selecting a small subset of individuals from the population, and then the individual with the highest fitness is chosen as the parent. It provides a balance between exploration and exploitation, as it allows both strong and weaker individuals to participate in the selection process.

We use linear combination crossover. It is a crossover method without any randomness. The offspring's genes are a simple linear combination of parents' genes as Eq. (2) states and in the end, the one with the better fitness value is accepted from the two. In our experiment, we use the $\alpha$ parameter fixed with 0.2. We apply a crossover probability of 0.8 through our tests.

$$\begin{cases} \textit{Offspring}_1 = (1 - \alpha) \cdot \textit{Parent}_1 + \alpha \cdot \textit{Parent}_2 \\ \textit{Offspring}_2 = \alpha \cdot \textit{Parent}_1 + (1 - \alpha) \cdot \textit{Parent}_2 \end{cases} \quad (2)$$

where $\alpha$ is the parameter that controls the degree of combination.

### 3.2 Used Benchmark Functions

We used four multidimensional continuous benchmark functions to evaluate our proposed methods. We made experiments with 3, 10 and 50 dimensions. The individuals are real coded, where the genes are values in the corresponding dimension of the problem. We have restricted the search space to the interval of $[-5; 5]$ for each benchmark function.

First, we used the Rastrigin function. The function is described according to Eq. (3). It has a complex multimodal structure and a large number of local minima.

$$f(\mathbf{x}) = A \cdot n + \sum_{i=1}^{n} \left( x_i^2 - A \cdot \cos(2\pi x_i) \right) \quad (3)$$

where $n$ is the number of dimensions. We denote the gene vector of individuals as $\mathbf{x} = [x_1, x_2, \ldots, x_n]$. The global minimum can be found at f(0, ..., 0) = 0, and the (usual) search domain is $-5.12 \leq x \leq 5.12$.

We used the Ackley function as Eq. (4) shows, which has a nearly flat outer region and a large hole at the center with local minima in-between. The function is usually evaluated in the domain $x_i \in [-32.768; 32.768]$, although it may be restricted to a smaller domain. The global minimum value is 0 at $x = (0, ..., 0)$.

$$f(\mathbf{x}) = -a \cdot exp\left(-b \cdot \sqrt{\frac{1}{d} \cdot \sum_{i=1}^{d} x_i^2}\right) - exp\left(\frac{1}{d} \cdot \sum_{i=1}^{d} \cos(cx_i)\right) + a + exp(1) \quad (4)$$

where $d$ is the number of dimensions of the problem, and $a, b, c$ are parameters with usual values $20, 0.2, 2\pi$ respectively.

We also used the Sphere function for the evaluation of the performance. It is a convex function with a smooth and symmetric shape. The function is described according to Eq. (5).

$$f(\mathbf{x}) = \sum_{i=1}^{n} \left( x_i^2 \right) \tag{5}$$

where $n$ is the number of dimensions. The global minimum can be found at f(0, ..., 0) = 0 and its search domain is $-\infty \le x \le \infty$.

In addition, we used the Rosenbrock function to evaluate the performance of the proposed methods. The function is described according to Eq. (6). It is a smooth, non-convex function with an elongated valley shape, resembling the outline of a banana.

$$f(\mathbf{x}) = \sum_{i=1}^{n-1} [100 \left( x_{i+1} - x_i^2 \right)^2 + (x_i - 1)^2] \tag{6}$$

where $n$ is the dimension of the problem. The global minimum can be found at $f(\mathbf{x}^*) = 0$, at $\mathbf{x}^* = (1, ..., 1)$.

### 3.3   Used Classical Control Problems

Classic control problems in Gymnasium [17] served as a benchmark for testing and comparing the effectiveness of our algorithms. They offer a diverse experimental environment for assessing the performance of different algorithms in control tasks. It can give back the reward gained based on multiple experiments conducted, called episodes. We used in our experiments 5 episodes of randomly initialized simulations for robust evaluation. We focused on the CartPole and MountainCar problems. The problems are described below, and the working principle of the agents can be seen in Fig. 2 (left) and in Fig. 3 (left).

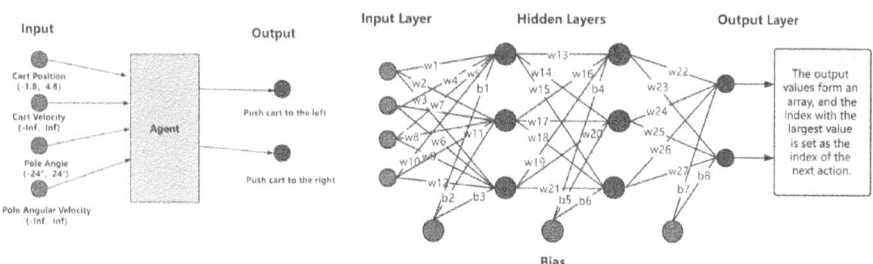

**Fig. 2.** Visualization of the agent and the neural network used in the CartPole control problem.

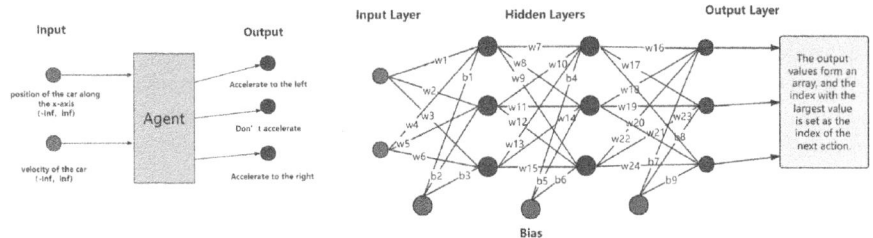

**Fig. 3.** Visualization of the agent and the neural network used in the MountainCar control problem.

– **MountainCar:** It involves a car positioned at the local minimum of a sinusoidal-shaped valley, where the sole available actions are accelerations in either direction or doing nothing. The objective is to tactically apply accelerations to guide the car towards the desired state (position) atop the right hill. The agent gets $-1$ reward for each timestep until it reaches the goal for 200 steps at most. Upon reaching the goal or running out of time the simulation concludes.

– **CartPole:** A pole is affixed to a cart via a passive joint, allowing a frictionless rotation, and a frictionless track for the cart. The inverse pendulum is initially positioned in an upright manner on the cart, and the objective is to maintain balance by exerting forces to the left or right on the cart. The agent gets $+1$ reward for each timestep for which it maintains the pole in a $\pm12$ deg rotation range from the vertical position and the car remains on the pole between $[-2.4;\ 2.4]$. The simulation concludes upon breaking any of the conditions above or after at most 500 steps.

For the problem solver agent, we use a Multi-Layer Perceptron (MLP) presented in Fig. 2 (right) and Fig. 3 (right). The training of the agent is conducted through the evolutionary process. We use two hidden layers in the MLP's structure which are visible in Fig. 2 (right) and Fig. 3 (right). This concludes, that the weights to optimize in the CartPole case are $15 + 12 + 8 = 35$ parameters while with MountainCar this is $9 + 12 + 12 = 33$. We restrict the weight values to the domain $w_i \in [-10; 10]$. The goal is to maximize the reward values.

## 4   Results

### 4.1   Preliminary Results

The experimental outcome of the preliminary tests can be seen in Fig. 4. We can see that the SAM compared to the RDM has a much higher convergence in the case of continuous problems. However, it turns out to be too greedy and probably stuck in a local minimum. This is highly visible in the results in cases of Rastrigin function 10D with configuration 2 and Ackley 10D with configuration

1 (and in some other cases as well), where SAM gets stuck in a local optimum, where in the long run RDM wins.

The case is similar with the control problems visible in Fig. 5. The evolution of the weights with SAM has an inferior result, even in the case of the MountainCar the majority of the methods are unable to finish the task, hence the $-200$ value of the reward.

### 4.2   Results Obtained on the Continuous Benchmarks

Because of the greediness of our SAM method, we combine the SAM with the RDM to achieve a better solution, resulting the Combined SAM method (CSAM). We do this by defining a change point in the generations to change between the SAM and RDM methods. We use a value of 15, meaning that in the first 15 generations we use SAM and then we change to RDM. The value of the parameter was determined heuristically based on preliminary experiments.

The experimental outcomes are depicted in the subsequent Fig. 6. We show in this figure only the 50D and configuration 1 results as the other results are similarly arranged and have similar forms as well.

The numerical outcome of all experiments (average of the last generation's best individuals' fitness value over each run) can be observed in Table 2 on the continuous benchmarks. We can observe that in most cases the combined SAM method with the different elite ratio parameters is superior to the traditional Random Deviation. From this categorical statement, only the Rosenbrock function with 3 dimensions and only with the larger population size (configuration 2) is an exception, where the results are worse than the RDM. Also, the Sphere function with 3 dimensions gives a mixed winner in the evolution. The results also suggest that there is an optimal range to the elite ratio parameter. Having a too large value such as 40% usually results in weaker results than smaller ones. On the other hand, a too-small ratio like 5% is occasionally undesirable.

### 4.3   Results Obtained on the Classical Control Problems

We can see the results obtained from the classical control problems visually in Fig. 7 and numerically in Table 3. We use the average reward of the agent at the end of a simulation as the metrics to compare the effectiveness of the methods. We compute this average over both the 10 runs and 5 simulation episodes together.

From Table 3 we can see that the results obtained with the modified versions are better both in the MountainCar problem and the CartPole problem. Although with the first we can see mixed results. This situation can be because of the problem's complexity.

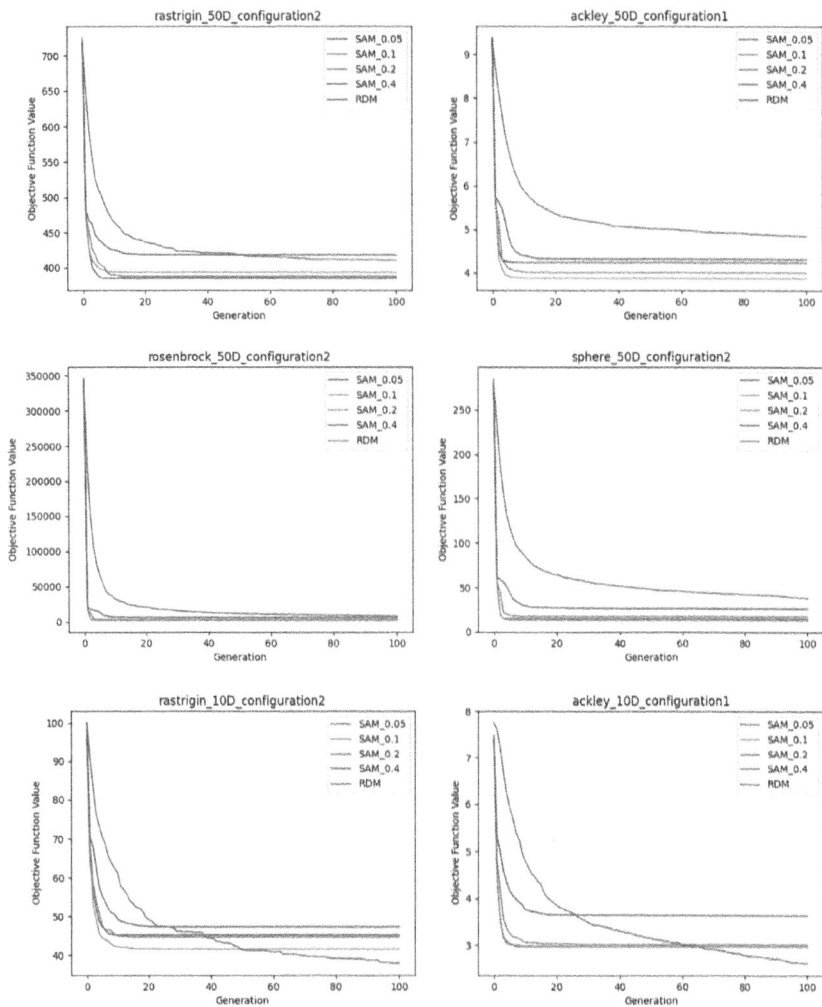

**Fig. 4.** Plot of the average best individual's objective function value along the generations on the continuous benchmark functions with 50 dimensions and with Config1, and two additional typical results.

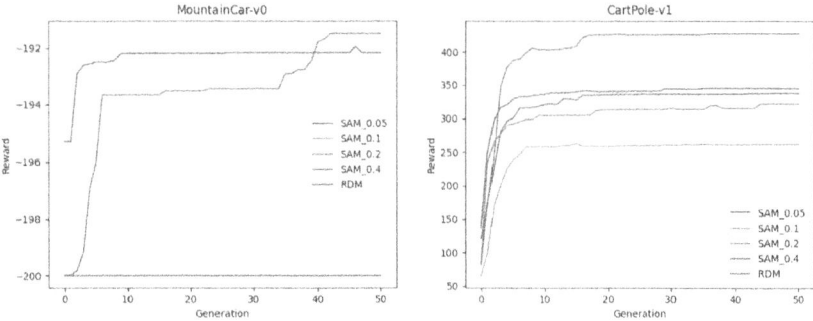

**Fig. 5.** Plot of the average best individual's reward along the generations on the control problems.

**Table 2.** Objective function values of the best individuals average of the last generation on the continuous problems.

| Name | Rastrigin | | Ackley | | Rosenbrock | | Sphere | |
|------|-----------|---------|---------|---------|------------|---------|---------|---------|
| | Config1 | Config2 | Config1 | Config2 | Config1 | Config2 | Config1 | Config2 |
| 3 dimensional case | | | | | | | | |
| CSAM_0.05 | 2.0127 | 1.276 | **0.1556** | **0.1018** | **0.5482** | 0.7127 | **0.0023** | 0.015 |
| CSAM_0.1 | 1.7032 | 1.2195 | 0.1604 | 0.1452 | 0.8112 | 0.7793 | 0.0027 | 0.02 |
| CSAM_0.2 | **1.5658** | **1.0366** | 0.1772 | 0.1338 | 0.623 | 0.7453 | 0.003 | 0.022 |
| CSAM_0.4 | 1.7529 | 1.3269 | 0.2078 | 0.2346 | 0.9631 | 0.6516 | 0.0039 | 0.036 |
| RDM | 2.6796 | 2.1856 | 0.2661 | 0.2232 | 1.1698 | **0.606** | 0.004 | 0.026 |
| 10 dimensional case | | | | | | | | |
| CSAM_0.05 | 35.1599 | **28.5759** | **1.8768** | **1.6526** | **46.1729** | 39.8044 | **0.2747** | 0.279 |
| CSAM_0.1 | 34.6565 | 30.7473 | 1.9842 | 1.8748 | 48.7194 | **37.0504** | 0.3088 | 0.279 |
| CSAM_0.2 | **33.4152** | 31.5854 | 2.1494 | 2.0657 | 48.9154 | 47.0035 | 0.3721 | 0.324 |
| CSAM_0.4 | 35.6967 | 33.1342 | 2.4721 | 2.3666 | 71.6023 | 72.4412 | 0.6596 | 0.5281 |
| RDM | 44.733 | 40.8161 | 2.6076 | 2.3298 | 109.6157 | 75.1374 | 0.6673 | 0.6304 |
| 50 dimensional case | | | | | | | | |
| CSAM_0.05 | 372.4921 | 366.4863 | 3.9572 | 3.5527 | 2586.5016 | **1797.6944** | 16.7775 | 13.1336 |
| CSAM_0.1 | 369.2041 | **358.7559** | 3.826 | **3.4887** | **2585.1576** | 1886.0143 | 16.9048 | **12.037** |
| CSAM_0.2 | **368.3327** | 365.0151 | 3.8708 | 3.7034 | 2714.3369 | 2438.0386 | 18.071 | 14.19 |
| CSAM_0.4 | 378.6553 | 373.5548 | 4.0545 | 4.1719 | 5079.4381 | 3971.4788 | 25.1691 | 23.9221 |
| RDM | 417.3542 | 402.3034 | 4.8439 | 4.7316 | 10707.9637 | 9866.8463 | 36.8251 | 35.8344 |

**Table 3.** Outcomes of the methods implemented for the classical control problems. The average reward of the best individuals of the last generation. The higher, the better.

| Name | MountainCar | Cartpole |
|------|-------------|----------|
| CSAM_0.05 | **−176.98** | 344.32 |
| CSAM_0.1 | −191.42 | 396.76 |
| CSAM_0.2 | −194.9 | **417.02** |
| CSAM_0.4 | −188.12 | 400.18 |
| RDM | −190.44 | 273.98 |

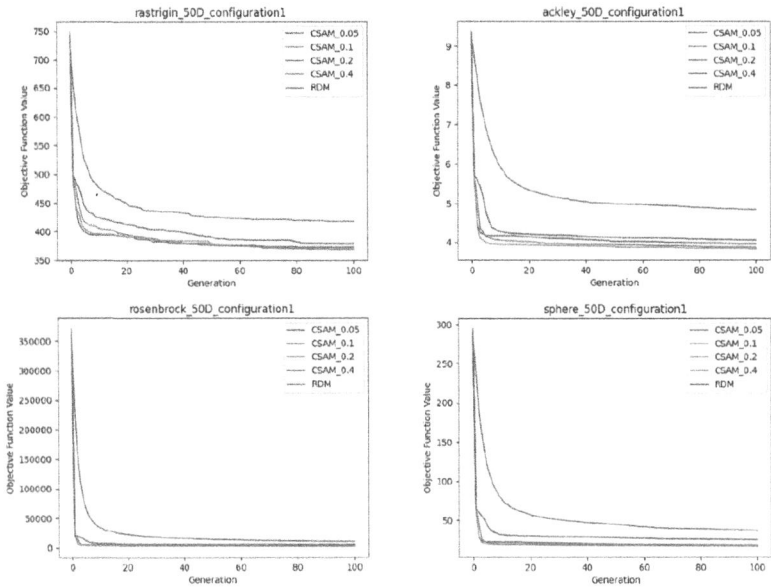

**Fig. 6.** Plot of the average best individual's objective function value along the generations on the continuous benchmark functions with 50 dimensions and with Config1.

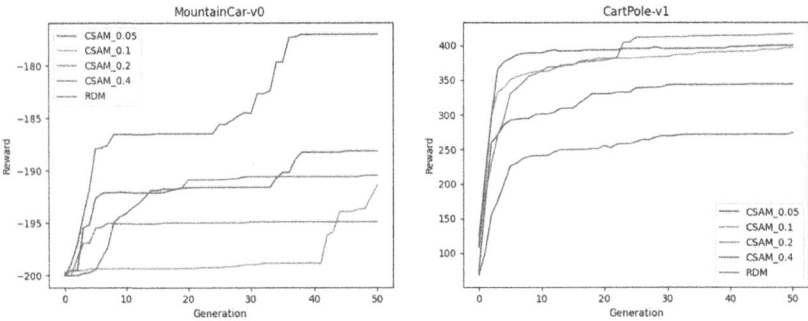

**Fig. 7.** Plot of the average best individual's rewards along the generations on the classical control problems.

## 5  Summary

In this paper we introduced our novel type of mutation operator called Social Attraction Mutation which is based on Random Deviation Mutation and inspired by the social attraction to a leader group. The new method uses an elite subset of the population and their single-cluster centroid for the mutation change.

We carried out experiments on three continuous problems with the problems having dimensions of 3, 10, and 50. The results suggest that the SAM has the advantage in convergence, but is too greedy. We experimented with CSAM, the combination of the SAM with RDM, and in most cases compared to the RDM we obtained better solutions with a change at generation 15 out of 100. A deeper investigation of the optimal change point is a matter of future research.

We experimented with two classical control problems, where we obtained mixed results for the SAM, but with the combination, we could also achieve better solutions consistently.

We would like to continue experimenting with different functionalities added to the attraction's parameter.

# References

1. Aviles, M., Sánchez-Reyes, L.M., Fuentes-Aguilar, R.Q., Toledo-Pérez, D.C., Rodríguez-Reséndiz, J.: A novel methodology for classifying EMG movements based on SVM and genetic algorithms. Micromachines **13**(12), 2108 (2022)
2. Aziz, R.M., Mahto, R., Goel, K., Das, A., Kumar, P., Saxena, A.: Modified genetic algorithm with deep learning for fraud transactions of ethereum smart contract. Appl. Sci. **13**(2), 697 (2023)
3. Baumeister, R.F., Leary, M.R.: The need to belong: desire for interpersonal attachments as a fundamental human motivation. Psychol. Bull. **117 3**, 497–529 (1995). https://api.semanticscholar.org/CorpusID:13559932
4. Blanchard, A.E., et al.: Automating genetic algorithm mutations for molecules using a masked language model. IEEE Trans. Evol. Comput. **26**(4), 793–799 (2022)
5. Fountas, N.A., Kechagias, J.D., Vaxevanidis, N.M.: Optimization of selective laser sintering/melting operations by using a virus-evolutionary genetic algorithm. Machines **11**(1), 95 (2023)
6. Hassan, A., Bass, O., Masoum, M.A.: An improved genetic algorithm-based fractional open circuit voltage MPPT for Solar PV systems. Energy Rep. **9**, 1535–1548 (2023)
7. Holland, J.H.: Adaptation in natural and artificial systems. Ann Arbor (1975)
8. Holland, J.H.: Adaptation in Natural and Artificial Systems: An Introductory Analysis with Applications to Biology, Control, and Artificial Intelligence. MIT press (1992)
9. Jalali Varnamkhasti, M., Vali, M., et al.: A special mutation operator in the genetic algorithm for fixed point problems. J. Function Spaces **2023** (2023)
10. Katoch, S., Chauhan, S.S., Kumar, V.: A review on genetic algorithm: past, present, and future. Multimedia Tools Appli. **80**, 8091–8126 (2021)
11. Li, H., Jiang, X., Wei, X.: Initialization method of genetic algorithm based on improved clustering algorithm. In: Proceedings of the Genetic and Evolutionary Computation Conference Companion, pp. 447–450 (2022)
12. Ma, T., Wang, T., Yan, D., Hu, J.: Improved genetic algorithm based on k-means to solve path planning problem. In: 2020 International Conference on Information Science, Parallel and Distributed Systems (ISPDS), pp. 283–286. IEEE (2020)
13. Miller, B.L., Goldberg, D.E.: Genetic algorithms, tournament selection, and the effects of noise. Complex Syst. **9** (1995). https://api.semanticscholar.org/CorpusID:6491320
14. Moroz, O., Stepashko, V.: New two-parametric mutation operator for inductive modelling using combinatorial-genetic algorithm. In: 2022 12th International Conference on Advanced Computer Information Technologies (ACIT), pp. 76–79. IEEE (2022)

15. Neto, A.D.A., Pereira, R.L., De Oliveira, R.C.L.: Application of a genetic algorithm with social interaction to search based testing. In: 2022 IEEE Latin American Conference on Computational Intelligence (LA-CCI), pp. 1–6. IEEE (2022)
16. Pereira, R.L., et al.: Game theory and social interaction for selection and crossover pressure control in genetic algorithms: An empirical analysis to real-valued constrained optimization. IEEE Access **8**, 144839–144865 (2020)
17. Towers, M., et al.: Gymnasium (Mar 2023). https://doi.org/10.5281/zenodo.8127026, https://zenodo.org/record/8127025
18. Tsai, C.F., Lu, S.L.: A novel mechanism for efficient the search optimization of genetic algorithm. Inter. J. Comput. Intell. Syst. **9**(1), 57–64 (2016)

# Extracting Common DNA Segments from the Complete Genomes of 7538 Viruses and Five Selected Mammals

Jing-Doo Wang[1]($\boxtimes$) and Yi-Chun Wang[2]

[1] Department of Computer Science and Information Engineering, Asia University, No. 500, Lioufeng Road Wufeng Division, Taichung, Taiwan
jdwang@asia.edu.tw
[2] Department of Biomedical Sciences, Chung Shan Medical University, No. 110, Sec.1, Jianguo North Road, Taichung, Taiwan
ycw@csmu.edu.tw

**Abstract.** This paper aims to extract common DNA consecutive sequences appearing in both of the complete genomes of viruses and that of mammals. With these common DNA sequences as genomic fossils, biologists or virologists can trace possible pathway of genomic evolution across species. To meet the requirement of huge computation to extract common DNA sequences from complete genomes of all viruses and mammals selected, this study adopts one previously developed approach that was based on MapReduce programming model. This study has experiments for extracting common DNA sequences run on a Hadoop cluster containing ten computing nodes. Experimental resources includes the whole genomic sequences of 7,538 viruses and five selected mammals, including *Homo sapiens* (Human), *Pan troglodytes* (Chimpanzee), *Mus musculus* (House Mouse), *Rattus norvegicus* (Brown Rat) and *Sus scrofa* (Wild Boar). There are a huge amount of common DNA consecutive sequences extracted and, for simplicity, there are only 26 ones whose lengths are longer than 50 base-pair (bp) selected for illustration. Among above 26, there are 13 reverified as no repetitive sequences that could be seemed as the clues to reinspect the relationship of viruses and mammals. Via cloud computing that can provide with more computing nodes then ten used in this study, it is believed that this approach can handle with more complete genomes of species, and then provide more common genomic fossils to biologists or virologist to verify the potential connections among diverse species in the future.

**Keywords:** Comparative genomics · Complete Genome · Maximal Repeat · Mammal · Virus · Hadoop · MapReduce

## 1 Introduction

### 1.1 Viruses Vs. Genomic Fossils

Viruses are able to infect all types of organisms, including animals, plants, bacteria and archaea. Due to the types of interactions between viruses and their

N.-T. Nguyen et al. (Eds.): ICCCI 2024, CCIS 2165, pp. 371–383, 2024.
https://doi.org/10.1007/978-3-031-70248-8_29

corresponding hosts, the range of hosts one virus can infect is limited [10]. According to the combination of viruses' nucleic acid (DNA or RNA), strand-edness (single-stranded or double-stranded), sense, and method of replication, Baltimore classification system [9] divided the viruses into seven groups. The International Committee on Taxonomy of Viruses (ICTV) [2], however, takes the responsibility of formal taxonomic classification of viruses according to their phenotypic characteristics, such as morphology, the types of nucleic acid, mode of replication, hosts infected or diseases caused. Due to high diversity of viruses and lack of conserved genomic sequences, however, it is not easy to construct a suitable and convinced structure for the taxonomy of all viruses [26]. So far there are still many viruses remaining unclassified in ICTV or NCBI(National Center for Biotechnology Information) [3].

It is well known that some viruses, e.g. *endogenous retrovirus* (ERV) [8], can replicate by inserting its genome into that of an infected cell. Once one virus might infect germ line cells of its host and if these cells survive and go on to create a new organism, that new organism will contain the genome of that virus as an inherent part within its own genome. After generations of that infected organism, the embedded genomes of that virus might become genomic fossils due to mutations. Furthermore, some of these ERVs, e.g. *human endogenous retrovirus type K* (HERV-K) [13], may remain active with biological functions. So far, there exists a huge number of common genomic DNA sequences appearing in both of viruses and organisms [20]. These existing common DNA sequences appearing in both of viruses and organisms may be helpful to trace the pathway of evolution across different species.

### 1.2   The Limitations of Multiple Sequence Alignment for Large and Diverse Genome Comparisons

Sequence alignment, based on dynamic programming [12], plays an important role in sequence and genome analysis in bioinformatics [22]. As more whole sequenced genomes is becoming publicly available, it is highly desired to have novel approach to make large genomes comparison possible and its computation efficient. There are well known online web sites for genomic sequence comparison based on sequence alignment. For example, Basic Local Alignment Search Tool (BLAST) [4], the UCSC (University of California Santa Cruz) Genome Browser [6] and the EMBL-EBI (European Molecular Biology Laboratory-European Bioinformatics Institute) search [19]. Because the computation of an accurate multiple sequence alignment (MSA) has long been known to be an NP-complete problem [12], there are heuristic approaches [11,23] proposed to tackle this problem in the past decades. To find the conserved regions, if existed, across species via multiple sequence alignment (MSA), however, it is essential to select correlated sequences whose lengths are similar. When some of these sequences are unrelated or diverse and their lengths are different, it will count on the knowledge of domain experts to rearrange the order of input sequences or narrow down the ranges of target sequences such that one can repeat the processes of MSA and then identify conserved regions if possible. Above MSA processes to handle diverse and large genome sequences will need a lot of manual efforts.

### 1.3  A Scalable Maximal Repeat Extraction Approach

There are many previous studies using the scalable approach developed in [25, 27] to extract DNA maximal repeats [15]. In [32], Wang et al. extracted and compared the common DNA sequences of 2,712 viruses and *Homo sapiens* (Human) from their complete genomes. However, all of common DNA sequences extracted were tandem repeats [24]. It is highly attractive and expected to extract nontandem repeats that appear in both of viruses and organisms. One can identify some distinctive repeats for bioinformatic researches [28, 29].

In [25, 27], Wang proposed a scalable maximal repeat extraction approach based on MapReduce programming model [18]. That approach can extract maximal repeats from tagged sequential data and meanwhile compute class frequency distributions of those repeats, where the tags are given or specified by domain experts in advance and the types of class can be any combination of tags attached with each of sequential data. The purpose of using maximal repeats as the units for computing frequency distribution in this approach, instead of using $k$-mers, is to reduce the candidate number of features. Intuitively, a maximal repeat [15] is one repeat that appears at least twice and can't always be a substring of another longer repeat in all of sequential data. Indeed, this scalable approach had been applied to extract repeat patterns and meanwhile compute the class frequency distributions of those patterns from a huge amount of tagged sequential data to analyze traffic flow [16, 17, 30, 31] or to improve quality control of product traceability [33].

## 2  Method

There are three steps in this study for identifying and analyzing genomic fossils appearing in both of viruses and selected mammals. The first step is to assign a unique "ClassID" as one tag to each of the complete genomes of viruses and mammals. Because these complete genomes may consist of many segments (viruses) or several chromosomes (mammals), it needs preprocesses to pack all of DNA sequences belonging to the complete genomes of one species into files, and then assign a unique "ClassID" to each of species. The second step is to use previous scalable approach [25, 27] to extract maximal repeats from tagged genomic sequences with ClassID in the first step and their frequency distribution of these repeats across distinctive ClassIDs. The third step is to scan all of maximal repeats and remove tandem repeats. Then, for illustration, find the common longest maximal repeats within each of viruses. Furthermore, check the taxonomy of these viruses and their known hosts in "Virus-Host DB" [7, 21]. The details of these three steps are described in the Sects. 2.1, 2.2 and 2.3, respectively.

## 2.1   Packing and Tagging Complete Genomic Sequences of Mammals and Viruses

To meet with the input format of the scalable maximal repeat extraction approach developed in [25,27], these fragmented DNA sequences of one species need to be packed, concatenated as one whole sequence and that whole sequence is tagged with one unique ClassID. After packing and concatenating all of DNA sequences of one species into one file, each of species is assigned with one class identifier ClassID and represented with one file in which contains all of DNA genomic sequences and their corresponding reverse complement sequences.

## 2.2   Extracting Common DNA Segments Appearing in both of Viruses and Mammals

This paper adopted one scalable approach developed in [25,27] to extract common consecutive DNA sequences that are contained in both of the complete genomes of viruses and at least one of selected mammals. Due to the limitation of the number of pages in this paper, the processes of above scalable extraction approach are illustrated using Table 1, Table 2, Table 3 and Table 4. The details are described in the following.

Table 1 gives ten 50 bp (base pair) DNA sequences, including five (M_1,M_2,M_3,M_4,M_5) from "hs_ref_GRCh38.p7_chrMT" of *Homo sapiens* and the others (V_1,V_2,V_3,V_4,V_5) from five viruses. Table 2 shows four DNA segments created manually as biomarkers (BM1,BM2,BM3,BM4) for further common sequences mining. In Table 3, one can carefully inspect those ten DNA sequences (C1,C2,C3,...,C10) in which they are embedded with some of four biomarkers (BM1,BM2,BM3,BM4) in Table 3. According to their original DNA sequences in Table 1, five DNA sequences (C1,C2,C3,C4,C5) are tagged with "Mammals" and the others (C6,C7,C8,C9,C10) are tagged with "Viruses"

This study is to extract these four biomarkers, (BM1,BM2,BM3,BM4), in Table 3 from those ten (C1,C2,C3,...,C10) in Table 3, and then identify the connections among "Mammals" and "Viruses". Using the input of ten DNA sequences (C1,C2,C3,...,C10) with tags, "Mammals" or "Viruses", as describe in Table 3, one can adopt the maximal repeat extraction approach in [25,27] to extract eighteen maximal repeats (S1, S2,S3,...,S18) whose lengths are greater than 7 bp and have their class frequency distributions of those repeats. as shown in Table 4. Table 4 show the statistic of maximal repeats whose length are longer than six base pair (bp) and their class frequency distributions among these ten tagged sequences, where "DF" (resp. "CF") of one maximal repeat. e.g. "S1", is the number of sequences (resp. "classes") containing that repeat, and "TF" is the total number of that repeat appearing in sequences. Note that the values of "DF" is the same with that of "CF" because each of sequences is estimated as one class in this study.

Due to the huge number of common maximal repeats (DNA segments) existed within selected mammals and viruses, for simplicity, this study only choose the

**Table 1.** Ten original DNA sequences.

| Sequence ID | Species: Reference (Position:Start..End) | Sample DNA Sequences (50 bp) |
|---|---|---|
| M_1 | Homo sapiens: hs_ref_GRCh38.p7_chrMT(0..49) | GATCACAGGTCTATCACCCTATTAACCACTCACGGGAGCTCTCCATGCAT |
| M_2 | Homo sapiens: hs_ref_GRCh38.p7_chrMT(50..99) | TTGGTATTTTCGTCTGGGGGGTATGCACGCGATAGCATTGCGAGACGCTG |
| M_3 | Homo sapiens: hs_ref_GRCh38.p7_chrMT(100..49) | GAGCCGGAGCACCCTATGTCGCAGTATCTGTCTTTGATTCCTGCCTCATC |
| M_4 | Homo sapiens: hs_ref_GRCh38.p7_chrMT(150..199) | CTATTATTTATCGCACCTACGTTCAATATTACAGGCGAACATACTTACTA |
| M_5 | Homo sapiens: hs_ref_GRCh38.p7_chrMT(200..249) | AAGTGTGTTAATTAATTAATGCTTGTAGGACATAATAATAACAATTGAAT |
| V_1 | Abaca bunchy top virus: NC_010314 (0..49) | AGCAGGGGGGCTTATTATTACCCCCCCTGCTCGGGGCGGGACATTCTGTG |
| V_2 | Abalone herpesvirus Victoria/AUS/2009: NC_018874 (0..49) | ACTCGTATGAACTTTGACTGGTTTTTGGGGCGCGAGAGTTTGGTTTGGAT |
| V_3 | Abalone shriveling syndrome-associated virus: NC_011646 (0..49) | CTATTTAACTAATTTAGTATTGTTTGTTGTTTTCGGTTGAGTCAATTGTT |
| V_4 | Abelson murine leukemia virus: NC_001499 (0..49) | GCGCCAGTCCTCCGAGTGACTGAGTCGCCCGGGTACCCGTGTATCCAATA |
| V_5 | Abisko virus: NC_035470 (0..49) | TCAATTTATTATTATAAATATACCATTTCGTGTACCATTACCATGTACCT |

**Table 2.** Four DNA BioMarkers created manually

| BioMarker ID | Length (bp) | Manual DNA Segment | Note |
|---|---|---|---|
| BM1 | 10 | ACCGGGTTTT | |
| BM2 | 10 | AAAAGGGCCT | |
| BM3 | 12 | ACGTACGTACGT | Tandem Repeat (ACGT)3 |
| BM4 | 8 | TTCCGGAA | |

**Table 3.** Ten DNA sequences in Table 1 embedded with four biomarkers in Table 2.

| Tag | ClassID | Manual DNA Segment | Sample DNA Sequences (50 bp) embedded with manual DNA Segments |
|---|---|---|---|
| Mammals | C1 | BM1 | GATCACAGACCGGGTTTTGTCTATCACCCTATTAACCACTCACGGGAGCTCTCCATGCAT |
| | C2 | BM2 | TTGGTATTTTCGTCTGGGGGGTATGCAAAAGGGCCTCGCGATAGCATTGCGAGACGCTG |
| | C3 | BM3,BM4 | GAGCCGGAGCACGTACGTACGTACCCTATGTCGCAGTATCTGTCTTTCCGGAATTGATTCCTGCCTCATC |
| | C4 | BM4 | CTATTATTTATCGCACCTACGTTCAATATTACAGGCGAACATATTCCGGAACTTACTA |
| | C5 | BM1,BM1 | AAGTGTGTTAATTAATACCGGGTTTTTAATGCTTGTAGGACATAATAATAACAACCGGGTTTTATTGAAT |
| Viruses | C6 | BM1 | AGCAGGGGGGCTTATTATTACCCCCCCTGCTACCGGGTTTTCGGGGCGGGACATTCTGTG |
| | C7 | BM2 | ACTCAAAAGGGCCTGTATGAACTTTGACTGGTTTTTGGGGCGCGAGAGTTTGGTTTGGAT |
| | C8 | BM3 | CTATTTAACTAATTTAGTATTGTTTACGTACGTACGTGTTGTTTTCGGTTGAGTCAATTGTT |
| | C9 | BM4 | GCGCCAGTCCTCCTTCCGGAAGAGTGACTGAGTCGCCCGGGTACCCGTGTATCCAATA |
| | C10 | BM4 | TCAATTTATTTCCGGAATATTATAAATATACCATTTCGTGTACCATTACCATGTACCT |

longest common maximal repeats within each of viruses as the clues (genomic fossils) for further discussion. For example, in Table 4 there are 18 maximal repeats, S1...S18, whose lengths are greater than or equal to seven bp. For each of viruses sequences (C6, C7, C8, C9, C10), Table 5 only selects five common DNA segments (S12, S2, S7, S17, S18) in which each is the longest common DNA segment within corresponding virus sequences (C6, C7, C8, C9, C10) and also appear in mammals sequences (C1, C2, C3, C4, C5). Based on these selected

**Table 4.** Maximal repeats (Length >= 7 bp) extracted from the tagged sequences in Table 3 and their class frequency distributions.

| Segment ID | Maximal Repeat (DNA Segment) | DF | TF | CF | Length | Class Frequency Distribution (ClassID#DF#TF) | Notes |
|---|---|---|---|---|---|---|---|
| S1 | AAAAGGGCCT | 2 | 2 | 2 | 10 | (C2#1#1)(C7#1#1) | BM2 |
| S2 | AATATTA | 2 | 2 | 2 | 7 | (C10#1#1)(C4#1#1) | |
| S3 | ACCCTAT | 2 | 2 | 2 | 7 | (C1#1#1)(C3#1#1) | |
| S4 | ACCGGGTTTT | 3 | 4 | 3 | 10 | (C1#1#1)(C5#1#2)(C6#1#1) | BM1 |
| S5 | ACGTACGT | 2 | 4 | 2 | 8 | (C3#1#2)(C8#1#2) | substring(BM3) |
| S6 | ACGTACGTAC | 2 | 3 | 2 | 10 | (C3#1#2)(C8#1#1) | substring(BM3) |
| S7 | ACGTACGTACGT | 2 | 2 | 2 | 12 | (C3#1#1)(C8#1#1) | BM3 |
| S8 | ATAATAA | 1 | 2 | 1 | 7 | (C5#1#2) | |
| S9 | GGTTTTT | 2 | 2 | 2 | 7 | (C5#1#1)(C7#1#1) | |
| S10 | GTTTTCGG | 2 | 2 | 2 | 8 | (C6#1#1)(C8#1#1) | |
| S11 | TACCATT | 1 | 2 | 1 | 7 | (C10#1#2) | |
| S12 | TACCGGGTTTT | 2 | 2 | 2 | 11 | (C5#1#1)(C6#1#1) | T[BM1] |
| S13 | TACGTAC | 2 | 4 | 2 | 7 | (C3#1#2)(C8#1#2) | substring(BM3) |
| S14 | TACGTACGT | 2 | 3 | 2 | 9 | (C3#1#1)(C8#1#2) | substring(BM3) |
| S15 | TACGTACGTAC | 2 | 2 | 2 | 11 | (C3#1#1)(C8#1#1) | substring(BM3) |
| S16 | TATTATT | 2 | 2 | 2 | 7 | (C4#1#1)(C6#1#1) | |
| S17 | TTCCGGAA | 4 | 4 | 4 | 8 | (C10#1#1)(C3#1#1)(C4#1#1)(C9#1#1) | BM4 |
| S18 | TTTCCGGAAT | 2 | 2 | 2 | 10 | (C10#1#1)(C3#1#1) | T[BM4]T |

five common DNA segments (S12, S2, S7, S17, S18) from virus sequences (C6, C7, C8, C9, C10), Table 6 shows the corresponding frequency distribution of these DNA segments appearing in mammals sequences (C1, C2, C3, C4, C5). It can be found that those four manual biomarkers (genomic fossils) are extracted precisely or hidden within these five selected DNA segments.

**Table 5.** The longest common DNA segment for each of viruses (C1, C2, C3, C4, C5)

| | ClassID | Segment ID (length:bp) | The Longest DNA Segemnt |
|---|---|---|---|
| | C6 | S4(10), S10(8), **S12(11)**, S16(7) | S12 |
| | C7 | **S1(10)**, S9(7) | S1 |
| Viruses | C8 | S5(8), S6(10), **S7(12)**,S10(8), S13(7), S14(9), S15(11) | S7 |
| | C9 | **S17(8)** | S17 |
| | C10 | S2(7), S11(7), S17(8), **S18(10)** | S18 |

**Table 6.** Frequency distributions of common DNA segments extracted from the sequences in Table 1

| The longest common DNA segment of each virus | Note | | | Mammals | | | | |
|---|---|---|---|---|---|---|---|---|
| | | | | C1 | C2 | C3 | C4 | C5 |
| S12 | T[BM1] | | C6 | | | | | 1 |
| S1 | BM2 | | C7 | 1 | | | | |
| S7 | BM3 | Virus | C8 | | | 1 | | |
| S17 | BM4 | | C9 | | | 1 | 1 | |
| S18 | T[BM4]T | | C10 | | | 1 | | |

## 2.3 Analyzing the Taxonomy and Hosts of Selected Viruses with Five Selected Mammals

Due to genetic mutation, one original DNA sequence would generate numerous DNA maximal repeats after generations. To focus on discussing with the significant common DNA sequences that appeared in both of mammals and viruses in this study, only the longest common DNA maximal repeats within each of those viruses that had common DNA maximal repeats were selected. Then, these selected DNA maximal repeats were submitted to the "Repeat-Masker" [5] to check the types of interspersed repeats and low complexity DNA sequences. RepeatMasker [5] can scan these selected common sequences to detect the existence of complex repeats, e.g. "Simple-repeat" [14], "LTR" (Long Terminal Repeat), "LTR/ERVK" (Long Terminal Repeat/Endogenous RetroVirus-K), and "SINEs/ALUs" (Short Interspersed Nuclear Elements/Arthrobacter Luteus). For viruses that contained the longest common DNA maximal repeats, their taxonomy and hosts could be obtained from "Virus-Host DB" [7,21]. One can observe these taxonomy and hosts of these viruses, and then try to trace back the location of the common longest DNA maximal repeat in corresponding viruses and mammals. That might provide some clues for biologists or virologist for their further researches.

## 3   Experimental Results

### 3.1   Tagged Whole Genomic Sequences for Each of Species

Experimental resources included the whole genomes of five selected mammals and 7,538 viruses whose complete genomic DNA sequences were available at NCBI FTP [1] when this work was launched at 2018/3/1. These mammals included *Homo sapiens* (Human), *Pan troglodytes* (Chimpanzee), *Mus musculus* (House Mouse), *Rattus norvegicus* (Brown Rat) and *Sus scrofa* (Wild Boar). Table 7 shows the locations (subdirectories) of complete genomes for five mammals at NCBI FTP site [1] and the mapping of "ClassID" to each of these mammals and viruses. One can observe that five mammals were assigned in the front with CIDs as "C1", "C2", "C3", "C4" and "C5", respectively, and the ClassIDs of 7,538 viruses were ranged from "C6" to "C7543".

**Table 7.** The whole genomic sequences of five selected mammals and 7,538 viruses.

|  | ClassID | Species | Genomes in NCBI FTP Site |
|---|---|---|---|
| Five Mammals | C1 | Homo sapiens | hs_ref_GRCh38.p7_chr*.fa.gz |
|  | C2 | Pan troglodytes | ptr_ref_Pan_tro_3.0_chr*.fa.gz |
|  | C3 | Mus musculus | mm_ref_GRCm38.p4_chr*.fa.gz |
|  | C4 | Rattus norvegicus | rn_ref_Rnor_6.0_chr*.fa.gz |
|  | C5 | Sus scrofa | ssc_ref_Sscrofa11.1_chr*.fa.gz |
| 7538 Viruses | C6 | Abaca bunchy top virus | NC_010314.gbk~NC_010319.gbk |
|  | C7 | Abalone herpesvirus Victoria_AUS_2009 | NC_018874.gbk |
|  | C8 | Abalone shriveling syndrome-associated virus | NC_011646.gbk |
|  | C9 | Abelson murine leukemia virus | NC_001499.gbk |
|  | C10 | Abisko virus | NC_035470.gbk |
|  | ... | ... | ... |
|  | C3341 | Human endogenous retrovirus K113 | NC_022518.gbk |
|  | ... | ... | ... |
|  | C3464 | Influenza A virus | NC_004905.gbk~NC_004912.gbk |
|  | ... | ... | ... |
|  | C7540 | uncultured phage WW-nAnB | NC_030449.gbk |
|  | C7541 | unidentified adenovirus | NC_030116.gbk |
|  | C7542 | unidentified circular ssDNA virus | NC_030449.gbk |
|  | C7543 | unidentified human coronavirus | NC_036584.gbk |

## 3.2    The Longest Common DNA Segments in Each of Viruses

From common DNA segments appearing in both of viruses and mammals, for simplicity, this paper chose the longest common maximal repeat appearing in one virus and their lengths are longer than 50 bp for further illustration.

Due to genetic mutations happened after generations, there might be a number of variant DNA segments derived from each of common maximal repeats that appear in both the whole genome of one virus and that of one mammal. Table 8 shows the DNA sequences of these selected 26 DNA segments, named from "S1" to "S26", Due to the limitation of the page length, some of DNA segments whose lengths are longer than 100 bp are truncated and the detail of these 26 DNA sequences can be obtained from additional files. Among these 26 DNA segments in Table 8, reported by RepeatMasker [5], there were 13 marked as "no repetitive sequences", nine as "LTR", two as "Simple repeat", one as "LTR/ERVK" and one as "SINEs/ALUs".

Above 26 common DNA segments whose lengths are all longer than 50 bp can reveal some footprints in genomic revolution. To further inspect the connections, linked by those 26 common DNA segments, between these corresponding viruses and mammals, first of all, one can check the known hosts of these 26 viruses to verify whether those connections (common DNA segments) were derived from their hosts or not. These 26 DNA segments can be used as genetic fossils for further researches about genomic revolution in this future.

**Table 8.** 26 common DNA segments that appears in both of one virus and one mammal.

| | | The longest common DNA sequences appearing in both of one virus and one mammal | |
|---|---|---|---|
| Segemnt ID | Length (bp) | RepeatMasker | Selected Common DNA Maximal Repeats |
| S1 | 805 | no repetitive sequences | GCCACCTGCTGGAGCAGCATCAACTGGCCCGCCAATTGTTCAAGACCATCAATCGCTGGCTGGCCGAAGCAGGCC |
| S2 | 185 | no repetitive sequences | ACGGCGACGGAGAGCACGGCCACCACGCCAGACCAGATATAAGTGCCGGCCCGGCCAGCACCCACAATCTTGTG |
| S3 | 63 | no repetitive sequences | AGGCGGTTTTTTCGAGGGTTCAGTAAGTTGGGGAACTTCTGAACCGTGGTAACAGGGTGTACA |
| S4 | 116 | no repetitive sequences | CGATGTAGAGCTGTCAATGCCTGCGAGCCAATTTTAAGCCGTGTAACATTTGTCTAGCTGTCTATAGCCACTATAG |
| S5 | 79 | no repetitive sequences | ATGAACTGGCAAAGCAAGGTCATAAGGTGATGTGCTTCAACTTTGACGGAGCCGATCACGGAGATTACGCAAAG |
| S6 | 78 | no repetitive sequences | AAAAAAGTACTGTATATATAAACATGCAGATAGAAAATATGCCAACAGTACGCTAGACAATGTAAGACAGCTATA |
| S7 | 53 | Simple_repeat:(ATACAC) | ATACACATACACATACATATACATATACACATACACATACATATACACATACA |
| S8 | 80 | no repetitive sequences | AGGGAGGTTTGAAATGGACCGGGAACCTAAGAGTGCCAGATACTGTGCTGAGTGTAATAGGCTGCATCCTGCTGAC |
| S9 | 80 | no repetitive sequences | AGGGAGGTTTGAAATGGACCGGGAACCTAAGAGTGCCAGATACTGTGCTGAGTGTAATAGGCTGCATCCTGCTGAC |
| S10 | 51 | Simple_repeat:(AGAA)n | AGAGAAAGAAAGAGAGAAAGAGAGAGAGAGAGAAAGAAAGAGAGAAAGAAA |
| S11 | 90 | SINEs/ALUs | GCCTGTAATCCCAGCACTTTGGGAGGCCGAGGCAGGTGGATCACATGAGGTCAGGAGTTCGAGACCAGCCTGGCC |
| S12 | 65 | no repetitive sequences | CTCAATGTGAAGCACTACAAGATCCGCAAGCTGGACAGCGGCGGCTTCTACATCACCTCACGCAC |
| S13 | 306 | LTR | GATCTCCTATTACAAGAAAAGGAAGATTGGCCTATTTCTTTATTAGGGTTCTTGGGAGAGGTTCATTTCCATCTTCC |
| S14 | 983 | LTR | GCTCCTGATCCCTCGAAGTGGATTTGGGCTTTTAGTTTAGTCAGCAAGTCTCTTCCTAGCAGAGGATAGGGGCAGT |
| S15 | 969 | LTR | GAGAGGTACTGGTAGGCCTTCCTACAGGGGGCAAAAGCTGGGGCTAAAAAGCCCACTAATATAGCTAAGATCAG |
| S16 | 860 | LTR | GCAACCTCTACCCCCGTGTCCATAAAACAATACCCCATGTCACAAGAAGCCAGACTGGGGATCAAGCCCCACATA |
| S17 | 770 | LTR | ACGTGGTTCTTTTAGGGAGCAGGAGGTCCAAGCCCTCGCCGCCTCCATCTGAATTTTTGCTTTCGGTTTTTCGCCGA |
| S18 | 647 | no repetitive sequences | ACCTAGTGGCTGTTCCTTTCTATGCAGACCTTTGAAATGCTCAAAAACACCAGAAGGCTGGATCCTTGTCATCAAC |
| S19 | 327 | LTR | ATCGGAGAGAACCTGAGGTTTAGAAGGTCTGGGTTTTATAGAGGGGGTAAGGGCAGGGTAAAGGGCAGATCGGC |
| S20 | 308 | LTR | GATGACTTACTGCTGGCCGCTACTTCCGAACTAGACTGCCAACAAGGTACTCGGGCCCTACTACAAACCCTAGGG |
| S21 | 294 | LTR | CGCGTCCAATCAGTCCGTAGATGTCAAGAAGAGACGCTGGGTCACCTTCTGCTCTGCCGAGTGGCCAACTTTCGG |
| S22 | 249 | LTR | AAACAGAGTCCCCGTTTTGGTGAGAGGGTACAAGGGGGCTGCCATTTCTGCAAACCCAGGGATCCAGAGGCGAC |
| S23 | 176 | no repetitive sequences | GACTCGCGGTCCTGCAGCATTCCTTTGGTGGCCCCGAGGAAGGCAGGGAAGCTCTTCCTGGGGACCACTCCTCCT |
| S24 | 115 | no repetitive sequences | CATAAAGAAATAGGCAGTACCACTAGCTGTATGCCCAACACAGAAGAGAACGCAGGATCCAGCATGGATCTCTC |
| S25 | 50 | no repetitive sequences | TCTGACCCGATTACCTTGGGAAGTCCAAATCTGGGGAAGATTTCTTCTAA |
| S26 | 927 | LTR/ERVK | CATTACCCAAAAACAAAGATCTTCCAGTTCTTAAAATTGACTACTTGGATTCTACCTAAAATTACCAGACGTGAAC |

## 3.3 Inspecting the Connections Linked by These Common DNA Segments

Table 9 shows the co-occurrence 26 DNA segment mapping among 26 viruses and five mammals, where each of cells is the frequency of one segment appearing in one of five selected mammals. There are 12 out of 26 viruses whose known hosts, reported in Virus-Host DB [7,21], are the same with the mammals that shared the same DNA segment with corresponding viruses. This observation may coincide with the statements about "retrovirus" in Sect. 1.1. However, the hosts of the others whose frequencies are attached with an extra symbol ⋆ and marked in red color, are not among those five mammals.

For example, the no-repetitive sequence "S1" appears in both of virus *Escherichia virus P1* and *Rattus norvegicus* (three times). However, as shown in Fig. 10, the known host of *Escherichia virus P1* in Virus-Host DB is reported as *Escherichia coli* . It is interesting to further analyze this fact that both of *Escherichia virus P1* and *Rattus norvegicus* contain such a long and no-repetitive DNS segment "S1" whose length is 805 bp, as shown in Table 8. Taking another DNA segment "S7", one simple tandem repeat "(TATCTA)n" with 53 bp as shown in Table 9, for example, one can find that the host of *Human betaherpesvirus 6A* is *Homo sapiens* as shown in Table 10. However, *Rattus norvegicus* contain that segment "S7", but *Homo sapiens* doesn't. According to above observation, the ancestors of *Human betaherpesvirus 6A* may be concerned with *Rattus norvegicus* in the past, and the "S7" could be a clue for tracking genetic revolution between *Homo sapiens* and *Rattus norvegicus*.

**Table 9.** 26 selected viruses and their corresponding mammals

| Segment ID | Virus (ClassID) | Five Selected Mammals (ClassID) | | | | |
|---|---|---|---|---|---|---|
| | | Homo sapiens (C1) | Pan troglodytes (C2) | Mus musculus (C3) | Rattus norvegicus (C4) | Sus scrofa (C5) |
| S1 | Escherichia virus P1 (C2446) | | | | *3 | |
| S2 | Delftia phage RG-2014 (C1944) | | | | *1 | |
| S3 | Salmonella phage SE2 (C5720) | | | | *1 | |
| S4 | Shigella phage pSf-2 (C6014) | | | | *1 | |
| S5 | Escherichia phage ADB-2 (C2323) | | | | *1 | |
| S6 | Shigella virus Shfl1 (C6016) | | | | *1 | |
| S7 | Human betaherpesvirus 6A (C3327) | | | | *1 | |
| S8 | Bovine viral diarrhea virus 1 (C1120) | *1 | | | | |
| S9 | Pestivirus giraffe-1 H138 (C4963) | *1 | | | | |
| S10 | Glypta fumiferanae ichnovirus (C2682) | | | | | *1 |
| S11 | BeAn 58058 virus (C622) | *2 | | | | |
| S12 | Y73 sarcoma virus (C7468) | | | | | *1 |
| S13 | Mouse mammary tumor virus (C4199) | | | 1 | | |
| S14 | Mus musculus mobilized endogenous polytropic provirus (C4231) | | | 9 | | |
| S15 | Porcine endogenous retrovirus E (C5076) | | | | | 1 |
| S16 | Moloney murine leukemia virus (C4166) | | | 2 | | |
| S17 | Murine type C retrovirus (C4227) | | | 1 | | |
| S18 | Abelson murine leukemia virus (C9) | | | 1 | | |
| S19 | Murine osteosarcoma virus (C4223) | | | 2 | | |
| S20 | PreXMRV-1 (C5143) | | | 2 | | |
| S21 | Spleen focus-forming virus (C6201) | | | 1 | | |
| S22 | Friend murine leukemia virus (C2593) | | | 19 | | |
| S23 | Moloney murine sarcoma virus (C4167) | | | 1 | | |
| S24 | Woolly monkey sarcoma virus (C7340) | | | | *1 | |
| S25 | RD114 retrovirus (C5432) | | *1 | | | |
| S26 | Human endogenous retrovirus K113 (C334) | 1 | | | | |

**Table 10.** The host of virus *Escherichia virus P1* reported by Virus-Host DB [7,21].

### Escherichia virus P1

| Scientific Name | Escherichia virus P1 [TAX:10678] |
|---|---|
| Lineage | Viruses; Caudovirales; Myoviridae; Punavirus |
| Genome type | Non-segmented |
| RefSeq | NC_005856  Enterobacteria phage P1, complete genome. |
| DBLINKS | KEGG BRITE:  NC_005856 |
| | ViralZone:  family, genus |

### Known hosts (2)

| Scientific Name | Enterobacteriaceae [TAX:543] |
|---|---|
| Lineage | Bacteria; Proteobacteria; Gammaproteobacteria; Enterobacterales |
| Evidence | UniProt |

| Scientific Name | Escherichia coli [TAX:562] |
|---|---|
| Lineage | Bacteria; Proteobacteria; Gammaproteobacteria; Enterobacterales; Enterobacteriaceae; Escherichia |
| Evidence | RefSeq |
| Evidence | NCBI Virus |

Above observations about the relationship among these viruses, their known hosts and corresponding mammals provides an interesting topic of genetic revolution for biologists or virologists to track why such long consecutive DNA sequences occurred simultaneously within these viruses and mammals.

## 4    Conclusion

This paper shows a novel method for whole genome comparison by extracting and analyzing co-occurrence DNA sequences (maximal repeats) from the whole genomes of five mammals and 7,538 viruses where their whole genomes were available and downloaded from NCBI FTP site at 2019/3/1. From these co-occurrence DNA sequences that appear in both of these mammals and viruses, one can find some factors or hidden clues for further research by inspecting the frequency distribution of DNA sequences among these mammals and viruses.

There are 26 viruses and their common and consecutive DNA sequences selected for illustration in this paper. 13 out of above 26 DNA sequences whose lengths are 50 bp at least were reported as "no repetitive sequences" via "Virus-Host DB". This study also provided the taxonomy of these 23 selected viruses and, for example, found two viruses, *Bovine viral diarrhea virus 1* and *Pestivirus giraffe-1 H138*, that were noted with "IV: (+)ssRNA viruses" in Baltimore classification of viruses. Their known hosts of that two viruses were *Bos taurus* and *Giraffa camelopardalis*, respectively, but they shared the same common DNA sequence with *Homo sapiens*. Indeed, there are still many interesting phenomena existed if one can further analyze or inspect those class frequency distributions of those extracted common DNA sequences.

Via those significant common DNA sequences and their class frequency distribution computed by the scalable maximal repeat extraction approach [25, 27]. It would be desired and very attractive to have whole genomic DNA sequences comparison among all viruses and organisms. This work shows a novel approach to address the problems in comparative genomics via extracting DNA maximal repeats from whole genomic sequences of viruses and mammals. Using distinctive DNA as clues to track the relationship among viruses and organisms, it is interesting in the future to discover the unknown track or pathway between viruses and organisms via these distinctive DNA repeats.

**Acknowledgement.** The project is funded in part by the Ministry of Science and Technology (MOST) under Grant No. 107-2632-E-468-002. Thanks for Prof. Tsung-Chi Chen discussing plant viruses. Thanks for Mr. Ren-Der Huang for maintaining Hadoop cluster computing environment.

## References

1. FTP Site for Genomes in NCBI. https://ftp.ncbi.nih.gov/genomes
2. International Committee on Taxonomy of Viruses (ICTV). https://talk.ictvonline. org/

3. National Center for Biotechnology Information(NCBI). http://www.ncbi.nlm.nih. gov/
4. NCBI Web BLAST (Basic Local Alignment Search Tool. https://blast.ncbi.nlm. nih.gov/Blast.cgi
5. RepeatMasker. http://www.repeatmasker.org/
6. The UCSC Genome Browser. https://genome.ucsc.edu/index.html
7. Virus-Host DB. https://www.genome.jp/virushostdb/
8. Aiewsakun, P., Katzourakis, A.: Marine origin of retroviruses in the early palaeozoic era. Nat. Commun. **8**, 13954–13954 (2017). https://doi.org/10.1038/ ncomms13954
9. Baltimore, D.: Animal Virology, vol. 4. Elsevier Science (1976)
10. Bandín, I., Dopazo, C.P.: Host range, host specificity and hypothesized host shift events among viruses of lower vertebrates. Veterinary Res. **42**(1), 67–67 (2011). https://doi.org/10.1186/1297-9716-42-67
11. Chatzou, M., et al.: Multiple sequence alignment modeling: methods and applications. Briefings Bioinform. **17**(6), 1009–1023 (2015). https://doi.org/10.1093/bib/ bbv099
12. Cormen, T.H., Leiserson, C.E., Rivest, R.L., Stein, C.: Introduction to Algorithms, 3rd edn. The MIT Press (2009)
13. Garson, J.A., Usher, L., Al-Chalabi, A., Huggett, J., Day, E.F., McCormick, A.L.: Quantitative analysis of human endogenous retrovirus-k transcripts in postmortem premotor cortex fails to confirm elevated expression of herv-k rna in amyotrophic lateral sclerosis. Acta Neuropathol. Commun. **7**(1), 45 (2019). https://doi.org/10. 1186/s40478-019-0698-2
14. Gulcher, J.: Microsatellite markers for linkage and association studies. Cold Spring Harbor Protocols **2012**(4), pdb.top068510 (2012). https://doi.org/10.1101/pdb. top068510
15. Gusfield, D.: Algorithms on Strings, Trees, and Sequences : computer science and computational biology. Cambridge University Press (1997)
16. Wang, J.-D., Noto Susanto, C.O.: Traffic flow prediction with heterogenous data using a hybrid cnn-lstm model. Comput. Mater. Continua **76**(3), 3097–3112 (2023). https://doi.org/10.32604/cmc.2023.040914, http://www.techscience.com/ cmc/v76n3/54369
17. Wang, J.-D., Noto Susanto, C.O.: Traffic flow prediction with heterogeneous spatiotemporal data based on a hybrid deep learning model using attention-mechanism. Comput. Model. Eng. Sci. **140**(2), 1711–1728 (2024) https://doi.org/ 10.32604/cmes.2024.048955, http://www.techscience.com/CMES/v140n2/56559
18. Li, F., Ooi, B.C., Özsu, M.T., Wu, S.: Distributed data management using mapreduce. ACM Comput. Surv. **46**(3), 31:1–31:42 (2014).https://doi.org/10.1145/ 2503009
19. Madeira, F., et al.: The embl-ebi search and sequence analysis tools apis in 2019. Nucleic Acids Res. **47**(W1), W636–W641 (2019). https://doi.org/10.1093/nar/ gkz268
20. Meyer, T.J., Rosenkrantz, J.L., Carbone, L., Chavez, S.L.: Endogenous retroviruses: with us and against us. Front. Chem. **5**, 23 (2017). https://doi.org/10. 3389/fchem.2017.00023
21. Mihara, T., et al.: Linking virus genomes with host taxonomy. Viruses **8**, 66 (2016)https://doi.org/10.3390/v8030066
22. Mount, D.W.: Bioinformatics: Sequence and Genome Analysis, 2 edn. Cold Spring Harbor Laboratory Press (2004)

23. Pérez-Wohlfeil, E., Diaz-del Pino, S., Trelles, O.: Ultra-fast genome comparison for large-scale genomic experiments. Sci. Rep. **9**(1), 10274 (2019). https://doi.org/10. 1038/s41598-019-46773-w
24. Usdin, K.: The biological effects of simple tandem repeats: lessons from the repeat expansion diseases. Genome Res. **18**(7), 1011–1019 (2008). https://doi.org/10. 1101/gr.070409.107
25. Wang, C.T.: Method for extracting maximal repeat patterns and computing frequency distribution tables, U.S. Patent No. 10,409,844 (Sep 2019)
26. Wang, J.D.: A study of comparing the ambiguity of existing virus taxonomy structures using protein's region names in the vector space model. In: 2015 IEEE Conference on Computational Intelligence in Bioinformatics and Computational Biology (CIBCB 2015), pp. C–15004 (2015)
27. Wang, J.D.: Extracting significant pattern histories from timestamped texts using mapreduce. J. Supercomput., 1–25 (2016)
28. Wang, J.D.: A novel approach to mine for genetic markers via comparing class frequency distributions of maximal repeats extracted from tagged whole genomic sequences. In: Abdurakhmonov, I.Y. (ed.) Bioinformatics in the Era of Post Genomics and Big Data, chap. 5. IntechOpen, Rijeka (2018). https://doi.org/10. 5772/intechopen.75113
29. Wang, J.D.: Reducing the gap between phenotypes and genotypes via comparing tagged whole genomic sequences. In: The 12th International Conference on Advancements in Bioinformatics and Drug Discovery; J. Proteomics Bioinform. (2018)
30. Wang, J.D., Hwang, M.C.: A novel approach to extract significant patterns of travel time intervals of vehicles from freeway gantry timestamp sequences. Appli. Sci. **7**(9) (2017). https://doi.org/10.3390/app7090878
31. Wang, J.D., Pan, S.H., Ho, C.Y., Chuan Liao, S., Lien, Y.N., Nurmandi, A.: Online web query system for various frequency distributions of bus passengers in taichung city of taiwan. IET Smart Cities **2**(3), 135–145 (2020)
32. Wang, J.D., Wang, Y.C., Hu, R.M., Tsai, J.: Extracting the co-occurrences of dna maximal repeats in both human and viruses. In: The 17th annual IEEE International Conference on Bioinformatics and Bioengineering (BIBE 2017) (2017)
33. Wang, J.-D.: A novel approach to improve quality control by comparing the tagged sequences of product traceability. MATEC Web Conf. **201**, 05002 (2018)https:// doi.org/10.1051/matecconf/201820105002

# Strategies to Use Harvesters in Trustworthy Fake News Detection Systems

Krzysztof Cabaj[1(✉)] 🆔, Marcin Kowalczyk[1] 🆔, Marcin Gregorczyk[1] 🆔,
Michał Choraś[2] 🆔, Rafał Kozik[2] 🆔, and Wojciech Mazurczyk[1] 🆔

[1] Warsaw University of Technology, Warsaw, Poland
{krzysztof.cabaj,marcin.kowalczyk2,marcin.gregorczyk,
wojciech.mazurczyk}@pw.edu.pl
[2] Bydgoszcz University of Science and Technology, Bydgoszcz, Poland
{chorasm,rkozik}@pbs.edu.pl

**Abstract.** Today, detecting fake news is an important challenge facing societies and scientific communities using AI and IT solutions. Many previous works tackle the problem of visual content or text analysis. Still, those seldom tackle the real-life data collection and harvesting problem, mainly focusing on well-crafted and prepared old datasets. That is why, in this paper, we focus on evaluating the performance of the distributed harvester in the context of fake news detection. To this aim, we developed a test-bed that simulated a real-life scenario and conducted experiments to determine the efficacy of the harvester system. The results obtained show that the investigated problem can become a crucial issue, and thus, various aspects must be considered to address it.

**Keywords:** Disinformation detection · Web harvesting · Performance evaluation

## 1 Introduction

During the last decade, disinformation, a.k.a. fake news, has become a very important issue, which has recently been throttled by the development of Artificial Intelligence (AI) [6]. Internet users want to receive news as soon as possible, leading to situations in which fake news can appear even on trusted media. In addition, controversial fake news can be echoed and/or cited by other media platforms, making them very popular. For this reason, automatic detection of disinformation news is crucial [14]. However, detecting fake news is not a simple process that can be easily solved by AI techniques alone. To build a trustworthy detection system, some human interaction is needed. Such a solution could consist of a few steps, presented below:

– Collect the content of a set of sample web pages using the web harvester,

N.-T. Nguyen et al. (Eds.): ICCCI 2024, CCIS 2165, pp. 384–394, 2024.
https://doi.org/10.1007/978-3-031-70248-8_30

– Human-based annotation of gathered web pages, for example: the addition of information concerning; how many sources the article contains; could it mislead; if content concerns religion, politics, etc.,
– train AI-based algorithms using gathered text and additional annotations,
– Gather suspicious pages collected by the harvester and decide if the content contains fake news.

From the steps presented above, it is visible that for the system to be effective, at least two fake news detection harvester systems are needed. Since training AI models requires large amounts of data, the performance of the used harvester cannot be neglected.

In the existing literature, various approaches to web harvesting have been proposed [5,12] that focus on various objects of the web pages, e.g., digital images, text, videos, or a mix of those. However, only a few works include the disinformation scenario and evaluate the system developed in real-life environments.

Taking into account the above, in this paper, we present our results concerning the performance evaluation of the distributed harvester system used by our team in the last few projects [7,15]. In contrast to previously conducted research, this work is devoted solely to investigating performance issues. We believe that our research could be helpful to other researchers in this field.

The rest of the paper is structured as follows. Section 2 contains related work on web harvester systems and associated research. Next, Sect. 3 presents the testbed utilized and details of the experiments performed. Next, Sect. 4 contains an exhaustive discussion of the results achieved. Finally, Sect. 5 concludes our work.

## 2   State of the Art

Various types of web harvesters (also known as web spiders, scrapers, or crawlers) have been under development since the late 1990 s (for a historical context, see e.g., [2,3]). Such solutions typically fall into one of three categories: specialized browser extensions, dedicated standalone systems, and auxiliary libraries usable in dedicated solutions (see [5] for an in-depth exploration).

During the last decades, the research on harvesters has revolved around search engines, focusing on the analysis of different parts of HTML web pages, e.g., digital images, text, videos, or a mix of those [12]. For example, the authors of [13] detail various malicious activities observed in the wild since 2016. They also introduce a framework using statistical tests and expert-based knowledge (e.g., Portable Executable file header detection) to identify malicious images. Conversely, [4] presents research on detecting malicious images using machine learning methods. The authors experiment with state-of-the-art techniques such as Decision Trees, Random Forest, Gradient Boosting, and K-Nearest Neighbors.

Regarding textual information, several approaches have been proposed for different applications. In [18], a new technique is proposed to retrieve the content by employing string methods, and supplementary information is proposed without the need to construct a DOM Tree. This approach is demonstrated to be approx. 60 times faster than extraction using DOM-based methods. Moreover, including extra information improves the extraction time by 2.35 times compared to relying solely on the string methods. On the other hand, [17] focuses on utilizing web harvesting to retrieve information from web journals. This enables the collection of information that is spread across different sources, it is always up-to-date. The harvesting technique was successfully developed on the basis of the web bootstrap framework.

Furthermore, some notable research also focused on detecting harvester activities, especially instances of misbehavior, such as ignoring the *robots.txt* file. For example, in [16], the authors propose rules, including monitoring access to *robots.txt*, tracking downloaded PDF files, and assessing the ratio of downloaded HTML files to images. A J48 classifier is used to differentiate between malicious and benign activity. Next, in [1], similar tasks are addressed using statistical tests. Specific connection-related features are identified for detecting undesired or disruptive crawlers. The research employs features such as visit duration, IP address, and visit frequency for detection purposes.

Finally, web harvesters have recently started to be utilized for fake news or disinformation detection purposes. To this aim, various parts of the web pages are captured and analyzed to detect any abnormalities, signs of modifications, or misuse. In [7], we evaluated the harvester developed for the DISSIMILAR fake news platform [15] to determine whether digital images from popular websites do not contain unwanted information in the metadata or other parts of the content.

## 3    Test-Bed and Experimental Evaluation

As presented in the introduction and state-of-the-art sections, efficient harvesters are needed to fight fake news. In the previous research, we developed a harvester system and conducted various research concerning the detection of fake news. For example, in the SWAROG project [8], we analyzed web page text using advanced NLP-based and artificial intelligence methods (e.g. [9–11] ), and in the DISSIMILAR [15] project, we analyzed the content of media files. However, in this work, we focus on the performance aspects of our harvester.

In this section, we present, with details, the test-bed utilized during conducted experiments as well as experimental scenarios. Later, we briefly describe the architecture of the developed harvester system, which was evaluated during these experiments. At the end of the section, we provide information concerning conducted experiments.

### 3.1    Test-Bed Design and Experimental Setup

Our test-bed was hosted in the Institute of Computer Science, Warsaw University of Technology private cloud. The cloud was built with OpenStack software with the latest stable release: 2023.2[1] (at the time of writing the paper). The physical

---

[1] https://releases.openstack.org/.

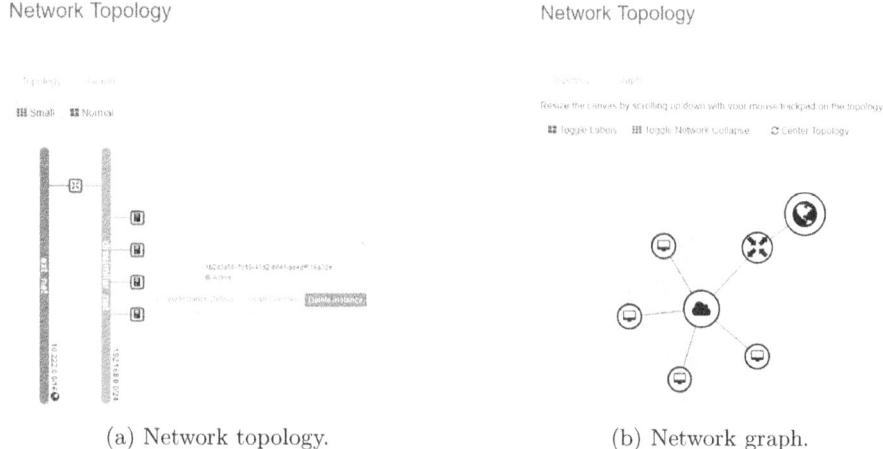

(a) Network topology.                    (b) Network graph.

**Fig. 1.** Test-bed topology as seen in OpenStack management interface.

layer consisted of four equal Intel S2600WFT servers (2 sockets x 20 cores x 2 threads; 1TB RAM; Ceph cluster for Software Defined Storage as backend storage). They were interconnected with $4 \times 10$GbE and bonded together using the Link Aggregation Control Protocol (LACP). Ubuntu 22.04 LTS was used for both physical hosts and OpenStack instances. It is worth noting that during our research, no other workload was running on our OpenStack cluster to exclude possible interference from other instances.

In the OpenStack environment, we create a dedicated project for our research. It contains one virtual LAN, one virtual router, and five virtual machines. The virtual machines contain:

- test web-server which contains copies of harvested web pages (see Sect. 3.3 for more details),
- one management node (see Sect. 3.2 for more details),
- three worker nodes with various number of vCPUs (see Sect. 3.2 and Sect. 4 for more details).

All experimental traffic, to reduce any congestion issues, is forwarded only in the LAN, and the virtual router is used only for remote access to the test-bed. The logical view of our test-bed, as is presented in the OpenStack management dashboard, is shown in Fig. 1a and 1b.

### 3.2  Developed Harvester

During previous studies, we developed a distributed harvester to allow for a massive collection of scalable web pages. The architecture of the harvester system is presented in Fig. 2. The system consists of one management node and multiple workers. The management node hosts databases (Postgres and Redis) used by

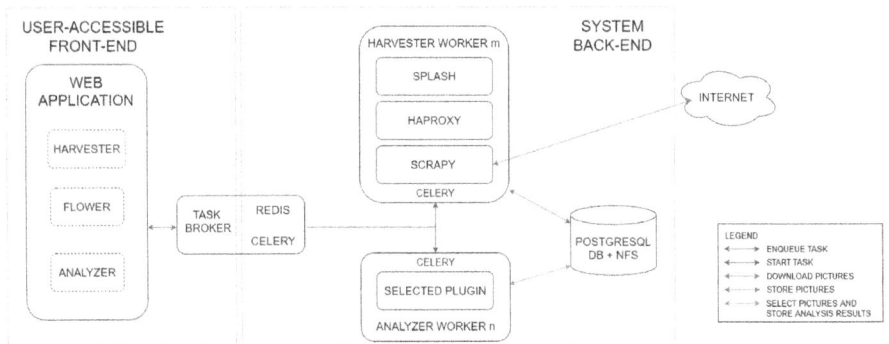

**Fig. 2.** A harvester architecture (source: [7]).

system elements for task registration and storing gained results. Multiple workers using the Celery task queue system perform web page downloading. This architecture allows easy scalability by adding additional virtual machines containing the next workers. Additionally, the management node provides a management interface that controls and monitors the workers running using Celery Flower software. More details on the harvester architecture can be found in [7].

In previous research, we mainly focused on harvesting media files. However, in this work, we focus on text presented on the web pages. To this aim, we integrate Scrapy Spiders with our harvester. Scrapy Spiders[2] play a pivotal role in the functionality of the Scrapy framework[3], a robust and open-source web crawling and scraping library designed for Python. They serve as the architects of data extraction, defining the rules and logic needed to navigate through websites and retrieve valuable information. Using XPath or CSS selectors, Scrapy Spiders precisely identify the location of data on webpages, ensuring accuracy in the scraping process. Their asynchronous capabilities enable them to handle multiple pages simultaneously, enhancing efficiency and speed. With a modular and extensible design, they can tailor their behavior to diverse websites' unique structures and content, making them a versatile and indispensable tool for web scraping projects.

### 3.3 Experiments Description

The main purpose of this research concerns the performance evaluation of the distributed harvester, which was briefly described in the previous section. However, we could not conduct such experiments using web pages hosted, for example, on social networks or news portals, for a few reasons. Firstly, most commercial web servers enable a counter-DoS (Denial of Service) system, and when the harvester connects to the web server with a high frequency, further connections are blocked.

---

[2] https://docs.scrapy.org/en/latest/topics/spiders.html.
[3] https://docs.scrapy.org/en/latest/intro/tutorial.html.

**Table 1.** Detailed information concerning harvested test data (January 2024).

| Topic | Website address | Count |
|---|---|---|
| News | *wiadomosci.wp.pl* | 481 |
| Business | *finanse.wp.pl* | 1321 |
| Weather | *pogoda.wp.pl* | 1454 |
| Tourism | *turystyka.wp.pl* | 142 |
| Sport | *sportowefakty.wp.pl* | 319 |
| Movie Stars | *gwiazdy.wp.pl* | 88 |
| Women | *kobieta.wp.pl* | 7 |
| All URLs | – | 7689 |
| Unique URLs | – | 3811 |

Second, in consecutive tests, the content of the news web pages could change rapidly, especially when some 'hot topic' appears. Thirdly, we could not predict network congestion, which could have a major impact on conducted tests. Due to these facts, we performed our test in a controlled, simulated test-bed environment.

In our test-bed, we would like to utilize representative data similar to what is used in real web portal data nowadays. To this aim, before our main experiments, we harvested real news pages from one of the main Polish media groups – WP (Wirtualna Polska in Polish, Virtual Poland in English, https://www.wp.pl/). The harvesting has been carried out through January 2024. The statistics of the collected data are presented in Table 1.

During the main part of our research, we host parts of original web pages (actually only HTML files) in our web server provided in the test-bed environment. In effect, we are able to conduct repeatable experiments using the full system performance.

## 4   Results

The main aim of the conducted research is to investigate the best strategy for configuration tuning of the harvester software, as well as virtual machines that host harvester workers. In the harvester software, we could tune, among others, such parameters as:

- the number of workers,
- enable/disable auto throttle feature, which detects anti-DoS systems in the harvested web servers and automatically reduces the frequency of concurrent connections ($AUTOTHROTTLE\_ENABLED$ = True/False option in Scrapy Spiders library),
- additional delay after each harvested page ($DOWNLOAD\_DELAY$ parameter in Scrapy Spiders library),

– number of concurrently working jobs in each worker (*concurrency* parameter of the Celery system).

The main parameter for the virtual machines that host workers is the number of virtual CPUs (vCPUs), and available RAM memory.

**Table 2.** Harvester runtime for 500 start pages, with various initial parameters of Scrapy Spiders

| Parameter | Time |
|---|---|
| No 'auto-throttle' | 23 [s] |
| With 'auto-throttle' | 1:31 [min:s] |
| With 'auto-throttle' (delay 1 s) | 35:37 [min:s] |
| With 'auto-throttle' (delay 2 s) | 1:10:45 [h:min:s] |

In the first step, we evaluate the most realistic and useful settings for the Scrapy Spiders. We evaluate enabling the auto throttle feature as well as some constant delay. For evaluation purposes, we chose the first 500 pages from the WP Business (*finanse.wp.pl*) website as the starting pages for our harvester. The obtained results are presented in the Table 2. For simplicity in this experiment, we use one worker with one vCPU, and all links are harvested in one task.

The very long times in the last two columns of Table 2 are directly associated with the set download delay. Even though we started harvesting 500 start pages, we actually downloaded 575 pages and tried to download 1173 other pages, which were not stored in our test web server, and ended with the HTTP 404 error. Such statically configured delays are a good idea for harvesting data from real web portals. Still, as presented in Table 2, they drastically reduce the possible level of performance of the harvester. After analysis of the first-step results, we conducted further experiments using no download delay based only on the auto-throttle.

The main research concerns the number of virtual CPUs used in the worker virtual machine and the number of concurrency tasks that could be executed in the particular worker. In the test-bed cloud used for hosting our harvester infrastructure, each worker virtual machine receives 1, 2, or 4 vCPU during creation. In our configuration, each vCPU is equivalent to one processor - with one core and one thread, in comparison to real hardware. A number of vCPUs define how many instruction streams can be executed concurrently at one time on the given virtual machine. Of course, modern operating systems using preemptive techniques could stop the execution of one process/software thread and give CPU to another. In situations when a concurrent process/software thread waits for some I/O operations, this gives some speed up. However, when we run too many process/software threads, this could lead to performance degradation.

The concurrency parameter in the Celery system defines how many concurrent tasks could be executed in the worker. We investigate 1, 2, or 4 (only for machines with 4 vCPUs) concurrent tasks executed on each worker.

During the experiments, we added two tasks to our harvester. The first contains 1,321 start pages from the Business topic (*finanse.wp.pl*, and the second contains 481 start pages from the News topic (*wiadomosci.wp.pl*). To gather reliable data, we conducted ten harvester experimental executions for each parameter's set (vCPUs and concurrency). Table 3 presents the results obtained, the average harvesting time for each task, with standard deviation.

**Table 3.** Detailed results showing harvesting times for various configurations of vCPUs number and concurrency of Celery tasks.

|               | 1 vCPU            | 2 vCPUs           | 4 vCPUs           |
| ------------- | ----------------- | ----------------- | ----------------- |
| 1 concurrency | wp_fin 147.26 s   | wp_fin 171.59 s   | wp_fin 163.52 s   |
|               | std_dev 18.44 s   | std_dev 24.81 s   | std_dev 18.41 s   |
|               | wp_wiad 83.12 s   | wp_wiad 94.50 s   | wp_wiad 96.56 s   |
|               | std_dev 14.23 s   | std_dev 14.65 s   | std_dev 15.82 s   |
|               | SUM **230.38 s**  | SUM **266.09 s**  | SUM **260.09 s**  |
| 2 concurrency | wp_fin 179.30 s   | wp_fin 174.12 s   | wp_fin 162.12 s   |
|               | std_dev 22.29 s   | std_dev 25.92 s   | std_dev 19.55 s   |
|               | wp_wiad 106.78 s  | wp_wiad 104.32 s  | wp_wiad 91.48 s   |
|               | std_dev 11.91 s   | std_dev 19.55 s   | std_dev 17.59 s   |
|               | SUM **286.09 s**  | SUM **278.45 s**  | SUM **253.61 s**  |
| 4 concurrency |                   |                   | wp_fin 153.49 s   |
|               |                   |                   | std_dev 16.13 s   |
|               |                   |                   | wp_wiad 84.34 s   |
|               |                   |                   | std_dev 9.17 s    |
|               |                   |                   | SUM **237.84 s**  |

The first conclusion from our research is that gathered data have a relatively large standard deviation. This is due to the existence of significant differences in harvesting time observed during experiments. For example, in an experiment using one vCPU and one concurrency, despite the average value of 147.26, we record a minimal time of 126.10 and a maximal of 187.13. Interestingly, during our experiments, we were the only user of the cloud, and no other user activity could temporarily reduce the performance of our machines, which was the first explanation of this behavior. However, as shown in the following results, the cause is probably a complication of the Harvester system, which utilizes various libraries that perform some additional tasks unpredictably.

The results seem to be as expected when we analyze the data in each column (for the same number of vCPUs). When we try to run more concurrency tasks and a given worker does not have sufficient performance (number of vCPUs), the harvesting time is longer. We could easily observe this phenomenon for one and two vCPUs. The situation for the four vCPUs is different. Because this machine has the highest number of vCPUs, more actions can be performed in parallel when we enable additional concurrency. In our opinion, this is the explanation of why, in this machine, adding an enlarging concurrency parameter reduces the calculation time.

## 5   Conclusions

Currently, the identification of disinformation stands as a significant hurdle facing both societies and scientific communities. Numerous previous studies have addressed the challenge of analyzing visual content or text, but few have delved into the practicalities of data collection and harvesting in real-life situations. That is why, in this research, our focus was directed towards assessing the effectiveness of a distributed harvester within the realm of fake news detection. For this purpose, we constructed a test environment that emulates real-life scenarios and conducted experiments to gauge the efficiency of the harvesting process. The obtained results indicate that the scrutinized problem has the potential to escalate, emphasizing the imperative need to consider various facets to mitigate it effectively.

As described in this paper, our research shows that the mismatch in software and hardware configuration could degrade the performance of the complicated distributed system in our research harvester system. This is especially important as more and more systems are moved to public or private clouds. As a future work, we would like to investigate performance issues when more machines in the cloud work cooperatively.

**Acknowledgments.** This research was funded by the National Center for Research and Development within the INFOSTRATEG program through the grant INFOSTRATEG-I/0019/2021-00, as well as within EIG CONCERT-Japan call to the project Detection of fake newS on SocIal MedIa pLAtfoRms "DISSIMILAR" through the grant EIG CONCERT-JAPAN/05/2021.

## References

1. Aghamohammadi, A., Eydgahi, A.: A novel defense mechanism against web crawlers intrusion. In: 2013 International Conference on Electronics, Computer and Computation (ICECCO), pp. 269–272 (2013). https://doi.org/10.1109/ICECCO. 2013.6718280
2. Bogonikolos, N., Fragoudis, D., Likothanassis, S.: "archimides": an intelligent agent for adaptive-personalized navigation within a web server. In: Proceedings of the 32nd Annual Hawaii International Conference on Systems Sciences. 1999. HICSS-32. Abstracts and CD-ROM of Full Papers. vol. Track5, pp. 9 pp.– (1999). https://doi.org/10.1109/HICSS.1999.772923
3. Bradford, C., Marshall, I.: A bandwidth friendly search engine. In: Proceedings IEEE International Conference on Multimedia Computing and Systems, vol. 2, pp. 720–724 (1999)https://doi.org/10.1109/MMCS.1999.778573

4. Cohen, A., Nissim, N., Elovici, Y.: Maljpeg: Machine learning based solution for the detection of malicious jpeg images. IEEE Access **8**, 19997–20011 (2020). https://doi.org/10.1109/ACCESS.2020.2969022

5. Diouf, R., Sarr, E.N., Sall, O., Birregah, B., Bousso, M., Mbaye, S.N.: Web scraping: State-of-the-art and areas of application. In: 2019 IEEE International Conference on Big Data (Big Data), pp. 6040–6042 (2019). https://doi.org/10.1109/BigData47090.2019.9005594

6. Kim, B., Xiong, A., Lee, D., Han, K.: A systematic review on fake news research through the lens of news creation and consumption: research efforts, challenges, and future directions. PLOS ONE **16**(12), 1–28 (2021). https://doi.org/10.1371/journal.pone.0260080

7. Kowalczyk, M., Malanowska, A., Mazurczyk, W., Cabaj, K.: Web page harvesting for automatized large-scale digital images anomaly detection. In: Proceedings of the 17th International Conference on Availability, Reliability and Security, ARES 2022. Association for Computing Machinery, New York (2022). https://doi.org/10.1145/3538969.3544471

8. Kozik, R., Komorniczak, J., Ksieniewicz, P., Pawlicka, A., Pawlicki, M., Choraś, M.: Swarog project approach to fake news detection problem. In: García Bringas, P., et al. (eds.) International Joint Conference 16th International Conference on Computational Intelligence in Security for Information Systems (CISIS 2023) 14th International Conference on EUropean Transnational Education (ICEUTE 2023), pp. 79–88. Springer Nature Switzerland, Cham (2023). https://doi.org/10.1007/978-3-031-42519-6_8

9. Ksieniewicz, P., Choraś, M., Kozik, R., Woźniak, M.: Machine learning methods for fake news classification. In: Yin, H., Camacho, D., Tino, P., Tallón-Ballesteros, A.J., Menezes, R., Allmendinger, R. (eds.) IDEAL 2019. LNCS, vol. 11872, pp. 332–339. Springer, Cham (2019). https://doi.org/10.1007/978-3-030-33617-2_34

10. Ksieniewicz, P., Zyblewski, P., Choraś, M., Kozik, R., Giełczyk, A., Woźniak, M.: Fake news detection from data streams. In: 2020 International Joint Conference on Neural Networks (IJCNN), pp. 1–8 (2020).https://doi.org/10.1109/IJCNN48605.2020.9207498

11. Kula, S., Choraś, M., Kozik, R., Ksieniewicz, P., Woźniak, M.: Sentiment analysis for fake news detection by means of neural networks. In: Krzhizhanovskaya, V.V., et al. (eds.) ICCS 2020. LNCS, vol. 12140, pp. 653–666. Springer, Cham (2020). https://doi.org/10.1007/978-3-030-50423-6_49

12. Kumar, M., Bhatia, R., Rattan, D.: A survey of web crawlers for information retrieval. WIREs Data Min. Knowl. Discovery **7**(6), e1218 (2017). https://doi.org/10.1002/widm.1218

13. Kunwar, R.S., Sharma, P.: Framework to detect malicious codes embedded with jpeg images over social networking sites. In: 2017 International Conference on Innovations in Information, Embedded and Communication Systems (ICIIECS), pp. 1–4 (2017)https://doi.org/10.1109/ICIIECS.2017.8276144

14. Mazurczyk, W., Lee, D., Vlachos, A.: Disinformation 2.0 in the age of ai: a cybersecurity perspective. Commun. ACM **67**(3), 36-39 (2024).https://doi.org/10.1145/3624721

15. Megías, D., Kuribayashi, M., Rosales, A., Cabaj, K., Mazurczyk, W.: Architecture of a fake news detection system combining digital watermarking, signal processing, and machine learning. J. Wireless Mobile Netw. Ubiquitous Comput. Dependable Appli. **13**(1), 33–55 (2022). https://doi.org/10.22667/JOWUA.2022.03.31.033

16. Narkar, N.G., Shekokar, N.M.: A rule based intrusion detection system to identify vindictive web spider. In: 2016 International Conference on Computing, Analytics and Security Trends (CAST). pp. 271–275 (2016).https://doi.org/10.1109/CAST. 2016.7914979
17. Rahayuda, I.G.S., Santiari, N.P.L.: Web harvesting for data retrieval on scientific journal sites. Jurnal Informatika Universitas Pamulang **6**(1), 202–209 (2021). https://doi.org/10.32493/informatika.v6i1.10077
18. Uzun, E.: A novel web scraping approach using the additional information obtained from web pages. IEEE Access **8**, 61726–61740 (2020). https://doi.org/10.1109/ ACCESS.2020.2984503

# Time Series Analysis of Sentiment Polarity Trends: A Case Study

Bernadetta Maleszka[✉][iD]

Faculty of Information and Communication Technology,
Department of Applied Informatics, Wroclaw University of Science and Technology,
Wybrzeze Wyspianskiego 27, 50-370 Wroclaw, Poland
Bernadetta.Maleszka@pwr.edu.pl

**Abstract.** The availability of a large number of opinion about the particular product or service may cause user confusion about final opinion. More and more methods for sentiment analysis are developed, that can judge the polarity of each opinion. The problem arises during the process of determining the final polarity and to predict future opinion. In the paper we analyze a set of opinions written by the group of users (e.g. familiar users in a social network) using time series analysis methods. The main aim is to analyze the trend of polarity in a group of people and to predict sentiment towards a particular topic in this group. We analyzed a dataset of real opinions using time series decomposition and prediction methods for determining the trend of sentiment score. The performed experimental evaluations have shown the efficiency of the proposed method. The results can be useful for future development of a complex model for opinion forecasting.

**Keywords:** Opinion Evolution · Sentiment Dynamics · Seasonal Decomposition · Holt-Winters Smoothing Exponential Model

## 1 Introduction

Nowadays, when a user wants to buy a product or use a service, he or she has an opportunity to check the opinion of other users about the interesting object. It is easy to find many user's reviews that describe product's or service's advantages and disadvantages. Usually, a problem arises due to information overload – there can be too many reviews to read or a lot of contradicting opinions. The user does not want to read all reviews and he or she bases on the opinion of his or her friends or on the best rated opinions. It can be an effective way to get the opinion but it may not be an objective one.

The task of opinion formation and mining is widely discussed in a literature and recent scientific papers. Many authors concentrate on the influence on the final user opinion by the opinion of other users e.g. in social networks [1,9]. There appear a lot of methods for analyzing sentiment polarity for any topic. Developed models are effective in determining the consensus opinion of the group of users.

© The Author(s), under exclusive license to Springer Nature Switzerland AG 2024
N.-T. Nguyen et al. (Eds.): ICCCI 2024, CCIS 2165, pp. 395–406, 2024.
https://doi.org/10.1007/978-3-031-70248-8_31

In some research areas there is still the problem of determining opinion of a single user based on the opinion of his or her familiar users, due to his or her level of knowledge or background [4]. There is also an important aspect of opinion dynamic of a single user and of a group of users.

The main idea of our current research is to analyze an evolution of users' opinions towards a selected topic. The contribution can be divided into tasks that cover the following problems:

- determining the clicks of users in social network and determining the features that can unambiguously characterize them;
- determining the collective sentiment of the group of users and its dynamic;
- determining the conditions that should be satisfied to rearrange the subgroups (recalculate the clicks of the users network based on current users' activities);
- predicting the future sentiment of a group of users taking into account e.g. technical parameters, network properties or outside factors that can influence the user opinions.

Accurate forecast of the future opinions can bring a profitable benefits for products' manufacturers or services' delivers and is a very important and current research problem.

In this paper we present a case study on analyzing the trend of sentiment polarity of a selected topic based on group opinion dynamics. The idea is to analyze a polarity trend for the selected topic. We use methods based on time series analysis to check the seasonal changes in sentiment and predict sentiment of users opinions.

In Sect. 2 we present a review of related works. Section 3 contains some definitions and problem formulation. There are also methods for analyzing opinion dynamics using time series approach and forecasting the trends. Information about methodology of experiments, its performance and results are presented in Sect. 4. Some final remarks and idea for future works are in Sect. 5.

## 2    Related Works

The aim of the paper is to analyze opinions dynamic of a single user and a group of users (e.g. in a social network). In this section we present some basic ideas of social changes and models of opinion changes that can be found in the subject literature.

It is not an easy task to discover the reasons why the person changes his or her mind. The authors of [18] divide mechanisms of social change into two general categories: the first is a one-directional change (accumulation, selection, and differentiation) and the second one is a curvilinear and cyclic change (saturation and exhaustion). The social change can be short-term transformations and long-term developments in society and it can be affected by conflicts, competitions, and cooperation.

Similar kinds of changes can be found in the area of social opinion dynamics. According to the theory of crowd, there exists an effect of crowd that can lead to conformity, deindividuation, and social loafing in a large group of people [22]. The similar situation is in a process of opinion formation – an individual bases his or her opinion on the opinions of other individuals (sometimes without checking what is the true). Such a strategy can not provide the desirable results as was shown e.g. by Frey and Van de Rijt [3].

### 2.1 Single User Opinion Evolution

An opinion of the individuals can change due to the fact of being a part of a society [10]. It can be influenced by others opinions, knowledge, behaviours or emotions. User can accept only a part of the opinions of others users (that confirm his or her opinion) [5]. It is hard to develop a model that takes into account all possible parameters that can have an impact on the final user opinion.

In real life, most people have their own opinion and it is hard to convince them to change it. A single user opinion can be influenced by many factors that can be conscious or unconscious. There are many attempts to model the opinion changes that take into accounts different aspects of user life. The control variables that are analyzed by Sobkowicz [14] are user knowledge and emotions. It is also popular approach that a user makes his or her opinion dependent on current opinion and the opinions of its neighbors and external influences. Liu and Fang [8] have proposed a non-Bayesian social learning model and they check the susceptibility to reach a consensus of a group of users.

To determine a final user opinion towards the particular topic, one can find in a literature a lot of approaches. The user can integrate his or her opinion based on the opinions collected from other users using e.g. average value, non-Bayesian model [8], probabilistic model [2], etc.

### 2.2 Opinion Dynamic in a Group of Users

There are two kinds of models of opinion dynamics that are the most popular [16]: models based on statistical physics and complex network (e.g. Ising model and the voter model) and models based on multi-agent systems [12] (e.g. the French-DeGroot model, the Friedkin-Johnsen model and the Hegselmann-Krause model). The first category focuses on the distribution of the opinion in a group of users (whole social network). The second category analyzes agent or user behaviours and its influence on the opinion of the whole group.

Hassani et al. [4] present a survey of existing opinion dynamics models. The aim of them is to model an opinion formation process that usually assume that all users updated theirs opinions to the average of all opinions of the neighbors. Su et al. [15] highlighted that many opinion evolution models are time consuming and to lower the complexity they proposed a nature-based heuristic model. Xie et al. [17] have used double-hazard scenarios to model the uncertainty and to model more factors that influence the results.

There are many papers that focus on public opinion fusion which could be interpreted in some contexts of opinion dynamic research.

Ianni [6] has developed a model of information fusion in a social network with different types of relationships between the users: positive and negative social relations that changes over time. The paper focuses on analyzing a possibility to obtain equilibrium state based on opinion configuration.

In the paper of Shen et al. [13], one can find an approach of consensus opinion determining that is based on trust aspects in the situation of incomplete information. Obtained results have shown that proposed methods are effective but there is still area to improve e.g. taking into account psychological aspects as e.g. user behaviour.

The role of a community structure in evolution process of opinion formation is analyzed by Peng et al. in the paper [11]. The authors claim that not only opinions can change over time but also the structure of relationships between users. There is a need to detect community from time to time and to analyze the final consensus opinion in new communities.

The above mentioned approaches focus on determining the sentiment polarity of a group of users towards a particular topic. Many effective algorithms are presented in this area.

The aim of this paper is to analyze the trend of users opinions. The task can be similar to opinion formation problem when we try to determine the consensus of the set of opinions for a small interval of time. The novelty of our approach is to analyze the dynamic of the opinions: we focus on the changes (evolution) of the opinion polarity over time.

## 3   A Method for Time Series Analysis of Sentiment Polarity

### 3.1   Time Series Analysis Methods

A prediction model of time series is formally defined as:

$$y'_t = f(t, y_{t-1}, \ldots, y_{t-p}, t, y'_{t-1}, \ldots, y'_{t-p}, \zeta_t) \tag{1}$$

where: $y_t$ is a value of variable $y$ in time $t$, $y'_t$ is a predicted value of $y$ in time $t$ and $\zeta$ is a random component of a time series.

The prediction of a variable $Y$ at time $t$ is calculated based on previous values of $y$ or previous predicted values of $y$.

The most popular models of time series analysis are additive and multiplicative models. They decompose the time series into the following parts:

- level $l(t)$ – is the average value of the time series;
- trend $tr(t)$ – is the value specifying the increase or decrease;
- seasonality $s(t)$ – is the value of cyclic repetition in series;
- noise $\zeta(t)$ – is the random value that influence the time series.

An additive model is defined as:

$$y'(t+1) = l(t) + tr(t) + s(t) + \zeta(t) \tag{2}$$

An multiplicative model is defined as:

$$y'(t+1) = l(t) \cdot tr(t) \cdot s(t) \cdot \zeta(t) \tag{3}$$

To analyze a trend part of time series, one can use many methods, e.g. linear regression or more sophisticated smoothing methods like moving average [7], Holt-Winters model [21], Prophet model [19], etc. All the models should be judged using standard metrics as e.g.: mean absolute error, root mean square error, mean absolute percentage error.

In the paper we analyzed the trend of group opinion using seasonal decomposition method and predicted the value of polarity average score using Holt-Winters Exponential Smoothing methods. This model takes into account all important aspects of time series: the value of polarity, changes over time, and it checks whether there are any cyclic changes that can influence the result.

The model of Holt-Winters Exponential Smoothing (HWES) is a method for calculating a forecast and capturing the seasonality. Similar to the decomposition approach, the time series is treated as a combination (additive or multiplicative model) of level, trend and seasonal components. In the additive model, the forecast is calculated as [21]:

$$y'(t+k) = l(t) + k \cdot tr(t) + s(t+k-m(h+1)) \tag{4}$$

The components are defined as:

$$l(t) = \alpha \cdot (y(t) - s(t-m)) + (1-\alpha) \cdot (l(t-1+tr(t-1))) \tag{5}$$

$$tr(t) = \beta \cdot (l(t) - l(t-1)) + (1-\beta) \cdot tr(t-1) \tag{6}$$

$$s(t) = \gamma \cdot (y(t) - l(t-1) - tr(t-1)) + (1-\gamma) \cdot (t-m) \tag{7}$$

where: $m$ is a frequency of the seasonality, $h = \lfloor (k-1)/m \rfloor$ specifies how wide is the window of past values that influence the forecast, and $\alpha, \beta$ and $\gamma$ are smoothing parameters.

## 3.2   Opinion Evolution Model of a Group of Users

The idea of analyzing group trend of polarity is presented as a list of steps in Algorithm 1.

To predict a sentiment of a group of users, it is necessary to determine a sentiment score for each single user opinion, and to calculate the average value of the score per day.

We propose to analyze the evolution of a group sentiment polarity and use forecasting methods using a time series approach for prediction. First, we analyze the cardinality of tweets sets with each polarity separately. We would like to know how the number of positive, neutral and negative tweets change over time. Similarly, we analyze the average score of sentiment polarity for the same tweets. Using the HWES method allows us to analyze different aspects of the time series, like trend, seasonal and residual components in both cases.

---

**Algorithm 1:** A method for forecasting the value of sentiment level for a group of users.

---

**Input:** $O$ – a set of users opinion towards a selected topic (with timestamps);
**Output:** predicted value of group opinion in time $t + k$

1. **foreach** *opinion* $o \in O$ **do**
   (a) Calculate the sentiment score $s_o(t)$ in timestamp $t$ ;
   (b) Determine the polarity of an opinion $o(t)$;

2. **foreach** *day* $\in T$ **do**
   Calculate the average value of the sentiment score $s_o(t)$ (average per day);

3. Decompose the time series data of a number of positive, neutral and negative sentiments into trend, seasonal and residual components;

4. Decompose the time series data of average score of sentiment into trend, seasonal and residual components;

5. Predict a value of time series in timestamp $t + k$ using prediction models.

---

## 4    Experimental Evaluation

In this section we present the data and methodology of the experiments. Next, we analyze the results of performed evaluations.

The data was taken from public available website [20]. It contains information about tweets of Apple Company. In our research, we analyze tweets from the year 2018. There are about 1,2 million tweets written by almost 30 thousand distinct users.

The aim of the research is to analyze a trend of sentiment polarity of the selected topic (Apple Company) in selected time period and forecast the value of a sentiment score for the future. To realize the defined contribution, we take into account the whole available set of tweets, calculate the sentiment score (and polarity) of each tweet using Vader Lexicon and SentimentIntensityAnalyzer package in Python. The range of sentiment score belongs to $[-1, 1]$ interval. The threshold for the polarity was proposed in [20]. A tweet has a positive sentiment when the score is greater than 0.33 and negative when the score is less than $-0.66$. Between these two thresholds the sentiment is neutral.

It is possible to calculate the cardinality of sets of positive, neutral and negative opinions. The number for tweets with neutral polarity is the biggest part of the dataset. There is more tweets with positive sentiment than negative ones. This is the reason for the fact that the average score over the year is greater then 0.

### 4.1    Sentiment Score Dynamics for a Single User

We have started the analysis with checking the changes of opinion polarity on a single user level. In this case, we can notice two kinds of users: the first group have a rather stable opinion – it can change slightly with time but at the global level (over the time of a year) it has the same average value; and the second group of users that changed their opinion diametrically (from negative to positive or vice versa). Due to this fact, it is not justified to develop any trend analysis methods on the single user level. The exemplary plots of both cases are presented in Fig. 1.

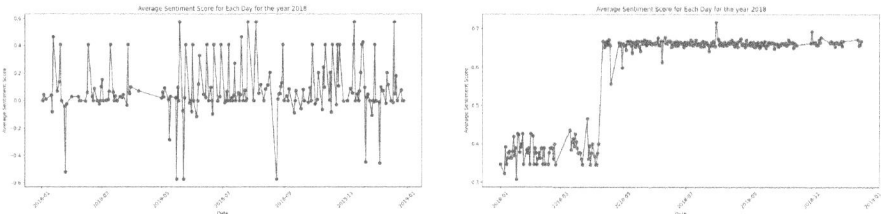

**Fig. 1.** Plots with examples of polarity series for a user with stable opinion (left part) and for a user that has changed the sentiment of opinion from neutral to positive.

In the case of the real user, it is rather not possible to notice the cyclic changes of sentiment polarity. When a user has his or her own opinion, it is really hard to convince him or her to stop thinking positively and start thinking negatively (or vice versa) about some product or service. It would be hard to find a situation when a user changes the polarity many times.

### 4.2    Seasonal Decomposition (Time Series of Tweets Cardinality)

The analysis of a whole set of tweets can provide some important and crucial information. The first aspect is a cardinality of positive, neutral and negative tweets.

Figure 2 presents the results for seasonal decomposition of the cardinality of all three values of polarity. We have tuned the parameters of the model—the most important factor was $frequency$ (the period of seasonality). We present the result for the tuned parameters. The best results were obtained for a weekly analysis ($frequency = 7$). The trend component of the time series shows that the overall number of tweets is rather stable over the year. We can observe the seasonality – there is a correlation between number of tweets and day of week. Analysis of the trend component of tweets cardinality time series should also focus on some external events during a year. We can observe a greater number of tweets with all values of polarity in 5 different period (Fig. 2). It can be related to some external factors e.g. financial results of the company, introducing a new technology or device into market, etc. The analysis of the correlation with external events is out of range of this paper.

The residual component of the time series shows the random fluctuations that can be independent from any outside factors.

Both kinds of decomposition models: the additive and multiplicative models obtain similar results (different values of variables but the same shape in the diagrams), so in the next analysis we present the results only for one model.

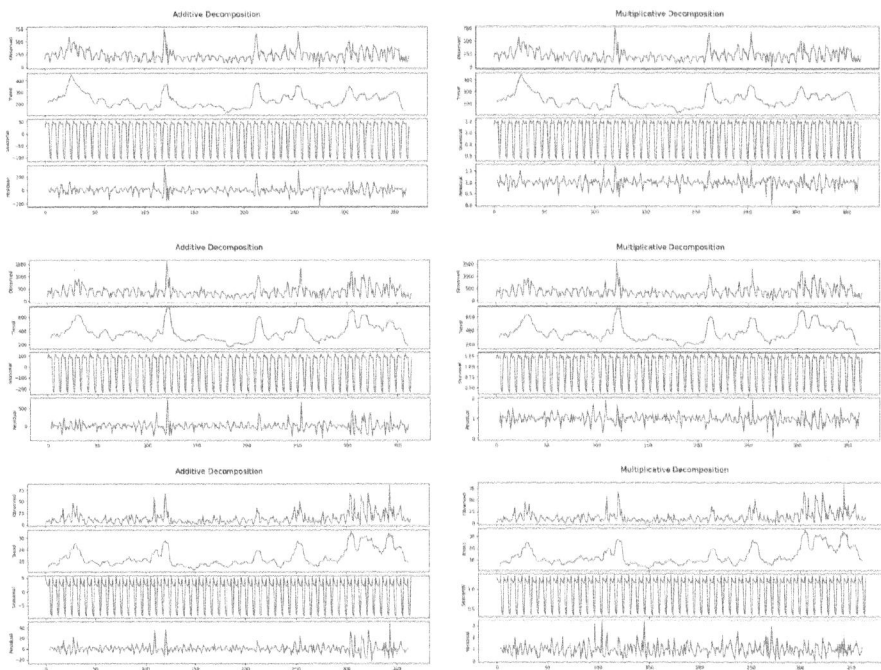

**Fig. 2.** Plots for seasonal decomposition of a number of tweets for different sentiment polarity: positive on the top, neutral in the middle and negative on the bottom. The left side diagrams are for additive decomposition model and the right side are for multiplicative model.

## 4.3   Seasonal Decomposition (Time Series of Average Sentiment Score of Tweets)

The crucial analysis of time series is determined for average sentiment score value. As it was shown in previous subsection, there are different numbers of tweets in different days, so the analysis focuses on the average score value.

Due to the seasonality of number of tweets, it was determined to analyze the average score of sentiment with the same value of parameter $frequency = 7$. The obtained results are presented in Fig. 3. The diagrams for different values of parameters were omitted due to lack of space.

The time series of average score of tweets sentiment is decomposed into three components: trend, seasonality and residual one.

Analyzing the trend component we can observe that the value starts with the level of 0.3 and systematically decreases during first 100 d of the year (over 3 months) and reaches the value of 0.15 (the range of the score is $[-1, 1]$). It can be interpreted as neutral polarity. Then we can notice a quick jump to the bigger value (about 0.3). Until the end of the year we can see also a decreasing trend but it is slower than in the first months in this year. At the end of the year we have a slight increase, which can be correlated with Christmas time.

Also, it is worth to notice that the increase of average score value are associated with a larger number of tweets in the same time periods in April. It shows users' interests in this company but decreasing trend of average score confirms that users were rather disappointed by the brand. The conclusions for the analysis in the second part of the year are different – there is a bigger number of tweets but the trend of average score slightly decreases. This can indicate that the sentiment of users' opinions are better.

The seasonal component plays in this case smaller role than in analyzing series of a number of tweets. The width of values interval for different frequency is small and it does not exceed the value of 0.07. It shows that the opinion of the analyzed company is not strongly correlated with the seasonality.

The last component in this model (a residual one) is a random value in range of ±0.1.

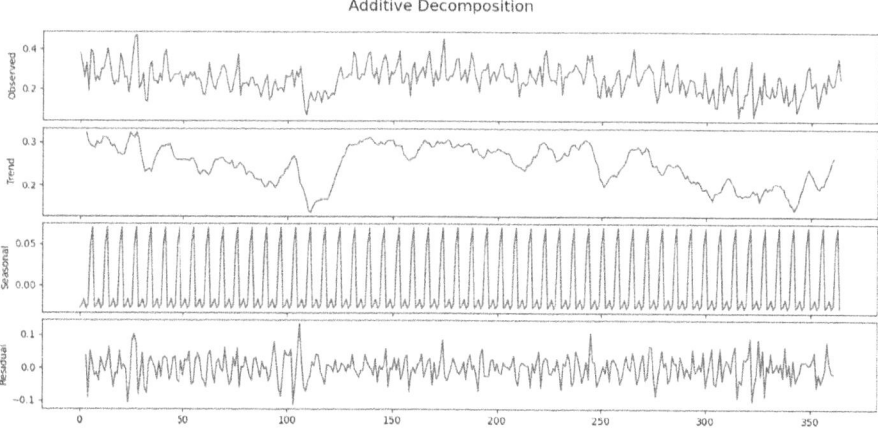

**Fig. 3.** The result of seasonal decomposition of time series of average sentiment score value using additive model. The time series is a sum of trend, seasonal and residual components.

### 4.4    Results for Forecast Model

To predict sentiment score based on time series we used the Holt-Winters model which belongs to a family of exponential smoothing methods. Results presented in Fig. 4 are divided into 3 cases: simple exponential smoothing which determine a trend of the series; double exponential smoothing that takes into account trend and seasonality (it was performed for both additive and multiplicative model) and triple exponential smoothing Holt-Winters model that allows us to fix the seasonality to the additive or multiplicative model.

The last figure in the Fig. 4 is a result of forecasting method based on triple exponential smoothing Holt-Winters model which was described in Sect. 3.1 and

Eqs. 4–7. We can observe that the more parameters are optimized in the model, the better the results fit to the time series.

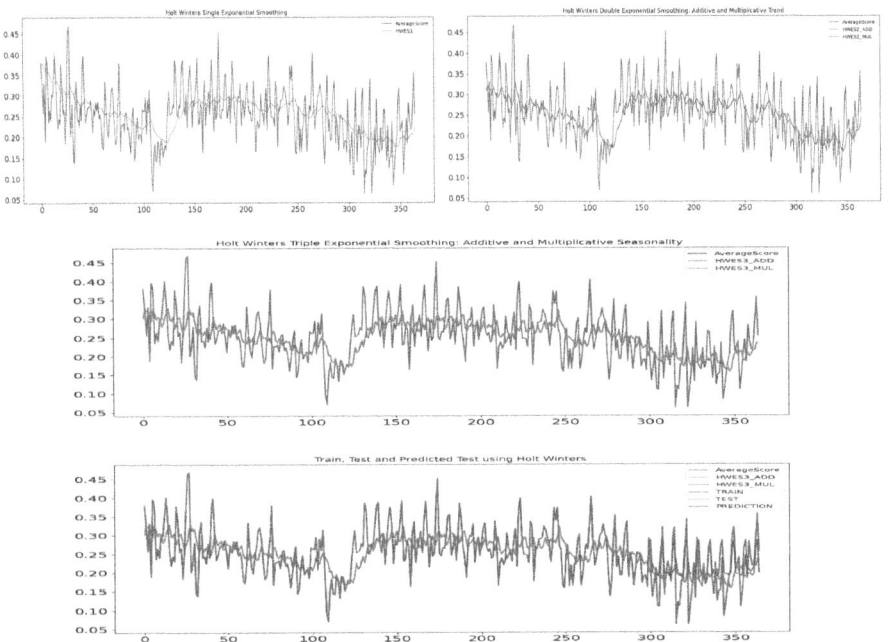

**Fig. 4.** The result of single, double and triple exponential smoothing using Holt-Winters method. The first model determines a trend of the series; the second one takes into account the trend and seasonality and the third one optimizes all three smoothing parameters from Holt-Winters model (Eqs. 4 – 7).

The most important result is presented in the last subfigure of Fig. 4 which presents a forecast of a time series. The data was divided into two parts: the training set (from the beginning of the year until 300 day – almost the end of the October) and the test set (the last two months). Visualized results allow us to confirm that the trend and seasonality were predicted successfully. To judge the results in a mathematical way, we analyzed the results with the classic effective measures.

The results of the following metrics: mean absolute error (MAE), mean square error (MAE), and mean absolute percentage error (MAPE) are presented in Table 1 for the different values of *frequency* parameter (*frequency* $\in \{7, 14, 21\}$ days) and for additive and multiplicative model of Holt-Winters forecasting method.

The obtained results show that the prediction of the Holt-Winters method for the analyzed time series is effective. The maximum values of MAE and MSE are not greater then 0.005 (in the range of $[-1, 1]$). The mean absolute percentage error varies from 30% to 33% and it means that the predicted values can be greater or less not more than 1/3 of the real value.

**Table 1.** Effectiveness of the Holt-Winters prediction model.

| Frequency | Additive model | | | Multiplicative model | | |
|---|---|---|---|---|---|---|
| | MAE | MSE | MAPE | MAE | MSE | MAPE |
| 7 | 0.04615 | 0.00315 | 33,19% | 0.04898 | 0.00353 | 30.99% |
| 14 | 0.04598 | 0.00316 | 33,40% | 0.04830 | 0.00343 | 30.91% |
| 21 | 0.04584 | 0.00309 | 32,65% | 0.04804 | 0.00346 | 30.13% |

## 5  Summary and Future Works

In this paper we developed the methodology to analyze the evolution of user opinions towards a fixed topic. The analysis focused on a technical parameters of trend and seasonality aspects of time series.

The obtained results have shown that seasonal decomposition model and Holt-Winters Exponential Smoothing model for predicting the time series trend and seasonality are very effective. The aim of performed analysis was to check the dynamic of average sentiment score towards a fixed topic in a group of users. This is a profitable approach for analyzing and predicting the evolution of users opinions.

Evolution of users opinions is a very important and up-to-date problem as it allows to predict the users' reaction and opinions which can bring measurable benefits for products manufacturers or services providers. In the future, we plan to continue analysis of the user group dynamics and enrich the presented approach with more aspects and factors that can influence the process of the user changing their opinion.

## References

1. Bernardo, C., Altafini, C., Proskurnikov, A., Vasca, F.: Bounded confidence opinion dynamics: a survey. Automatica **159**, 111302 (2024)
2. Deng, L., Liu, Y., Zeng, Q.-A.: How information influences an individual opinion evolution. Phys. A **391**, 6409–6417 (2012)
3. Frey, V., Van de Rijt, A.: Social influence undermines the wisdom of the crowd in sequential decision making. Manag. Sci. **67**(7), 4273–4286 (2021). https://doi.org/10.1287/mnsc.2020.3713
4. Hassani H., Razavi-Far R., Saif M., Chiclana F., Krejcar O., Herrera-ViedmaE.: Classical dynamic consensus and opinion dynamics models: a survey of recent trends and methodologies. Inform. Fusion **88**, 22–40 (2022)
5. Hou, J., Li, W., Jiang, M.: Opinion dynamics in modified expressed and private model with bounded confidence. Phys. A **574**, 125968 (2021)

6. Ianni, M.D.: Opinion evolution among friends and foes: the deterministic majority rule. Theoret. Comput. Sci. **959**, 113875 (2023)

7. Ibrahim, N.F., Wang, X.: Decoding the sentiment dynamics of online retailing customers: time series analysis of social media. Comput. Hum. Behav. **96**, 32–45 (2019)

8. Liu, Y., Fang, A.: Analysis of opinion evolution based on non-Bayesian social learning. Appl. Math. Comput. **464**, 128399 (2024)

9. Macias, R.C., Vera, J.M.R.: Dynamics of opinion polarization in a population. Math. Soc. Sci. **128**, 31–40 (2024)

10. Pedraza L., Pinasco J. P., Saintier N., Balenzuela P.: An analytical formulation for multidimensional continuous opinion models. Chaos, Solitons & Fractals **152**, 111368 (2021). https://doi.org/10.1016/j.chaos.2021.111368. ISSN 0960-0779

11. Peng, Y., Zhao, Y., Hu, J.: On the role of community structure in evolution of opinion formation: a new bounded confidence opinion dynamics. Inf. Sci. **621**, 672–690 (2023)

12. Proskurnikov, A.V., Tempo, R., Cao, M., Friedkin, N.E.: Opinion evolution in time-varying social influence networks with prejudiced agents. IFAC PapersOnLine **50–1**, 11896–11901 (2017)

13. Shen, Y., Ma, X., Xu, Z., Herrera-Viedma, E., Maresova, P., Zhan, J.: Opinion evolution and dynamic trust-driven consensus model in large-scale group decision-making under incomplete information. Inf. Sci. **657**, 119925 (2024)

14. Sobkowicz, P.: Discrete model of opinion changes using knowledge and emotions as control variables. PLoS ONE **7**(9), e44489 (2012). https://doi.org/10.1371/journal.pone.0044489

15. Su, Y., Li, Y., Xuan, S.: Prediction of complex public opinion evolution based on improved multi-objective grey wolf optimizer. Egyptian Inform. J. **24**, 149–160 (2023)

16. Tian, Y., Wang, L.: Dynamics of opinion formation, social power evolution, and naive learning in social networks. Annu. Rev. Control. **55**, 182–193 (2023)

17. Xie, Z., Weng, W., Pan, Y., Du, Z., Li, X., Duan, Y.: Public opinion changing patterns under the double-hazard scenario of natural disaster and public health event. Inf. Process. Manage. **60**, 103287 (2023)

18. Britannica Homepage. https://www.britannica.com/topic/social-change/Explanations-of-social-change, Accessed 09 Mar 2024

19. Hyndman, R.J., Athanasopoulos, G.: Forecasting: principles and practice, 3rd edition, OTexts: Melbourne, Australia. OTexts.com/fpp3 (2021), Accessed 09 Mar 2024

20. https://www.kaggle.com/code/kapilstp84/sentiment-analysis-with-twitter-data-for-apple/input, Accessed 09 Mar 2024

21. https://medium.com/analytics-vidhya/python-code-on-holt-winters-forecasting-3843808a9873, Accessed 12 Mar 2024

22. https://www.tutorchase.com/answers/a-level/psychology/what-is-the-effect-of-crowd-behavior-on-individual-decision-making, Accessed 09 Mar 2024

# Author Index

# GPSR Compliance

*The European Union's (EU) General Product Safety Regulation (GPSR) is a set of rules that requires consumer products to be safe and our obligations to ensure this.*

*If you have any concerns about our products, you can contact us on ProductSafety@springernature.com*

In case Publisher is established outside the EU, the EU authorized representative is:

Springer Nature Customer Service Center GmbH
Europaplatz 3
69115 Heidelberg, Germany

The manufacturer's authorised representative in the EU is Springer
Nature Customer Service Centre GmbH, Europaplatz 3, 69115 Heidelberg,
Germany. If you have any concerns regarding our products, please
contact ProductSafety@springernature.com

Printed and bound by CPI Group (UK) Ltd, Croydon, CR0 4YY
06/05/2026
02104369-0004